Ecologies of Photography in the American West

The Lieven Gevaert Series is a major series of substantial and innovative books on photography. Launched in 2004, the Series takes into account the ubiquitous presence of photography within modern culture and, in particular, the visual arts. At the forefront of contemporary thinking on photography, the books offer new insights on the position of the photographic medium within art historical, theoretical, social and institutional contexts. The Series is produced by the Lieven Gevaert Research Centre for Photography, Art and Visual Culture (www.lievengevaertcentre.be) and covers four types of approaches: publication of outstanding monographic studies, proceedings of international conferences, book length projects with artists, translations and republications of classic material. The Lieven Gevaert Series is published by Leuven University Press, and distributed in North America by Cornell University Press.

Series editors
Alexander Streitberger
Hilde Van Gelder

Lieven Gevaert Series vol. 36

Ecologies of Photography in the American West

Edited by Monica Bravo and Carolin Görgen

Leuven University Press

Table of Contents

7 Acknowledgments

9 Introduction
Monica Bravo and Carolin Görgen

Part 1: Past Made Present

23 Chapter 1: Rethinking the History of Western American Photography
Martha A. Sandweiss

37 Chapter 2: Mirrors with a Memory: War, Memory, and the Western Landscape
Binh Danh

47 Chapter 3: (Dis)Placing Edward Weston: The Politics of Shutter Release
Monica Bravo

Part 2: Natural Resource Extraction

65 Chapter 4: Missing the Forest for the Trees: Pulp, Pollution, and Landscape Photography in the Pacific Northwest
Katherine Mintie

79 Chapter 5: Looking at Cows, Reanimating Gelatin: The Biopolitics of Photographic Emulsion
Alan C. Braddock

97 Chapter 6: The Death of 7000 Songbirds: Photography's Toxic Affinities
Alice Cazenave

113 Chapter 7: The Nomadic Alchemist
Andrés Pardo Piccone

Part 3: Native Sovereignty

127 Chapter 8: Corn Pollen Futures: Photography, Environmental Memory, and Indigenous Sovereignty on the Navajo Nation
Will Wilson (Diné)

141 Chapter 9: Āina and Accession Photographs of Queen Kapiʻolani and Queen Liliʻuokalani
Emily Cornish

161 Chapter 10: Bedrock Mortar vs. Photograph: Nevada City Rancheria Nisenan Cosmovision and the Malakoff Hydraulic Mine
Jordan Reznick

183 Chapter 11: Everywhere is West
Mercedes Dorame (Tongva), with an essay by Emilia Mickevicius

Part 4: Against the Classical Landscape

199 Chapter 12: "Mightier than the Pen?" Amateur Photography and the Circulation of Environmental Narratives in California, 1880s–1920s
Yves Figueiredo

221 Chapter 13: Black Gold: Petroleumscapes in Southern California Pictorialism and Beyond
Carolin Görgen

241 Chapter 14: Documenting "Restless Earth": John S. Shelton's Photographs for *Geology Illustrated* (1966)
Jennifer Tucker

259 Chapter 15: Arctic Time: Ecology and Photography in the Ice Fields
Elizabeth Hutchinson

277 Afterword
Robin E. Kelsey

287 Notes on Contributors

Acknowledgments

This book began as a two-day research conference held at the Huntington Library, Art Museum, and Botanical Gardens in San Marino, California, in March 2024. Thanks to generous funding provided by the E. L. and Ruth B. Shannon Research Endowment for Western Programs at the Huntington, we were able to bring together a dozen scholars, artists, and curators who share our interest in the histories of photography in the American West. While the convening at the Huntington marks the formal starting point of this collection of essays, the intellectual groundwork for the project was laid much earlier.

We, Monica Bravo and Carolin Görgen, first met virtually through an informal exchange organized by Photography Network during the COVID-19 pandemic. United by a deep interest in the long and complicated histories of Californian photography, we were, from both sides of the Atlantic, shaped by the writings of Martha Sandweiss, Robin Kelsey, and François Brunet—scholars who engaged closely with what we call the "changing natures" of photography in the American West. As our conversations expanded to our own research on the transnational circulation of western American photographs in the nineteenth century and the extensive networks of amateur and professional photographers in the twentieth century, we found further common ground in the more recent ecocritical approaches to American art history articulated by Alan Braddock, Karl Kusserow, and others.

Recognizing the need to bring together both established and emerging voices in the field—and to do so in a place so central to the collecting and narrating of western American photographic history—we turned to the Huntington. There, we remain grateful to Susan Juster, who was a steadfast champion and host for this project; to Linde Lehtinen, who expertly guided us through the institution's significant holdings; and to staff including Juan Gomez, Catherine Wehrey-Miller, David Badesch, and Sandra Castro, who all helped execute the program flawlessly. Other conference speakers not represented in this volume, including moderator Jim Ganz, Cara Romero (Chemehuevi), and Christine Hult-Lewis—whose decades-long work on the trailblazer of Californian photography, Carleton Watkins, made so much of this intellectual work possible in the first place—have our particular gratitude, along with members of the audience who enriched our thinking with their insightful questions.

We are so pleased that this edited volume found a home at Leuven University Press, thanks to the tireless efforts and enthusiasm of Veerle De Laet, who understood the project and its importance from the start. We are proud that it stands alongside other titles in the Lieven Gevaert Series, coedited by Hilde Van Gelder and Alexander Streitberger. Our gratitude extends to the editorial board, the anonymous peer reviewers, the staff of Leuven University Press, and especially copyeditor Suzanne Needs-Howarth, who have brought this book to fruition. A generous press subvention from the Barr Ferree Foundation Fund for Publications, Department of Art and Archaeology, Princeton University made invaluable aesthetic enhancements to the publication possible.

Finally, our greatest thanks are to the contributors to *Ecologies of Photography in the American West*. We are so honored that the artists and authors have entrusted us with their exceptional work. It is especially fitting that the volume begins with an essay by Martha Sandweiss and concludes with an afterword by Robin Kelsey, both of whose contributions to the field have inspired our own academic journeys. We hope that this book will serve as inspiration to another generation of curious scholars, makers, and appreciators of photography of the American West.

Introduction

Monica Bravo and Carolin Görgen

Photography and the environmental imagination in the American West have long been intertwined. Almost since the medium's inception, photographers have not only documented but produced indelible imagery of the western imaginary. For example, early daguerreotypists whose names are now forgotten documented boom towns, the likes of Timothy O'Sullivan and William Henry Jackson accompanied the so-called Four Great Surveys of the American West, and Carleton E. Watkins produced mammoth prints of Yosemite celebrating a desirable, "untouched" landscape. By the late nineteenth century, throngs of outdoor Kodakers poured into the national parks, and, during the twentieth century, Ansel Adams's conservation aesthetics became the literal poster child of the Sierra Club. However, while the phrase "western American photography" may call to mind dramatic vistas of mountains and canyons, rivers and the Pacific, there is "another West" to be found in archives and collections, as well as in contemporary scholarship and photographic practice.

In an era of climate change and ever-louder calls for climate justice—and in a place such as the American West, where drought, flash floods, wildfires, and heat waves are undeniable signs of disasters both "natural" and "cultural"—we seek to explore the ecologies of photography beyond the frame. Exploring the ecological dimensions of photography forces us to consider the inseparable ecosystems of the region and the history of photography writ large. How did photography profit from, consume, or otherwise transform the environment? Just as importantly, how have various constituents used photographs to promote settlement and tourism to the region, as well as its preservation and conservation? Centering photographic ecologies—literally, the study of home—this volume explores the medium's interconnectedness with the earth, its organisms, as well as with economic structures and systems of power that shape the environment in which we live.

Ecologies of Photography in the American West brings together cutting-edge scholarship on nineteenth- and twentieth-century photographic practices with portfolios by important contemporary artists to embark on a conversation about the changing nature(s) of the medium in the region. Photographs have long been read as documenting self-evidently ecological sites—as wishfully environmentalist pictures of pristine nature. Instead, contributors to this volume shift the focus from representations of landscape toward photography's rootedness in the land.

This project had its origins with a symposium held at the Huntington Library, Art Museum, and Botanical Gardens in March 2024. That in-person experience modeled some of what we hope to achieve in this volume. We came together as interested, and invested, parties from around the world on the exquisite grounds of the Huntington—a site purchased as a ranch by Henry E. Huntington in 1903, with a fortune amassed through the railroads—to talk (in some ways paradoxically) about the environment. One hundred and twenty years after Huntington's purchase, we gathered around an incredible collection of hundreds of thousands of photographs, many of them California-based—objects ripe for aesthetic valuation as well as critical examination. We hope that this volume, too, will prompt an ongoing conversation. We love both photography and the American West, as much as we love sharing their complicated histories. We invite you to join us in exploring the surfaces, depths, and contradictions at the heart of these subjects.

The West Once Written

Western American photography has been a privileged site of practice as well as historical and theoretical examination, with the two realms often developing in tandem. This category has historically referred to photographs made in California, the Pacific Northwest, and parts of the Southwestern United States. Government survey photographers and the first producers of photographic views initiated a discourse on the production and meaning of American landscape that was quickly taken up by periodicals in California and on the East Coast, featuring writings by the likes of Oliver Wendell Holmes and Marcus Aurelius Root. A few generations later, curator Beaumont Newhall and his friend Ansel Adams gave pride of place to these western photographs—*Ancient Ruins in the Canyon de Chelly* (1873) by O'Sullivan, for example, now recognized as an important artist rather than a commissioned maker—in the inaugural exhibition of the Museum of Modern Art's (MoMA) Department of Photography in 1940. (fig. 1) In this and subsequent shows, Newhall and Adams were, in effect, plumbing America's "usable past" to discover in nineteenth-century expeditions and mining company commissions precedents for their own Modernist tastes and practices, and for those of their contemporaries, such as Walker Evans and Edward Weston.

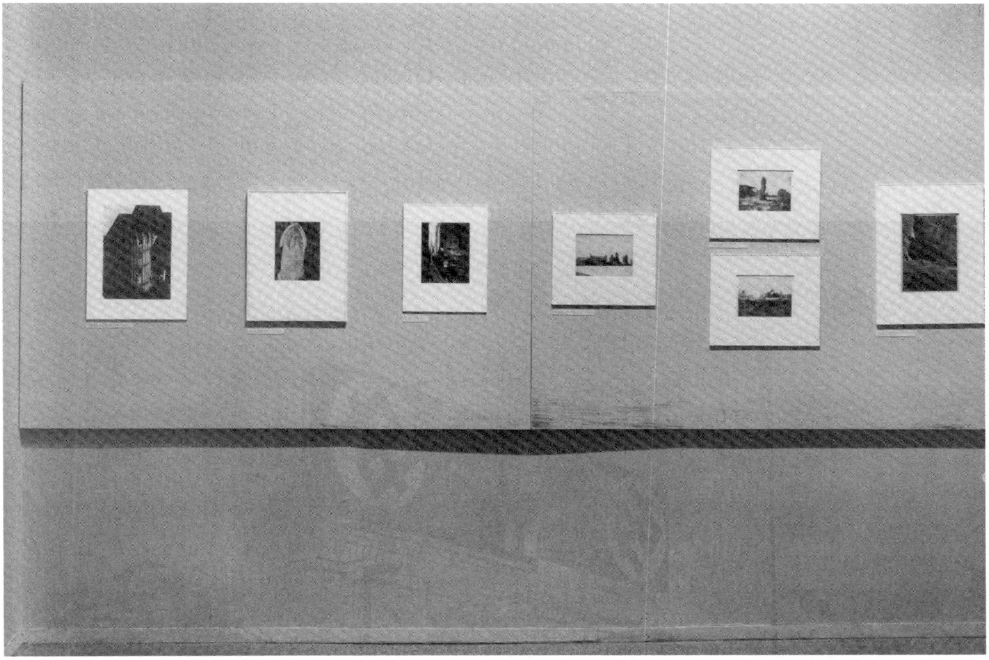

This emergent canon promoted a lineage of white male photographers that would become synonymous with western American photography into the twentieth century, culminating with the original—but nevertheless decidedly male—visual grammar of the New Topographics. So named for a 1975 exhibition curated by William Jenkins, this generation of photographers—many of them working in the American West but against the grand, romantic tradition of Adams and Weston—focused on the "man-altered landscape," inclusive of suburban tract homes, parking lots, and industrial warehouses. (Jenkins, 1975) (fig. 2) A decade later, motivated by the feminist turn, artist Deborah Bright, in her influential essay "Of Mother Nature and Marlboro Men," produced a more overt social critique of constructions of the American landscape. (Bright, 1985) She argued for a more capacious understanding of the category that would include nuclear sites and "feminine" spaces. This cogent analysis appeared contemporaneous with Postmodern photographer Richard Prince's appropriation of Marlboro cigarette advertisements. He selected for macho cowboys on horses and the arid, dramatic vistas that had already been critiqued by the New Topographics and Bright, but that still loomed large in the popular imagination.

Fig. 1.
Installation view of the exhibition *Sixty Photographs: A Survey of Camera Esthetics*, Museum of Modern Art, New York, December 31, 1940–January 12, 1941. Gelatin silver print, 17.8 × 24.1 cm. New York, Photographic Archive, Museum of Modern Art Archives. IN121.12.

By the late 1970s, both the study of photographic history as an academic pursuit and the use of social history as a critical methodology in art history had become acceptable, if not widespread. Social history holds that cultural, economic, and political (among other) contexts

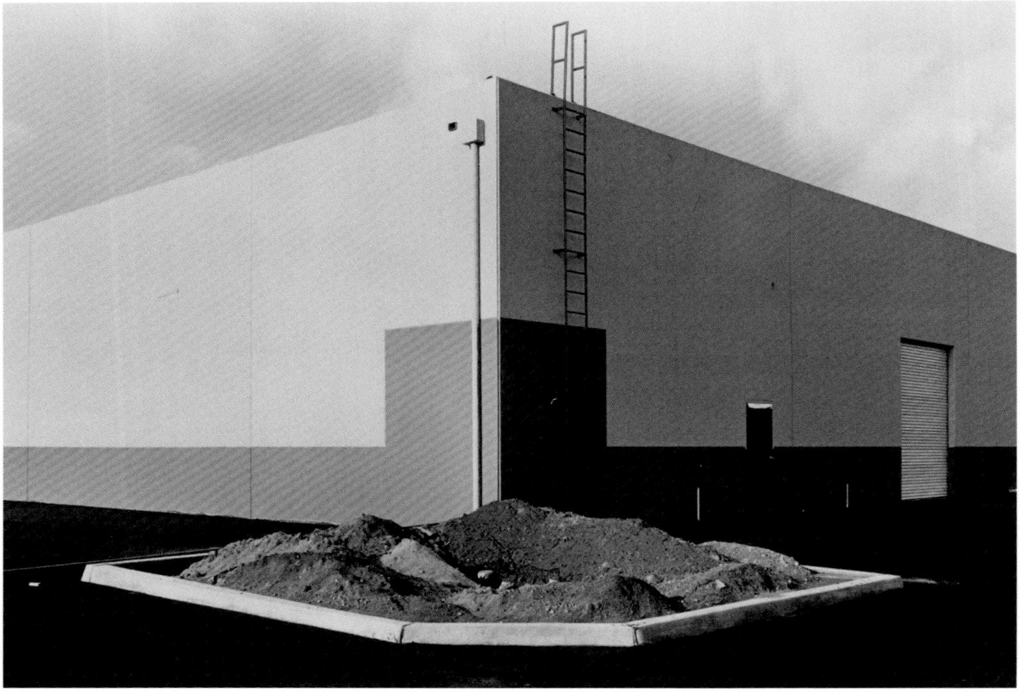

are central to the understanding of artistic production, and vice versa. Much of the scholarship fundamental to photographic history and theory derives from these principles. For example, Rosalind Krauss was among the most influential, in the 1980s, to draw attention to the museological, and sometimes market-based, construction of the American landscape photography canon's contours, with its notable exclusions. (Krauss, 1982) In the early 2000s, Martha A. Sandweiss (see her contribution to this volume) demonstrated how photography and the American West "came of age together" as "a new medium and a new place." (Sandweiss, 2002: 2) Ever since, other scholars and archivists have attempted to grapple with the large quantity of photographs produced by an ever-growing number of photographers in this vast territory. These social histories of the medium—which Alan Trachtenberg, Joel Snyder, Robin Kelsey (see his contribution to this volume), and François Brunet significantly shaped—have made us aware of the production, circulation, and long-lasting mythologies of western American photography. (Brunet, 2019; Snyder, 2002; Kelsey, 2007; Trachtenberg, 1989)

In almost all respects, the ecocritical approaches to American photography that characterize an emergent strain of scholarship are indebted to such social histories. As art historian Stephanie O'Rourke writes, "ecology supplies nothing less than an entirely new theorization of 'context.'" (O'Rourke, 2025) We continue to understand context—that capacious

Fig. 2.
Lewis Baltz, *South Corner, Riccar America Company, 3184 Pullman, Costa Mesa,* 1974. Gelatin silver print, 15.4 × 23.02 cm. San Francisco Museum of Modern Art, Gift of Carol Campbell Wenaas. © Estate of Lewis Baltz.

term—as central to artistic analysis, though now those contexts are more apt to be environmental, industrial, material, or decolonial. Examining the extractive practices of photo industries, the material components of photographs (light, water, minerals, paper, albumen, gelatin, etc.), and decolonial readings of well-known landscape photographs—to mention only a few of the contributions to this volume—all broaden our understanding of the political ecologies of art.

Revisionist Approaches: "There Goes Watkins!"

Ecologies of Photography in the American West builds on and expands what many have considered the favorable photographic climate of the West. While the book is rooted in the conceptual framework of a social history of American photography, our contributors push western ecosystems—and the ways in which photography transformed them—to the forefront. With the growing awareness that "ecology permeates all forms," we seek to examine the myriad ways in which the photographic medium is, and has always been, entangled with the environment. (Morton, 2010: 11; Braddock and Kusserow, 2018)

If the environment has always been present in pictures of the American West—in rocks, rivers, valleys, and prairies—so, too, have the machines (the cameras, trains, and drills) that surveyed, transformed, and partially destroyed this very environment. With these observations in mind, well-studied nineteenth-century western survey photographs by the likes of Alexander Gardner, O'Sullivan, and Watkins take on multiple new dimensions. Consider Watkins's *Rounding Cape Horn.* (fig. 3) The print shows not just "the hard fact" of the railroad, as contemporary observers termed it. When viewed through an ecocritical lens, it also gestures at how the camera assisted territorial destruction under the banner of progress, cutting through Native lands, detonating rocks, and elevating tracks on which coal-powered trains would carry raw materials extracted from the western earth. The infrastructure so gloriously depicted by Watkins—and the very materials used to make the photograph—lead to an investigation of mining practices that are not pictured in the frame, and that further complicate environmental and labor histories of the American West. While these new considerations might compromise the market value of a Watkins print, they are necessary to re-evaluating that work's capitalist entanglements. As one photo dealer overheard at the symposium put it, "there goes Watkins!"

These ecocritical ways of seeing are woven throughout the contributions to this volume. They broaden the scope of inquiry to the matter and material of western American photographs, and they explore

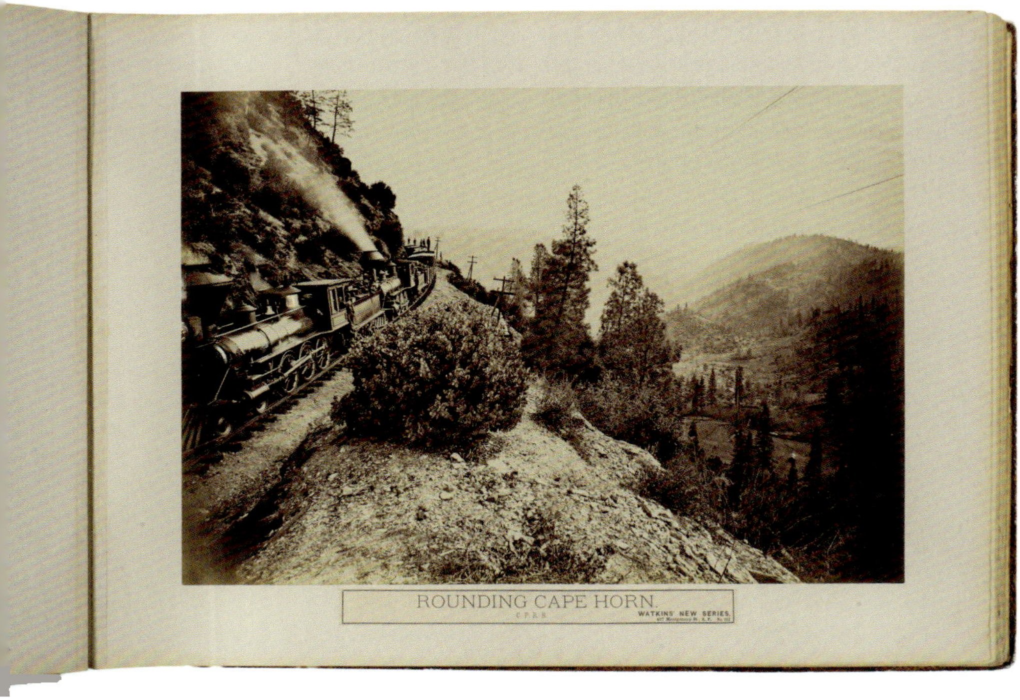

photography's extraction of value from the land, which had been home to Indigenous populations prior to and throughout Euro-American settlement. The same possessive and extractive impulses driving Manifest Destiny—the belief that white settlers were divinely ordained to push westward across the North American continent, from the Atlantic to the Pacific Ocean—also undergirded twentieth-century imperialism. However, the movement of people and ideas does not strictly follow the geopolitical boundaries imposed by colonization and governing structures. Consequently, the volume takes an expansive—but not comprehensive—view of geography to include the continental western United States, the Arctic, Hawai'i, and other places particularly impacted by US settler colonialism. We not only address an expanded western geography, but also burrow under the surface, both of the earth and of the photograph, looking at photographs not only as images, but as material objects. This includes—as in the case of Watkins's *Rounding Cape Horn*—questions ranging from the making and production of photographs to the infrastructures facilitating their circulation and consumption, as well as the far-reaching political ecologies of the medium's underlying resource extraction, land use, and colonial power structures. At a moment when the American West is reckoning with climate change, it is imperative to investigate what has always been at the heart of western American photography: the environment.

Fig. 3.

Carleton Watkins, *Rounding Cape Horn*, ca. 1889–1890. Albumen print, 50 × 68 cm. "Central Pacific Railroad and Views Adjacent. C. E. Watkins, Photo. San Francisco, Cal." San Marino, Calif., Huntington Library, Art Collections, and Botanical Gardens. RB 137501.

Contributors and Four Lines of Inquiry

The contributions to *Ecologies of Photography in the American West* develop four key themes: 1) critical evaluations of the canon of western American photography and its contemporary iterations; 2) the raw materials and labor that make up the photographic object and their roots in western ecosystems; 3) the various forms of Indigenous stewardship that have marked and continue to shape the American West; and 4) new readings of the "classical" western landscape and its practitioners.

The first lens addresses the canonical histories of western American photography. While it has much too often been told as the story of individual "master" photographers, our first line of inquiry challenges this premise. Martha Sandweiss—who was among the first to conceptualize the simultaneous development of photography and the American West—emphasizes the importance of biography when studying the iconography of expansionism. Focusing on the confrontation with the reality of place, she traces how photographs of white makers, such as Alexander Gardner, circulated back to Indigenous sitters, including Spotted Tail at Fort Laramie, thus unfolding more nuanced readings of photographic collaboration and reciprocity. Monica Bravo investigates the legacy of another canon maker—Edward Weston—and his relational ecologies. Considering Weston's oeuvre alongside that of his friend Toyo Miyatake and contemporary artists Sherrie Levine and Tarrah Krajnak, she questions notions of authorship and autonomy, thus bringing to the fore "the politics of exposure" of the twentieth-century western photographer. While the selective canon of western photo artists is constantly being re-evaluated, contemporary artists have reclaimed the techniques of these makers to tell their own stories. Binh Danh's portfolio uses the mirror-like surface of nineteenth-century daguerreotypes to reflect on his family's history as refugees in the American out-of-doors. Danh, who is Vietnamese American, recounts his experience making daguerreotypes in a mobile darkroom across the American West. Like the many Euro-American photographers before him, and like us, he is a guest on the land, exploring questions of home and belonging.

The second theme shifts the focus to photographic materiality. Artists and scholars alike have recently drawn attention to the stuff the medium is made from, which generates an ecological imprint of its own. (Levin et al., 2022; Angus, 2024) Contributors to this volume engage with the matter and work fundamental to the production of photographs and are asking: Where do the materials for photographic processes come from? How are they extracted and transformed? And whose labor is required to do so? Historically, the "dirty business" of extracting resources has been decoupled from the study of the medium—with crucial exceptions, such as Marxist photographer and writer Allan Sekula's scholarship—which we and others no longer feel is tenable. (Sekula, 1983; Coleman and James, 2021; Henning, 2025) Awareness of this alienation of labor troubles notions of authorships—human and non-human—that have long been taken as art-historical givens prior to the structuralist turn and the emergence of actor-network theory. (Latour, 2005) Over the course of the medium's almost two-century-long history, photographers have relied on a vast array of materials, including silver and gold, cotton, gelatin, and—more recently—rare earths, all of which are directly embedded in the earth's ecosystems.

Consider, for instance, the vast quantities of silver extracted through dynamite blasts in Nevada in the 1860s. To separate silver from its ore, mercury was needed, and both substances were used to develop photographs. (Bravo, 2025) Alice Cazenave's essay traces the afterlives of mercury contamination in Silver City, Nevada—a place where abandoned mining sites can still furnish the raw materials for photo developing, as her present-day landscapes show. The photographic industry also required large amounts of gelatin as the emulsion for gelatin silver prints, the most popular process of the twentieth century. Alan Braddock re-establishes the biopolitics of such emulsions through a case study of George Eastman's Kodak company, which obtained gelatin from cow bones. Paper, as the most common material support for photographs, was made from pulp derived from trees in the twentieth century, as Kappy Mintie explains in her essay on the Kodak paper mills that operated in the Pacific Northwest. Encouraging us to look at photographs in a holistic manner, her analysis links the material composition of photographs to the global impacts of deforestation and chemical contamination. Together, these mineral, animal, and vegetal components pose larger questions about supply chains and sustainability, as large-scale facilities have supported the growing demand for plates, film, papers, and chemicals, relying on vast transportation networks. Questioning these ever-growing extractive economies, Andrés Pardo Piccone explores how plants growing in the transnational American West—such as coffee—can be used as sustainable developers that are not reliant on international supply chains, nor on artificial chemistry. Collectively, the essays shift our understanding of photographs away from merely their images toward the components drawn from the earth that are fundamental to image production.

Our third lens reconsiders the supposedly environmentalist goals of many of the nineteenth- and twentieth-century photographs by recentering questions of home and belonging, of stewardship and sovereignty. While the environment of the American West had been tended to for thousands of years prior to the arrival of Euro-Americans, by the late nineteenth century, the traditional ecological knowledge of Indigenous communities would be largely sidelined by an aesthetic understanding of preservation. (Kelsey, 2013) Nevertheless, it is possible to productively re-read such photographs. Martin Berger argued that landscape views such as Watkins's actually consolidate white identity through their focus on an identical set of traits, names, and their construction of nature as a resource to be exploited. Far from positing an essentialist white gaze, he teaches us to recognize the settler-colonial ideologies embedded within apparently innocent, non-raced photographs. (Berger, 2003) Building upon this framework, Jordan Reznick asks how a Nisenan person might perceive the washed-up sediments of mining that were captured by Watkins's camera at Malakoff Diggins in the 1870s that, to them, inevitably link genocidal and ecocidal violence. Such scholarship defamiliarizes well-known landscape photographs and grants space to an often-marginalized Indigenous stewardship. Of course, the categories "Indigenous" and "Native American" are settler inventions: Contributors to this volume use the names of local communities when available and acknowledge the specificity of distinct social experiences.

Any historical understanding of the American landscape must be in conversation with contemporary artists and communities, who have inherited these traditions and are grappling with their legacies. In the same section, a broader reclamation of photographic processes is at the heart of Will Wilson's practice. The Diné photographer created an unconventional survey, using drone-based photography, to uncover the environmental degradation caused by more than 500 abandoned uranium mines on the Navajo Nation. Relying on survey photography—which historically enabled Indigenous dispossession—Wilson instead creates a platform for resilience. Tongva artist Mercedes Dorame pushes photography into an expanded field of artistic practice. Her "Everywhere is West" portfolio, here accompanied by an interpretive text by Emilia Mickevicius, explores the waters of the Los Angeles basin, highlighting the often-invisible traditional ecological knowledge embedded in the Pacific Ocean and its littoral surround. Sovereignty and care for the land are, likewise, at the heart of Emily Cornish's essay, in which she explores how royal Native Hawaiians used photographs showcasing *āina* (land) on their bodies as expressions of their pride for and responsibility to their homelands at a time of Euro-American annexation. Together, the scholarly essays and artist portfolios in this section center survivance and sovereignty.

The fourth dimension of *Ecologies of Photography* questions the "classical" western landscapes and practitioners. The term American West has tended to refer to California; the Pacific Coast; and, sometimes, the region west of the Rockies. It has often been examined from the perspective of the American East Coast, as the product of Manifest Destiny, the Oregon Trail, and the transcontinental railroad. And yet, the history of American photography looks different when told from the continent's western edge and with deference to the oceanic turn. Many mid-nineteenth century photographers arrived by sea, often arriving via such places as Panama and Chile. American photography's relationship, even indebtedness, to European histories has been over-represented in the literature because of scholarly bias and the medium's "invention" there. (Brunet, 2015) Yet attention to materials, industry, belonging, and other dimensions of photographic practice reveal much larger North–South relationships, trade and labor routes crisscrossing the Pacific Rim, as well as the importance of certain sites particularly affected by US imperialism, including Latin America, Hawai'i, and Alaska.

Challenging the received geographies of "the American West," Elizabeth Hutchinson examines photographs of the Arctic Ocean made off the coast of Alaska in the late nineteenth century. In the harsh environmental conditions of the Arctic, the environment became a co-producer of photographs, she argues. While explorations of the Arctic challenge the polished landscape aesthetics associated with the American West, so, too, do the aerial photographs made by geoscientist John Shelton that Jennifer Tucker discusses. Catering to popular audiences, Shelton's illustrated books of the American Southwest of the mid-twentieth century reflect the changing visual culture of the post-war landscape. Popular photographic practices are also at the heart of Yves Figueiredo's examination of Theodore Lukens—a largely unknown amateur photographer whose documentation of Southern California forestry introduced an ecological consciousness to the budding amateur photographer scene of the late nineteenth century. Carolin Görgen's essay similarly centers amateur networks active in Los Angeles, yet with an emphasis on how their outdoor practices became intertwined with and dependent on the growing petroleum infrastructure of the first decades of the twentieth century, expanding extractive visions of western American photography in the process. Taken together, the essays encourage us to question the well-established geographies and aestheticized stereotypes of "the West" to consider the layers of the land and the diversity of practices on it.

Finally, in the afterword, Robin Kelsey reflects on these myriad themes and the ways in which photographic history has evolved since his groundbreaking intervention in the field almost twenty years ago. (Kelsey, 2007) Through the lens of smoke, he weaves together meditations on Tohono O'odham poetry, western wildfires, and the ongoing question of photographic agency. In bringing together these perspectives, *Ecologies of Photography in the American West* stands at the forefront of a new wave of scholarship that interrogates the material, political, and ecological entanglements of photography. By centering de-canonization, non-human agency, Indigenous sovereignty, and new perspectives on landscape, this volume not only expands the scope of environmental humanities, but also redefines how we understand photography's role in shaping the American West and how this photography, in turn, was shaped by the West. In so doing, this volume challenges conventional narratives of both the medium and the region, offering a necessary and transformative intervention in photographic history.

Bibliography

Siobhan Angus, *Camera Geologica: An Elemental History of Photography* (Durham: Duke University Press, 2024).

Martin A. Berger, "Overexposed: Whiteness and the Landscape Photography of Carleton Watkins," *Oxford Art Journal* 26, no. 1 (2003): 3–23.

Alan C. Braddock and Karl Kusserow (eds), *Nature's Nation: American Art and Environment* (New Haven, CT: Yale University Press, 2018).

Monica Bravo, "Mercury is a Messenger: Photography's Dependence on Quicksilver and Mining Labor," *American Art* 39, no. 1 (2025): 16–25.

Deborah Bright, "Of Mother Nature and Marlboro Men: An Inquiry into the Cultural Meanings of Landscape Photography," in *The Contest of Meaning: Critical Histories of Photography*, ed. Richard Bolton (Cambridge, MA: MIT Press, 1992), 125–143.

François Brunet, "'An American Sun Shines Brighter,' or, Photography was (not) Invented in the United States," in *Photography and Its Origins*, eds Tanya Sheehan and Andrés Zervigón (New York: Routledge, 2015), 131–144.

François Brunet, *The Birth of the Idea of Photography*, trans. Shane B. Lillis (Toronto: Ryerson Image Centre; Cambridge, MA: MIT Press, 2019).

Kevin Coleman and Daniel James, eds, *Capitalism and the Camera: Essays on Photography and Extraction* (London: Verso, 2021).

Michelle Henning, *A Dirty History of Photography: Chemistry, Fog, and Empire* (Chicago: University of Chicago Press, 2025).

William Jenkins, *New Topographics: Photographs of a Man-altered Landscape* (Rochester: International Museum of Photography at the George Eastman House, 1975).

Robin Kelsey, *Archive Style: Photographs and Illustrations for U.S. Surveys, 1850–1890* (Berkeley: University of California Press, 2007).

Robin Kelsey, "Sierra Club Photography and the Exclusive Property of Vision," *Rachel Carson Center Perspectives* 1 (2013): 11–26, https://doi.org/10.5282/rcc/6205.

Rosalind Krauss, "Photography's Discursive Spaces: Landscape/View," *Art Journal* 42, no. 4 (1982): 311–319.

Bruno Latour, *Reassembling the Social: An Introduction to Actor-Network-Theory* (Oxford: Oxford University Press, 2005).

Boaz Levin, Esther Ruelfs, and Tulga Beyerle (eds), *Mining Photography: The Ecological Footprint of Image Production* (Leipzig: Spector Books, 2022).

Timothy Morton, *The Ecological Thought* (Cambridge, MA: Harvard University Press, 2010).

Stephanie O'Rourke, *Picturing Landscape in an Age of Extraction: Europe and Its Colonial Networks, 1780–1850* (Chicago: University of Chicago Press, 2025).

Martha A. Sandweiss, *Print the Legend: Photography and the American West* (New Haven, CT: Yale University Press, 2002).

Allan Sekula, "Photography Between Labor and Capital," in *Mining Photographs and Other Pictures: Selection from the Negative Archives of Shedden Studio, Glace Bay, Cape Breton, 1948–1968,* eds Benjamin H. D. Buchloh and Robert Wilkie (Halifax, NS: Nova Scotia College of Art and Design, 1983), 193–268.

Joel Snyder, "Territorial Photography," in *Landscape and Power*, ed. W. J. T. Mitchell (Chicago: University of Chicago Press, 1994/2002).

Alan Trachtenberg, *Reading American Photographs: Images as History. Mathew Brady to Walker Evans* (New York: Hill and Wang, 1989).

PART 1
Past Made Present

CHAPTER 1
Rethinking the History of Western American Photography

Martha A. Sandweiss

Like most histories, the history of western American photography itself has a history. To my surprise, I have lived through much of it. When I began my career as a Curator of Photographs at a museum of western American art in 1979, few people much cared about photographs of the American West, especially those made during the nineteenth century. Yet, as this book demonstrates, there is now a vibrant conversation among scholars, curators, and photographers, all asking how western photography can help us understand broader cultural questions. A body of work once considered marginal—because of the medium and the subject matter—now attracts a new kind of intellectual and creative energy.

The current conversations swirl around two main points: photographs as material objects and photographs as primary source documents that illuminate larger historical and social issues.

The focus on western photographs as material objects, with particular attention to the materials from which they were made, builds on the ecocritical thinking done in recent years about American painting and furniture. (Braddock and Kusserow, 2018) This is important work, long overdue for a medium that depends not just on wood-pulp paper, but on metals mined from the earth and a slew of heavily refined chemicals. Ironically, a focus on the environmental consequences of mining and manufacturing those materials draws us away from the distinctive "western" subject matter of the photographs and reminds us that, in terms of their material composition, western photographs are no different from any other photographs. Commodity flows transcend regional boundaries. So even western photographs, if made before 1865, have a likely link to slavery, the great moral stain at the heart of the nation. Just consider the cotton used for photographic polishing cloths and for the rag content of printing papers. (Mintie, 2020) Photographs made in remote parts of the western United States were bound up in larger economic forces and labor systems, none of them visible in the photographic image itself. Those mid-nineteenth-century photographs of western landscapes, Native peoples, and burgeoning railroad towns sit nestled inside local stories, regional stories, and stories that span the continent and the globe.

The second focus of the expanding conversations about western photography, on what some call the "political ecology" of western photographs, is really part of a broader movement to consider photographs as documents that can illuminate social, cultural, and economic questions that transcend the putative subject of the work. This sort of inquiry expands the sorts of questions photo historians have been asking since the turn away from more conventional art-historical analysis some 40 or 50 years ago. It hinges on the idea that photographs can be subject to the same sorts of rigorous thinking that scholars apply to other sorts of historical records. It takes as a given that photographs not only record visual *facts* but convey complex *ideas*. And it demands that scholars acknowledge that photographs might do more than simply visualize ideas gleaned from more conventional literary documents. In addition to being *illustrations*, historical photographs can be *sources*.

The ecocritical analysis of photographs and the consideration of photos as primary source documents decenter the old art-historical focus on the visual content and composition of the image, as well as the time-worn historical focus on photographs as straightforward sorts of visual evidence. They might take us in different directions—into a study of chemical factories or back into the archives—but both sorts of inquiry push us to consider how photographs function and circulate within networks of power. They push us to think about the inequitable power relationships that might exist between manufacturers and workers, between photographers and subjects, and between the institutions that paid to have photographs made and the people who later saw them. From still images come dynamic stories.

The pivotal break in the older ways of thinking about western American photographs as either straightforward illustrations or aesthetic objects occurred several decades ago.[1] That conceptual pivot hinged on three main events: first, the shift from art museum-based scholarship to a broader interest in photographic history; second, the expansion of the archives used in photo-historical studies, facilitated among other things by the expansion of computerized museum and library cataloging systems, digital photography, and the internet; and, finally, the turn toward visual studies within the field of American studies and the turn toward cultural studies with the broader field of art history.

In the late 1930s, as the centennial of Louis-Jacques-Mandé Daguerre's photographic invention neared, two American men were at work on groundbreaking histories of photography. The University of

Kansas chemistry professor Robert Taft was at work on *Photography and the American Scene*, a remarkable book published in 1938, written "not primarily from a technical viewpoint, but from that of social history." Taft explained that, "[s]ince this field has had no previous historian, my work in some respects has been easier than if there had been previous ones. On the other hand, first-hand facts (reliable source material, as the historian calls them) have had to be found and appraised." (Taft, 1938: viii) Taft worked hard to recover the most basic information about the early western photographers now deemed central to the canon. As he did so, he was corresponding with his competition, Beaumont Newhall, then at the Museum of Modern Art and working on his own, more international history of photography. (Robert Taft Papers)

Not surprisingly, Taft devoted far more attention to western photography than did Newhall, who folded his analysis of the work into a larger discussion of international documentary work in the 1860s and 1870s. Conversely, Newhall paid considerably more attention to the formal qualities of photographic images than did Taft, and as he brought his survey up to the present, he covered the self-consciously artistic efforts of the Pictorialists and the more abstract works of some of the photographers who followed. (Newhall, 1964)

In piecing together his story about nineteenth-century western photography, Taft delved deeper into the archives than did Newhall. But, remarkably, both men gleaned firsthand information from William Henry Jackson, then well into his nineties. (Taft, 1938: ix-x; Newhall, personal communication, 1982) Jackson had worked on some of the great post–Civil War photographic surveys of the American West and in 1871 famously photographed what would soon become Yellowstone National Park. Taft and Newhall wrote during an era of Kodaks and Graflexes; Jackson made vivid the era of wet-plate negatives, massive view cameras, and albumen photographs printed with the rays of the sun. When I had lunch with Newhall in the early 1980s, I sat awestruck as he related his conversations with Jackson from nearly half a century before. The history of western photography collapsed into the size of a soup bowl. The gourmand sitting across the table connected me to a photographer who had trekked across the American West more than a century earlier with a mammoth-plate view camera strapped to the side of a mule.

Taft's work embraced a social and cultural history of the medium; Newhall's seminal *Photography, 1839–1937* (1937), subsequently revised as *The History of Photography, from 1839 to the Present Day* (1964), pointed toward a more art-historical approach. (Brunet, 2012) For many reasons, the Newhall approach prevailed, in part because Newhall had the support of MoMA and Taft had the University of Kansas and the Kansas Historical Society. It was no contest; the influential New York art museum won.

Decades later, as the photo market began to boom in the 1970s, photo-historical scholarship remained located largely in art museums, where the institutional attention focused on exhibition-worthy "masterworks." Many of the curators then writing about western photographs took pains to emphasize that photographs were art and photographers were artists. That made sense given the agendas of the museums for which they worked and the collectors whom they courted. This emphasis on photographic "masterworks" stressed the aesthetic links between photographs and paintings and, inevitably, focused on a small handful of photographers. This approach left little space for the consideration of lesser-known photographers, including some of those included in this volume, and ignored the work produced in less "display-worthy" formats, such as stereographs or cartes de visite. These smaller-format pictures, however, hold great value to the newer generation of scholars, who recognize that aesthetics alone do not determine the potential value of a photograph as a cultural document.

The book *Era of Exploration: The Rise of Landscape Photography in the American West, 1860–1885*, published by the Albright Knox and Metropolitan Museums in 1975, exemplifies the more art-historical, museum-based approach that prevailed as western photography began to claim broader attention in the art market as well as within the scholarly community. Private collectors made up nearly one-third of the thirty-five sources for illustrations in the publication, their photographs long severed from their original historical context and selected for their aesthetic appeal. The institutional lenders came largely from the US East Coast. Although the curators acknowledged that the photographers were often in the employ of others—the railroads, the federal government—they focused on the museum-worthiness of the photographs, not on their potential value as sources illuminating the politics of those patrons (or of the photographers themselves). "The appreciation of nature as a work of art is essential to the rise of landscape photography," wrote Weston Naef, then curator at the Metropolitan Museum of Art and chief author of the catalog text. The photographers in his exhibition focused on "nature for its own sake" and produced pictures that offered "evidence of deep artistic conviction." (Naef, 1975: 16)

The 1980 National Gallery of Art catalog *American Light: The Luminist Movement, 1850–1875* included an essay on photography again written by Naef. Here, almost two-thirds of the essay illustrations came from private collectors, suggesting the ways in which curators had not yet delved into the many small libraries and historical societies where they might have found the work of other western photographers not already included in the canon. At the National Gallery, the focus on a kind of art-historical formalism remained strong as the curators—still fighting against the suspicion that photography was too mechanical to be a legitimate form of art—argued that photographers demanded serious consideration alongside their painter colleagues. As Naef wrote, "Carleton E. Watkins and Timothy H. O'Sullivan emerged simultaneously with the classic phase of luminist painting and stand parallel to Martin Johnson Heade, Sanford Gifford, and John E. Kensett in their tendency to simplify compositions, their fondness for low horizon lines, their attraction to solid forms in nature, their desire to model palpable space, and above all their belief in the transcendent beauty of nature." (Naef, 1980: 270) In art-historical arguments such as this, it is hard to discern any trace of Taft's focus on how nineteenth-century western photographers sought to meet the documentary demands of their government employers.

These projects sponsored by major art museums had an aesthetic agenda, a limited archive from which to draw key works, and an intimate connection with private collectors, who, in some cases, were acquiring important works—adjudged by aesthetics—before museums really got into the game. The limitations of the collections from which the curators drew their work, coupled with the art-historical agendas of the exhibitions and the focus on exhibition-worthy images, limited the scope of the work presented in museum exhibitions and catalogs in the 1970s and 1980s.

Things begin to change between the 1980s and early 2000s, with two roughly concurrent developments. One occurred within the academy, as visual culture became a more critical part of the disciplines of American studies and history, and as social history became a more integral part of art history. In the first case, scholars long focused on literary documents began to consider visual documents as fair subjects of inquiry. In the second, scholars traditionally focused on aesthetic analysis approached their traditional subject matter with a broader set of cultural and historical questions.[2]

These shifts within academia loosely corresponded with the other development, the growing use of computer cataloging in libraries, museums, and historical societies, as well as the expansion of the internet. Technology allowed researchers to access vast archives of heretofore difficult-to-find collections and expanded the pool of images they could consider in their work.[3] Computerized access to collection records came first; access to digitized images would not come till later, stimulating another revolution in research possibilities. Robert Taft had explicitly positioned his social history of photography against the older histories of photographic technologies. Yet technology is what made photo archives more widely available to those who wished to use the medium to write social history, and those online archives continue to grow. A quick search of the photography collections of History Colorado, for example, turns up online records and digital images for nearly 70,000 photographs; the Online Archive of California, which aggregates data from collections around the state, includes more than 85,000 photographic images. With only a small percentage of the corpus of western photographs reproduced in printed books, online digital resources such as these have transformed the ways in which scholars and curators do their work.

These two shifts—within the academy and within collecting institutions—laid the groundwork for a broader range of scholars to think about photographs as imbricated in a wide range of political, cultural, and environmental projects. It bears noting, however, that a reliance on digital archives mitigates against an interest in the materiality of photographs. A digital image is not the same as an albumen print mounted on board. In a sense, the current interest in the physical materials of photographic prints is a kind of push back against the very forces that make images more accessible on a computer screen.

I would situate my own book, *Print the Legend: Photography and the American West* (2002), within this fin de siècle moment of nascent change in the practice of photo-historical scholarship. While doing research in the 1980s and 1990s, I went to most of the western state historical societies, countless libraries, museums large and small. Illustrated computerized catalogs did not exist. It was not easy to know what was in a particular collection without showing up, but I felt committed to expanding the archive of western photographic images with which I could work. I recall the photo collection organized into "People," "Places," "Things"—the countless drawers of ill-arranged photographs arranged only by size. I had worked in an art museum for a decade; by the time I wrote *Print the Legend,* I no longer did, and the possibilities for thinking beyond the needs of an art museum felt expansive. It felt liberating to write about what seemed interesting (or even lost to view), rather than about what would work on a gallery wall; liberating to think about photographs as material, political, and commercial objects of interest beyond their mere subject matter; freeing to think about how photographs moved into, through, and sometimes out of the marketplace. (Sandweiss, 2002)

Print the Legend argued for photographs as rich primary sources that could address a broad range of historical questions extending well beyond the art-historical: about the federal government's involvement in westward migration and Native dispossession; about Native people's engagement with the new photographic technologies; about the uses of photography by the transcontinental railroads. If this approach seemed novel then, it now seems only an early iteration of today's conversations among scholars and artists who have a comfort with the analysis of visual images that an earlier generation of scholars lacked.

In recent years, I have been working on the Scottish-born photographer Alexander Gardner (1821–1882), a figure whose western work from the 1860s offers us a way to think about some of the larger questions raised in this volume about settler colonialism and Native dispossession. (Sandweiss, 2025) At the same time, his life and photographs offer up a caution about using nineteenth-century photographs to answer twenty-first-century questions. Setting aside for a moment the general advisability or utility of using a broad, sweeping phrase like "colonialism" to describe an immensely complicated historical process, I also worry about the continuing allure of using photographs to illustrate a particular point, rather than dealing more carefully with the broader archive of photographic evidence or the complexity of a single photographic image. Like their literary counterparts, photographs are complex sorts of documents.

Gardner went west twice. He photographed his way across Kansas for the Union Pacific Railroad, Eastern Division (later the Kansas Pacific), in 1867. The following year, he accompanied a federal peace commission to Fort Laramie, where the commissioners sought to persuade the Lakotas and other northern plains tribes to sign a treaty that would confine them to fixed reservations and set them on the road to cultural assimilation. I understand how easy and tempting it may be to use Gardner as an exemplar of a particular sort of thinking about westward expansion. He not only traveled under the protection of a railroad company and, later, a government commission, he also made photographs that seemed to endorse their intertwined "civilizing" missions.[4]

On October 19, 1867, while documenting the railroad's slow crawl across the southern plains, Gardner explained that he set up his cameras just west of Hays, Kansas, "at the end of the track [...] the locomotive and construction train upon the last laid rails." (fig. 1) He wrote, "I intend to call this 'westward the course of empire takes its way.'" (Gardner, 1867: 1) The phrase was in the air. It came from a 1728 poem by the Irish philosopher Bishop George Berkeley, who hoped to create a progressive new college on the British island of Bermuda. By the 1860s, however, it was the rallying cry for western expansionists in the United States. The painter Emanuel Leutze used the phrase as the title for his monumental 1862 mural in the US capital. A few years later, the College of California honored the poem's author by adopting his name for their school.

During the early 1860s in Washington DC, Gardner lived next door to Asa Whitney, perhaps the nation's greatest advocate for the idea that a cross-continental railroad would lead the nation to her Manifest Destiny. In their casual after-dinner conversations, the photographer would have come to understand the resonance of the phrase "westward the course of empire." His juxtaposition of a locomotive and some horse-drawn carts would serve as a visual metaphor for the dividing line between his

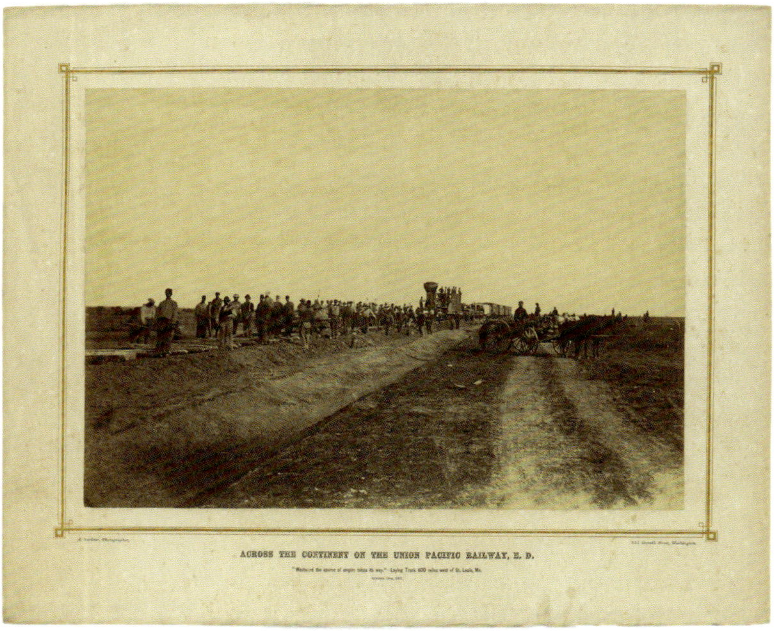

Fig. 1.

Alexander Gardner, *"Westward the Course of Empire Takes Its Way": Laying the Track 600 Miles West of St. Louis, Missouri*, 1867. Albumen silver print, 33.2×47.6 cm. Los Angeles, J. Paul Getty Museum.

https://www.getty.edu/art/collection/object/104GZY

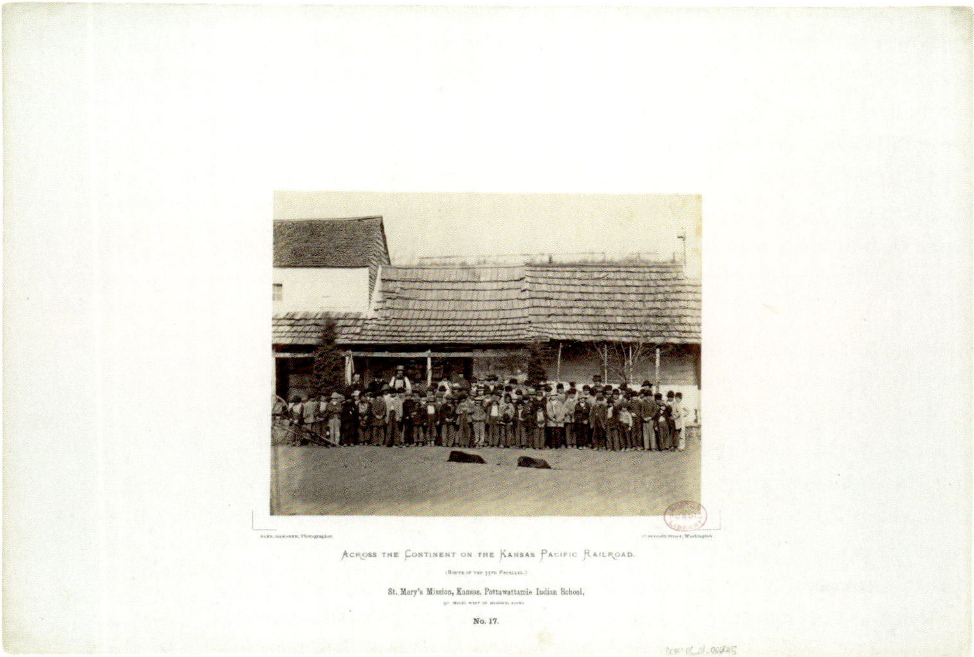

Fig. 2.

Alexander Gardner, *St. Mary's Mission, Kansas, Pottawatomie Indian School*, 1867. Albumen silver print, 15×20.2 cm. Boston Public Library.

https://ark.digitalcommonwealth.org/ark:/50959/70796d495

adopted nation's future and its past. Still, it is difficult to say whether the moralizing tone of the image reflected Gardner's personal views or those of his corporate patrons.

In that picture and in the photographs he made shortly afterward at the Pottawatomie Indian School at St. Mary's Mission, near Topeka, Kansas, he appears to advocate for the expansion of American settlements and the erasure of Native cultures. (fig. 2) As with his iconic photograph of railroad construction, he imagined a title for his picture—"Dawn of Civilization"—before he even made it. (Gardner, 1867: 1) Ultimately, he scrapped the words: The children standing stiffly in front of a schoolhouse conveyed the idea in purely visual terms. Shorn of their tribal dress and uncomfortably posed in western clothes, they exemplified what good Christian charity could accomplish.

And yet, it is important to resist the impulse to use Gardner's photographs as illustrations for particular ideas, based on their subject matter alone. Gardner was a complicated man, hardly a pawn of his corporate or government employers. He was a well-read socialist and an advocate for the poor and the dispossessed. He first set his sights on an American life in the late 1840s, when he joined with a group of relatives and fellow Scotsmen to plan a utopian colony somewhere in the American Midwest. They imagined the land empty and without history, theirs for the taking. Gardner, however, did not sail with the original colonists. By the time he came to the United States, in the mid-1850s, the so-called Clydesdale Colony was gone, its utopian aspirations washed away in the violent rains of an Iowa summer. The colony actually stood on land made available through the recent forced relocation of the Ho-Chunk (Winnebago) people. That was a history Gardner never reckoned with (and likely did not know). But as so often happens in American history, the dreams of one group arose from the wreckage of another's.

When Gardner finally went west in 1867 and 1868, he discovered for himself the profound disjuncture between his Scottish fantasies and an American reality. The American West was no empty place. It was filled with Native peoples and a long, deep history, evident at every turn.

The photographs Gardner made out west did not always correspond with what he saw and experienced, thought and felt. He took note of the darker underside of western life, so different from the idyllic world he once imagined. He observed the ugly anti-Native racism endemic in Kansas and the vigilante violence that left White miscreants hanging in the streets of a new railroad town. He turned down an offer to set up a studio in Hays, Kansas. He did not want to work in a "hard" place where in the mornings men could be found shot dead in the street. (Gardner, 1867: 1) Gardner's railroad photographs are no straightforward index of his personal beliefs. Think of the man, the photographs, and the interests of Gardner's corporate employer in separate buckets. What we learn from one is an unsteady guide to understanding the other. Like any photographer, Gardner worked in different registers, sometimes making photographs to please himself, sometimes making them to please others, always working with a visual vocabulary he hoped viewers would understand.

When Gardner got to Fort Laramie, on his second trip west, in late April 1868, he encountered thousands of Native people, their lodges and tipis stretched out across the plains. This was nothing like the work he had done earlier, photographing visiting Native delegations in his Washington, DC, studio. Out in the camps, he photographed people engaged in the activities of daily life, some with their hair hacked short in the traditional symbol of mourning that marked the persistent violence on the northern plains. These pictures of vernacular life were unlike anything Americans had seen. They lacked the romance of contemporary paintings and the studied formalism of studio portraits made back east. But let us not forget: Gardner's subjects had an interest in these pictures, too, and they imbued them with very different sorts of meanings.

Consider Spotted Tail, a Brulé Lakota leader well known to the peace commissioners with whom Gardner worked. He was a so-called friendly, committed to his people's values but also a realist who accepted the inevitability of federal power. Gardner photographed him repeatedly and also photographed one of his wives. Spotted Tail, it seems, was already familiar with photography, understood how to deploy photographs for his own purposes, and accepted Gardner as an ally.

Spotted Tail had been photographed before, by the ill-fated Ridgway Glover, the Philadelphia Quaker who photographed at Fort Laramie in 1866, just weeks before he himself was killed in a skirmish at a fort farther west. Glover initially encountered some reluctance on the part of Lakotas unfamiliar with the photographic process. But he patiently demonstrated how his camera worked, illustrating the principle with a small hand mirror. He also made tintypes to distribute as keepsakes, rendering what seemed an abstract process more concrete.

It is hard to know how Gardner might have fared had Spotted Tail's Brulé band not been at Fort Laramie when he first arrived. It was a stroke of almost unimaginable good luck. Spotted Tail and several other men already photographed by Glover became Gardner's first photographic subjects at this remote military fort. When they posed, they understood what they were doing. By modeling the process for others, they paved the way for Gardner's success. At Fort Laramie, a culture of making pictures and a newer culture of posing for them collided.

Gardner got something out of his pictures of Spotted Tail, but so did Spotted Tail. He and his wife posed again for Gardner in Washington in 1872. (fig. 3) In 1874, he urged some reluctant Brulé to sit for their portraits because his picture and others "had attracted so much attention in the east, that their condition interested all the land, the people of which must certainly have a limited idea of their location, and manner of living, and that it would be highly beneficial to spread true pictures of their people, as much as possible, throughout all the land." (Wilhelm, 1874: 82) In 1877, he posed for yet another photographer at Fort Robinson; later, he posed for photographers in Nebraska and in Carlisle, Pennsylvania.

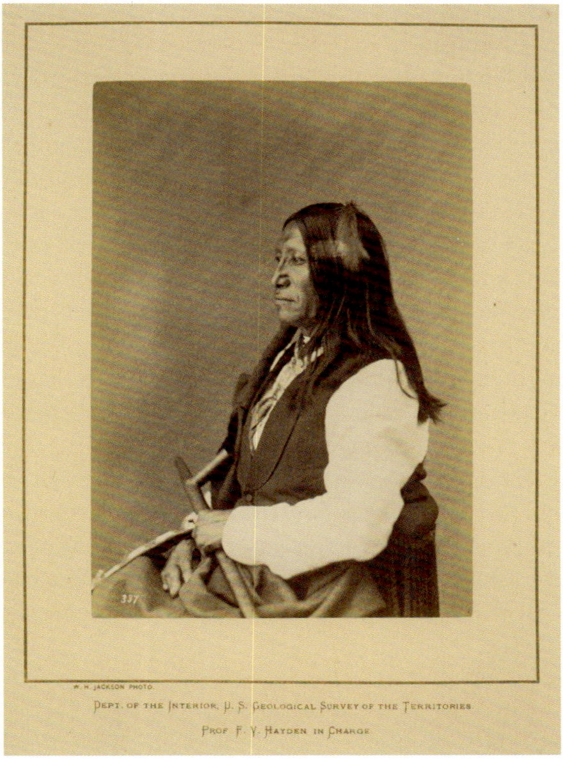

Fig. 3.
Alexander Gardner, *Spotted Tail*, 1872. Albumen silver print, 20.2×13.2 cm. Amon Carter Museum. P1970.54.10.
https://www.cartermuseum.org/collection/spotted-tail-p19705410

Some of Spotted Tail's contemporaries, likewise, expressed interest in the new technology. A group of Native people from the Upper Missouri region, photographed in Washington in 1867, requested that the federal government provide them with copies of their portraits to take home. (Statements of Disbursements, 1867: 20) The circulation of such photographs back west into Indian Country is a topic begging for more research. Photographers had ambitions for their work, but their subjects often did, too, sometimes seeking photographs for political or commercial reasons, sometimes seeking family mementos. (Sandweiss, 2002: 208–271)

When General William T. Sherman, a member of the 1868 peace commission, later thanked Gardner for a set of his Fort Laramie pictures, he wrote, they "all give details of Indian dress and physiognomy that will be valuable for all time to come." (Sherman to Gardner, 1869) He treasured them because they preserved what he thought time (and the United States military) would soon destroy. Spotted Tail understood the pictures differently, as tools of political power that could work to his own benefit. Meanings are not fixed in the visual content of the photographs; they are constituted and reconstituted by users. The pictures that for Sherman affirmed the inevitable disappearance of Native culture, for Spotted Tail affirmed cultural persistence, and for Gardner documented an experience that disrupted everything he once imagined about the far West and its peoples. That is the logical endpoint for conceding that the historical value of nineteenth-century photographs of the American West extends well beyond their apparent subject matter. The very concept of "meaning" becomes slippery.

It is too simplistic to argue that Gardner was a mouthpiece for the railroad company or the federal commission that took him West; wrong to imagine him an exemplar for the proponents of Manifest Destiny. We must dig deeper, into his patron's ambitions, his sitter's desires, even that most old-fashioned of concerns—the biographical facts of his life. In the era before digital photography, photographs almost always recorded what lay before the camera lens. But each photograph is also an act of imagination, and not just on the part of the photographer or their human subjects. Viewers—including the scholars, critics, and artists who use photographs as a subject for their own work—bring their own imaginations to photographic images. The pictures themselves might have been fixed in the chemical baths that rendered the images impervious to further exposure to light, but their meanings evolve as new viewers ask new questions.

Gardner posed as a mountain man some years before he went himself. (fig. 4) Would he have dressed this way after working in the Native camps around Fort Laramie, or after spending a night in a raw and violent Kansas railroad town? I think not. The romance was gone, despite what one might infer from some of his photographs.

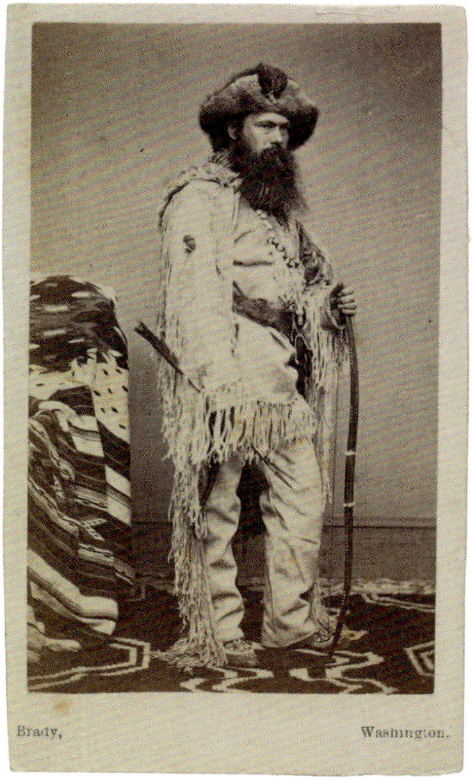

Fig. 4.

Mathew Brady Studio [Alexander Gardner], *Self-Portrait*, ca. 1861. Albumen silver print, 8.7×5.4 cm. National Portrait Gallery, Smithsonian Institution.

https://www.si.edu/object/alexander-gardner-self-portrait:npg_NPG.2002.86

Notes

1. For an example of how western photographs and prints were often used uncritically to illustrate arguments developed from literary sources, see Gabriel, 1929. For a dispiriting look at how textbooks throughout the twentieth century continued to use prints and photographs as illustrations, without interrogating them as sources themselves, see Masur, 1998.
2. A search for the term "visual culture" in the Google Books Ngram Viewer shows a dramatic spike in the use of the phrase beginning in 1990. For a general overview of the rise of cultural history during this period, see Cook, 2012, and Cook and Glickman, 2008.
3. The impact of computer cataloging and digitization on the study of historical photographs merits further research by historians and art historians. For a more general overview of the issues, see Mulligan, 2022.
4. All biographical information about Gardner referenced in this essay comes from Sandweiss, 2025. Information on the photographs of Spotted Tail comes from the same source.

Archival Collections

Robert Taft Papers, Kansas Historical Society, Topeka, KS.

W. T. Sherman to Alexander Gardner, 18 December 1869, collection of Philip D. Nathanson, Pasadena, CA.

"Statement of Disbursements for Indian Service, 1867," 1332 H. ex. doc 71 (20 December 1867), 20.

Thomas Wilhelm, *Memorandum from the Adjutant's Office, Eighth Infantry: A Diary Covering Events from Dec. 2, 1873 to October 5, 1874* (Fort Whipple, Arizona Territory: Army Press, 1874), 82, cited in Ephriam Dickson III, "Capturing the Lakota Spirit," *Nebraska History* 88, No. 1 (Spring–Summer 2007): 21.

Bibliography

Alan C. Braddock and Karl Kusserow (eds), *Nature's Nation: American Art and Environment* (New Haven, CT: Yale University Press, 2018).

François Brunet, "Robert Taft in Beaumont Newhall's Shadow: A Difficult Dialogue between Two American Histories of Photography," *Études Photographiques* 30 (2012). https://journals.openedition.org/etudesphotographiques/3481 (accessed April 28, 2026).

James W. Cook, "The Kids are All Right: On the 'Turning' of Cultural History," *American Historical Review* 117 (June 2012): 746–771.

James W. Cook and Lawrence B. Glickman, "Twelve Propositions for a History of U.S. Cultural History," in *The Cultural Turn in U.S. History: Past, Present, & Future*, eds James W. Cook, Lawrence Glickman, and Michael O'Malley (Chicago: University of Chicago Press, 2008), 3–57.

Ralph Henry Gabriel, *The Lure of the Frontier: A Story of Race Conflict* (New Haven, CT: Yale University Press, 1929).

Alexander Gardner, "Westward the Photographer Takes His Way," *Evening Star* (Washington, DC), October 28, 1867: 1.

Louis P. Masur, "'Pictures Have Now Become a Necessity': The Uses of Images in American History Textbooks," *Journal of American History* 84 (March 1998): 1409–1424.

Katherine Mintie, "Material Matters: The Transatlantic Trade in Photographic Materials during the Nineteenth Century," *Panorama* 6, no. 2 (2020). https://journalpanorama.org/article/re-reading-american-photographs/material-matters/ (accessed April 28, 2026).

Ian Mulligan, *The Transformation of Historical Research in the Digital Age* (Cambridge, MA: Cambridge University Press, 2022).

Weston J. Naef, "'New Eyes'—Luminism and Photography," in *American Light: The Luminist Movement, 1850–1875*, ed. John Wilmerding (Washington, DC: National Gallery of Art, 1980), 267–290.

Weston J. Naef, in collaboration with James N. Wood, *Era of Exploration: The Rise of Landscape Photography in the American West, 1860–1885* (New York: Albright Knox Art Gallery and the Metropolitan Museum of Art, 1975).

Beaumont Newhall, *Photography 1839–1937* (New York: Museum of Modern Art, 1937).

Beaumont Newhall, *The History of Photography from 1839 to the Present Day,* rev. and enlarged ed. (New York: Museum of Modern Art, 1964).

Beaumont Newhall, personal conversation with the author, Santa Fe, NM, 1982.

Martha A. Sandweiss, *The Girl in the Middle: A Recovered History of the American West* (Princeton: Princeton University Press, 2025).

Martha A. Sandweiss, *Print the Legend: Photography and the American West* (New Haven, CT: Yale University Press, 2002).

Robert Taft, *Photography and the American Scene: A Social History, 1839–1889* (New York: Dover Publications, 1938/1964).

CHAPTER 2
Mirrors with a Memory: War, Memory, and the Western Landscape

Binh Danh

I am an artist who reconfigures photographic techniques to explore the connections between history, identity, and place. My work is personal, rooted in the complex circumstances of my family's displacement and migration, shaped by war, and intertwined with foreign and familiar landscapes.

My journey begins in Vietnam, a country forever marked by the lasting impact of conflict and colonial history. I was born in Vietnam in 1977, after the Vietnam War had officially ended. (fig. 1) In late 1979, amid growing tensions, my family fled Vietnam as part of the wave of refugees later known as "boat people." Our journey eventually brought us to a refugee camp in Malaysia, where we lived for nearly a year before being granted asylum in the United States. My father, a technician who had worked alongside American forces in South Vietnam during the war, possessed valuable electronics skills that eased our transition into a new life. In America, we settled in Silicon Valley, California—a region undergoing its transformation at the time. My father opened a small television repair shop, where I grew up surrounded by flickering screens and humming electronics. My mother worked night shifts at a semiconductor plant, one of thousands of Southeast Asian women on high-tech assembly lines in the 1980s. On weekends, she joined the shop, where I helped with chores and tested repaired TVs—often by watching Vietnam War films for hours to make sure the sets worked.

Reflecting on the past, I now understand that I was searching for my reflection in those films—an image of my people and homeland. Yet it was nowhere to be found. The Vietnam War, in Hollywood's grasp, transformed into a theater of American trauma, where the

Vietnamese were either invisible or reduced to fleeting, voiceless figures. As I matured, I realized that, just as Westerns TV shows as a genre had dwindled in popularity by the 1970s, Vietnam War films had replaced the old frontier West with tropical jungles. American soldiers on screen now navigated rice paddies and tunnel networks instead of desert plains, battling an enemy as abstracted and othered as any outlaw or "savage" from earlier Westerns.

This realization led me in my early twenties to return to Vietnam. I was determined to witness the war's aftermath and understand how my family's past could shape my identity as an artist. During this period, I began building a photographic archive by collecting photographs related to the Vietnam War. The advent of the internet and eBay enabled me to access a wealth of material culture previously out of reach. This practice soon became central to my work, as I used these artifacts to explore the intersections of personal and collective memory.

One of the most ambitious expressions of this exploration emerged through my chlorophyll print series. In these works, I utilized the process of photosynthesis to embed historical images directly onto the surfaces of leaves. I contemplated what the trees and plants, silent witnesses to conflict, might recall if they could speak. This work draws upon the concept of "elemental transmigration," the idea that matter, in its constant state of decomposition and re-composition, carries memories of past events. The landscape itself, scarred by bomb craters and defoliation, bore witness to atrocities, retaining traces of trauma within its very soil and foliage. (fig. 2)

At the heart of my practice lies a love for photographic processes, learned during my teenage years in a high school darkroom. In particular, the daguerreotype has become a crucial medium for me. Created by exposing a silver plate inside a camera, daguerreotypes are unique, unreproducible objects. What fascinates me most is their mirrored surface: When viewers gaze at a daguerreotype, they see both the photographic image and their reflection. In the nineteenth century, daguerreotypes were called "mirrors with a memory." The concept is remarkably resonant in a contemporary context.

In recent years, my daguerreotypes have captured sites across the American West, places rich in histories of violence, displacement, and transformation. For example, my photograph of Shoshone Falls deliberately echoes Timothy H. O'Sullivan's 1868 image of the same site. O'Sullivan's work, created as part of the first federally funded survey of the American West after the Civil War, reflects an earlier wave of expansion and conquest. (fig. 3) Today, as I photograph these locations, I consider both their historical uses and their current realities—towns once driven by gold mining are now engaged in a "green rush" of cannabis cultivation, rivers still carrying the toxic legacy of hydraulic mining. (figs 4 and 5)

As a naturalized citizen living through the tumultuous years of the Trump administration, I have watched with alarm as environmental protections have been dismantled, public lands threatened, and historical amnesia encouraged. (figs 6 and 7) Places such as the Trinity Site in New Mexico, where the first atomic bomb was tested, now attract new waves of visitors in the wake of the 2023 movie *Oppenheimer*. These sites remain potent reminders of humanity's capacity for destruction. (fig. 8)

Ultimately, my project Daguerreotyping the Western Landscape seeks to reinsert marginalized histories and displaced narratives into the broader tapestry of American storytelling. I was drawn to these vantage points through the historical traces left by daguerreotypes created during the era of Manifest Destiny—many commissioned to depict the so-called barren landscapes, promoting their resources and justifying territorial claims for the US government and private interests. Today, many of these daguerreotypes exist only as drawings and engravings, such as those referenced in Martha A. Sandweiss's book *Print the Legend*. (2002) By returning to these sites with the daguerreotype, I invite viewers to reconsider their relationship with history, land, and nationhood. (fig. 9)

In this way, my work asserts that stories like mine—of war, migration, survival, and reclamation—are not peripheral but central to the American experience. From my birthplace in Vietnam to the valleys of Yosemite, my photographs become acts of remembering, forming a visual archive of landscapes marked by conflict and renewed through resilience. Though the Vietnam War remains ever-present in my consciousness, I offer spaces for reflection, belonging, and hope through these silvered plates, leaves, and images. (fig. 10)

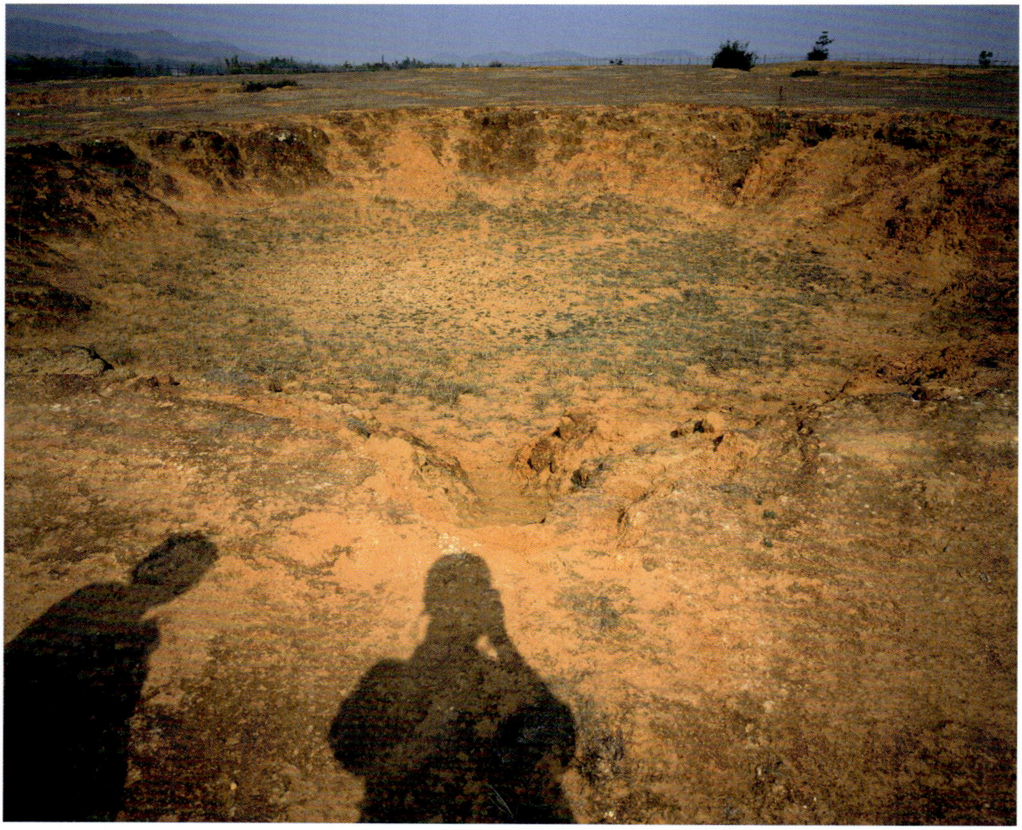

Fig. 1.

Binh Danh, *Bomb Crater of Phonsavan, Laos*, 2009. Archival pigmented print, 16 × 20 in.
Part of the series *Collecting Memories*.

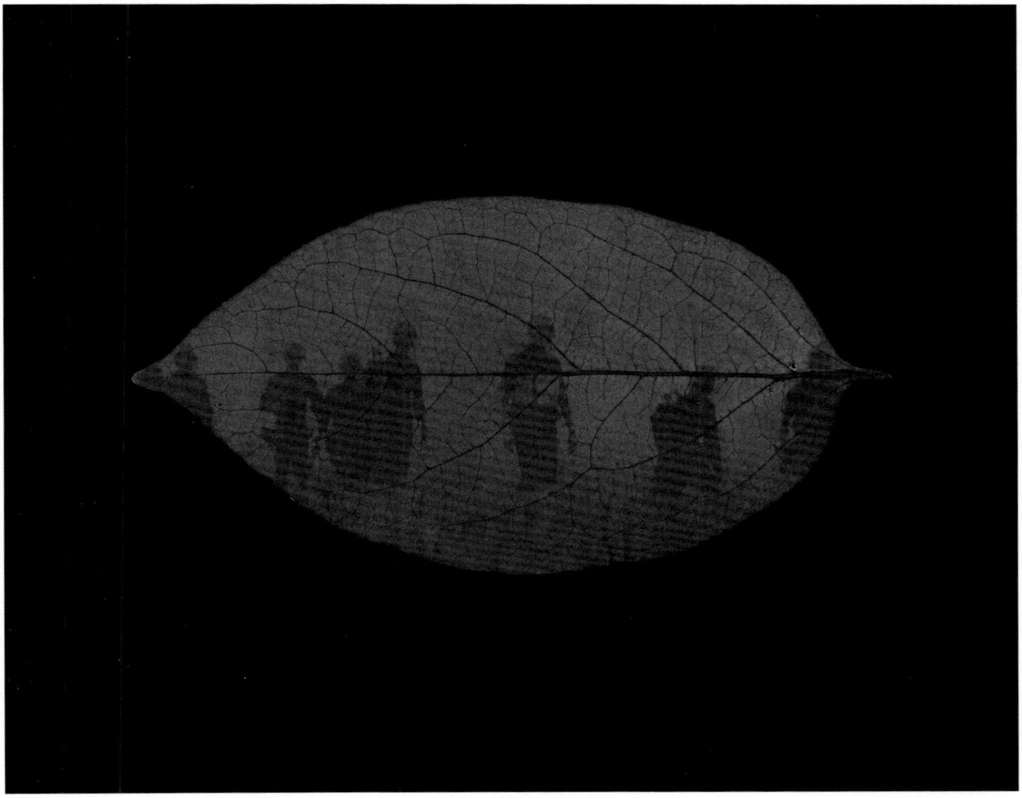

Fig. 2.

Binh Danh, *Drifting Souls*, 2002. Chlorophyll print and resin frame, 10 × 13 × 1 1/2 in.
Part of the series *Immortality: The Remnants of the Vietnam and American War.*

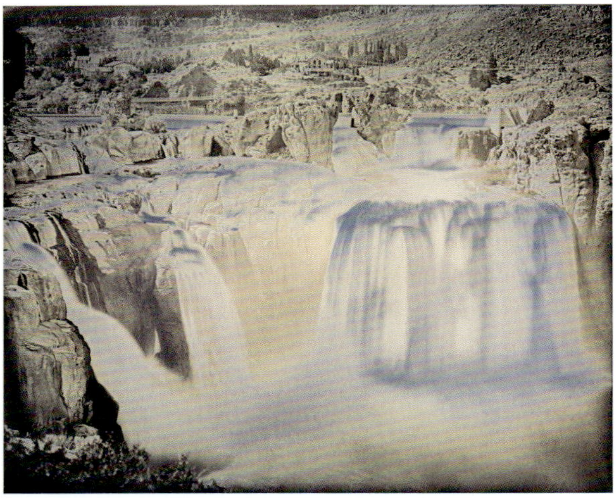

Fig. 3.

Binh Danh, *Shoshone Falls Park, Twin Falls, Idaho,* 2016. Daguerreotype plate, 8×10 in. / frame 13 1/8×14 3/4×1 1/2 in.

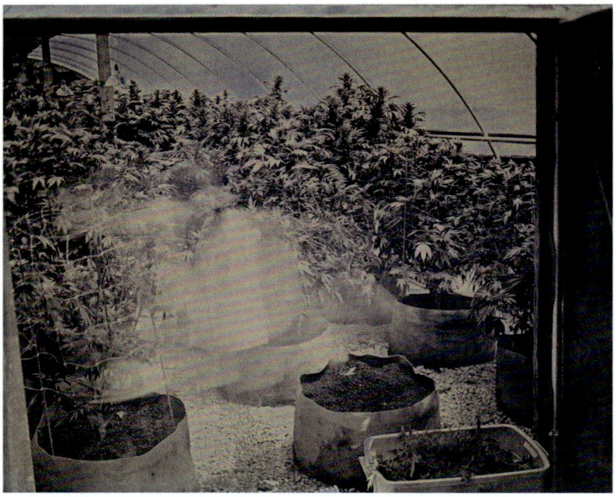

Fig. 4.

Binh Danh, *Cannabis Farm, Nevada City, CA,* 2019. Daguerreotype plate, 8×10 in. / frame 13 1/8×14 3/4×1 1/2 in.

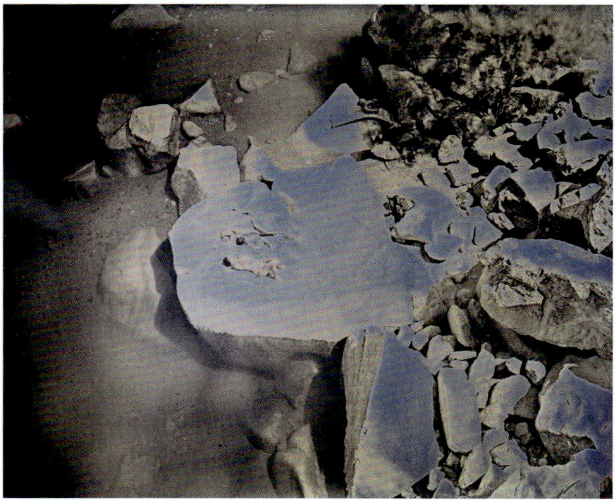

Fig. 5.

Binh Danh, *South Yuba River, Nevada City, CA,* 2019. Daguerreotype plate, 8×10 in. / frame 13 1/8×14 3/4×1 1/2 in.

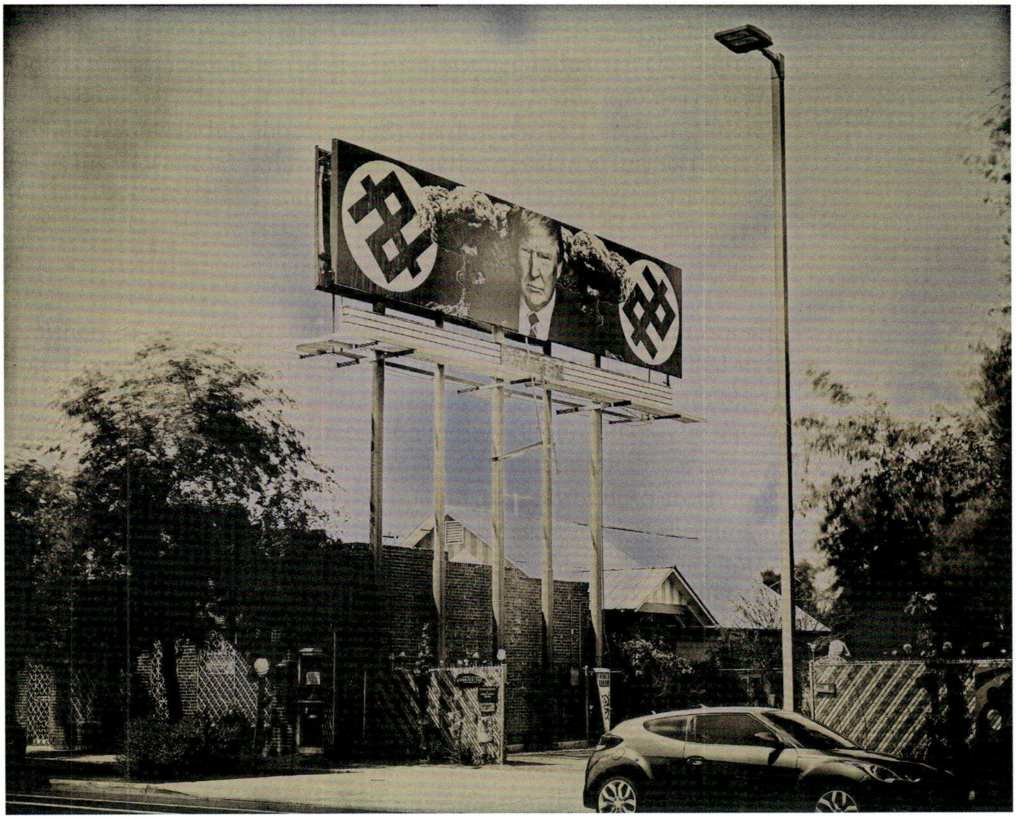

Fig. 6.

Binh Danh, *Trumpocalypse, Billboard by artist Karen Fiorito,* 2017.
Daguerreotype plate, 8 × 10 in. / frame 13 1/8 × 14 3/4 × 1 1/2 in.

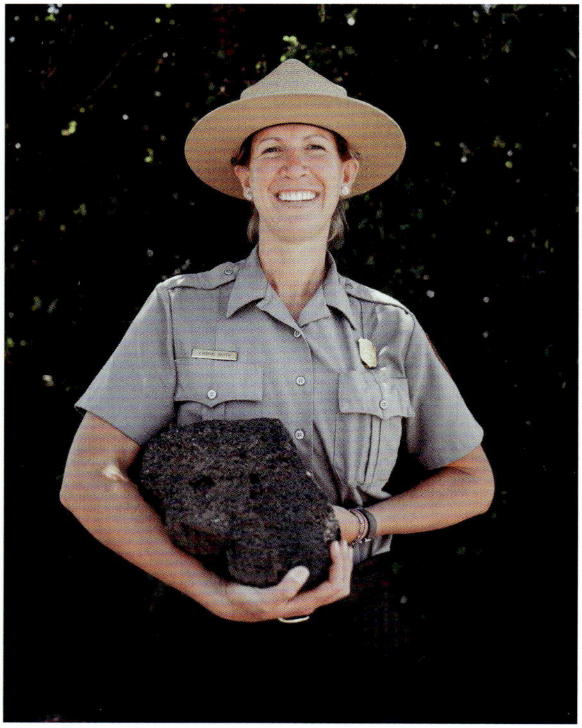

Fig. 7.

Binh Danh, *Park Ranger Lindsay Smith*, 2009. Archival pigment print, 40×30 in. From the series *Crater of the Moon.*

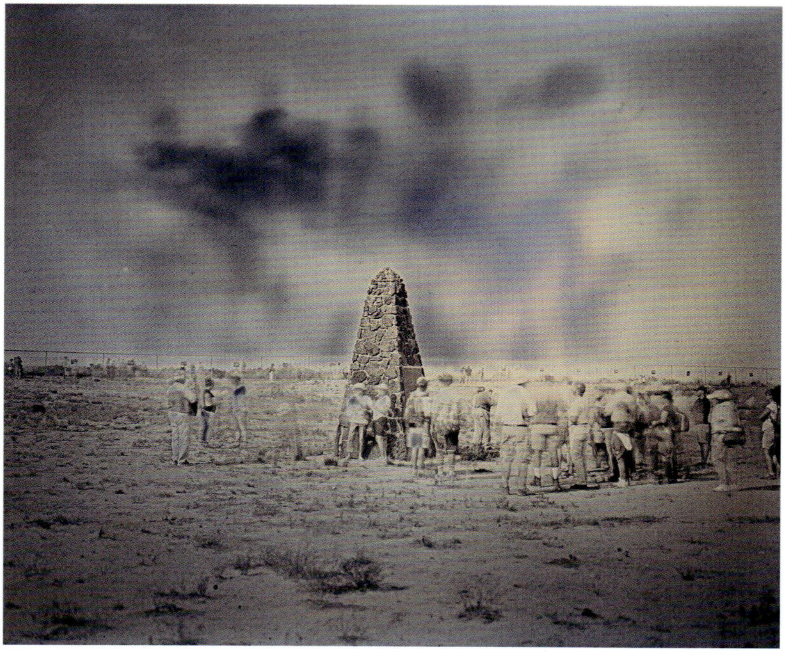

Fig. 8.

Binh Danh, *Trinity Site, White Sands National Monument,* 2016. Daguerreotype plate, 8 × 10 in. / frame 13 1/8 × 14 3/4 × 1 1/2 in.

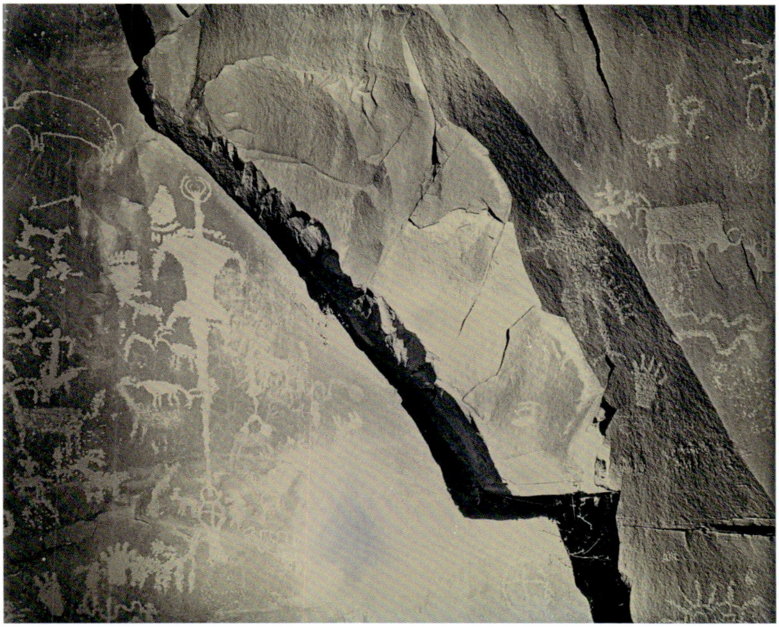

Fig. 9.

Binh Danh, *Newspaper Rock, Utah,* 2018. Daguerreotype plate, 8×10 in. / frame 13 1/8 × 14 3/4 × 1 1/2 in.

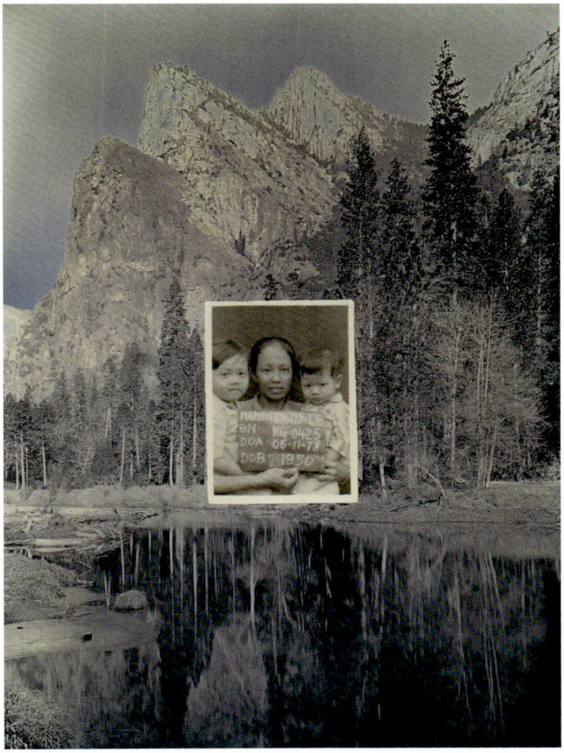

Fig. 10.

Binh Danh, *Two Brothers and a Mom (After Three Brothers)*, 2021. Archival pigment print, 12.5 × 14.75 in.

CHAPTER 3
(Dis)Placing Edward Weston: The Politics of Shutter Release

Monica Bravo

"Since then, I have never been able to decide whether Weston tried to kill me, or saved my life. For reasons more abstract, I suspect that many photographers, over the past thirty or forty years, have felt the same way." (Frampton, 1978: 48) Hollis Frampton, the avant-garde filmmaker, photographer, and theorist, wrote these words in 1978. He was reflecting on an experience nearly two decades prior, when, surprised by the large scale and, one might surmise, the impressive quality of a major retrospective of Edward Weston's photographs in the Midwest, he delayed his return flight to New York to experience the exhibit fully.

The flight for which Frampton was originally scheduled crashed in midair. The lone survivor of the 128 passengers and crew aboard the two planes was a young boy, Stephen Baltz, who fell 5,000 feet into a Staten Island snowbank. Frampton's memory is selective, sometimes faulty: The touring show *The World of Edward Weston* featured some 100, not 400, prints; the eleven-, not ten-year-old, boy died the next day, from fume inhalation; and I have been unable to locate the press photograph he vividly describes, of "the broken child [...] in front of an *Iglesia Pentecostal* called Pillar of Fire." (Frampton, 1978: 48; Swopes, 2018) (fig. 1) Nevertheless, the recollection *feels* true, as revelatory as the moments when Frampton must have pieced together the strange circumstances of his survival. He indirectly states that Weston's photographs were for him—as for other photographers, albeit likely "for reasons more abstract"—a matter of life and death. (Frampton, 1978: 48)

After this cold open, Frampton's essay "Impromptus on Edward Weston: Everything in Its Place" takes a decidedly irreverent turn. Weston's dogma of pre-visualization—in which the final image is imagined by the photographer before its "taking," thereby necessitating little darkroom manipulation—is presented as evidence of the artist's

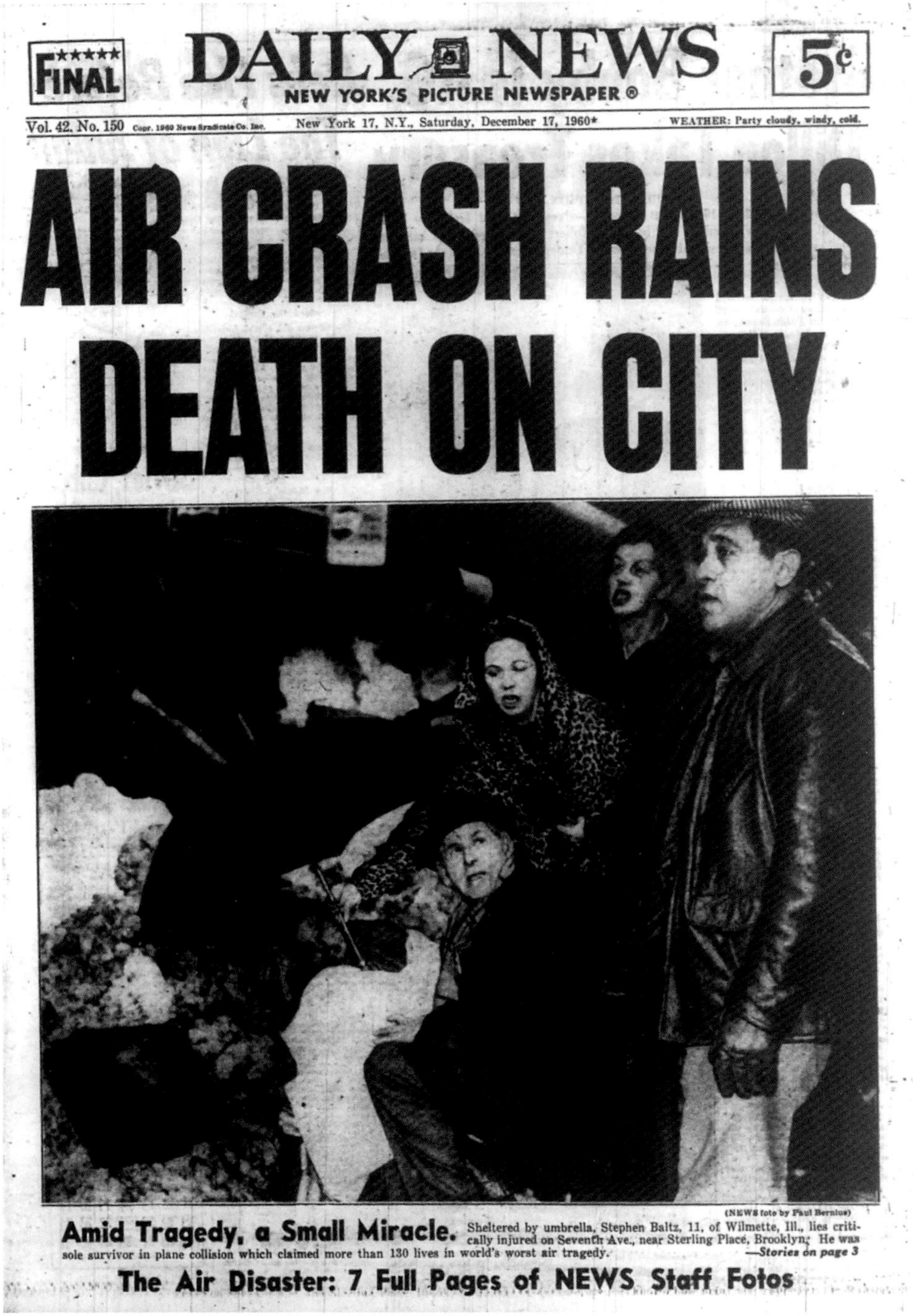

FINAL ★★★★★

DAILY NEWS

NEW YORK'S PICTURE NEWSPAPER ®

5¢

Vol. 42. No. 150 Copr. 1960 News Syndicate Co. Inc. New York 17, N.Y., Saturday, December 17, 1960★ WEATHER: Party cloudy, windy, cold.

AIR CRASH RAINS DEATH ON CITY

(NEWS foto by Paul Bernius)

Amid Tragedy, a Small Miracle. Sheltered by umbrella, Stephen Baltz, 11, of Wilmette, Ill., lies critically injured on Seventh Ave., near Sterling Place, Brooklyn. He was sole survivor in plane collision which claimed more than 130 lives in world's worst air tragedy. *—Stories on page 3*

The Air Disaster: 7 Full Pages of NEWS Staff Fotos

Fig. 1.

Front page of *Daily News: New York's Picture Newspaper,* December 17, 1960.

Modernist elevation to canonical sainthood. But the essay is no more "about Weston" than it is about the plane crash that haunted Frampton. The essay makes a largely structuralist argument about photography and its medium specificity. Yet it also seizes upon truths drawn mostly from Weston's writing, diagnosing the artist's attitude toward the medium above all as a logic of substitution. Frampton writes:

> The photographic image, for Weston, affirms the existence and enforces the persistence of its immediate pretextual object and thereby of its grand pretext, namely, the space in which that object subsists. The artist reaffirms his own existence through gradually replacing the space of the given world with the inventory of spaces of all the photographs he has made. It may be that Weston's refusal to emancipate his images from the patriarchal house of his own perceptions amounts to nothing more than the simple declaration of a territorial claim. (Frampton, 1978: 58)

The assertion—crudely put—is that, for Weston, the artist inheres within his photographs; through them, he claims space.

This essay is a riposte to Frampton's, though I respect his essay deeply and, in some respects, take it as my model. I do not disagree that Weston is prominently situated in the Modernist canon (a project to which my first monograph contributed) and that some of Weston's assertions may look (justly) ridiculous in this light of day. (Bravo, 2021) But here I want to attempt to decenter Weston by, like Frampton, writing an essay *around* but not *about* him and his photographs. Instead, I am interested in an ecology that coalesces around his (absent) presence. If, as Frampton indicates, Weston's perception inheres in his photographs as a (metaphorical) "territorial claim," I take seriously that claim to life and land. Here, I am taking a networked approach to the history of photography, investigating a practice profoundly rooted in the American West and specifically Northern California, though with tentacles reaching out among Weston's successors over several decades to New York and the midwestern city where Frampton's life was "saved."

Recent scholarship in the history of photography has displaced attention from the photographic object to the photographic event. (Azoulay, 2008) It acknowledges the multiple agents—human and non-human—involved both in the photographic encounter and in the creation of photographic meaning. Of the million micro-events within any photographic encounter, the triggering of the shutter must be the most consequential, marking the dramatic before and after of the action of "writing with light." The camera shutter is a window that allows variable durations of light to expose the photo-sensitive surface. Media theorist Ariella Aïsha Azoulay refers to the camera shutter as drawing at one and the same time three dividing lines: in time, space, and social relationships (the body politic). Co-extensive with a longer history of imperial violence, she refers to this action as the "imperial shutter," implicating larger social structures in the camera's material operations. (Azoulay, 2019: 5–10) At almost the opposite extreme, in terms of locating the shutter's agency exclusively with the photographer, is Modernist photographer Henri Cartier-Bresson's notion of "the decisive moment." Yet even this concept was mistranslated from *Images à la Sauvette* in 1952, leading to a fetishization of the fraction of a second, and clear—almost clairvoyant—seeing in photography.

Between these two poles of thinking on photographic exposure lies my attention to the *politics* of shutter release, examined through three vignettes surrounding Weston, revealing the action as ascribable neither to "imperial violence" nor to the photographer alone. Instead, I ask: What happens in the split second before a photograph has been recorded? Who gets to wield the shutter release? And how has variable access to or even appropriation of the shutter demonstrated broader sociopolitical dynamics in the American West? This geography is not incidental: In Weston's age, Northern California was the center of naturalist movements as much as a racist internment campaign (Macfarlane, 2025); by the 1970s, Southern California and particularly the California Institute of the Arts (CalArts) became a hub for feminist and Conceptual artistic practice. Today, the region continues to cultivate contemporary art and artists grappling with the "frontier" legacy of their photographic predecessors.

Frampton's essay demonstrates Weston's use of the shutter to be almost beside the point, but also paramount: Pre-visualization was everything, but so, too, was capturing that pre-visualized image with the camera and then (re)producing it in the darkroom. Weston is one of American photography's first auteurs, yet the politics of authoring a photograph are more complicated for those in his orbit. Weston's friend, the established studio photographer Toyo Miyatake, was interned at Manzanar under Executive Order 9066, which mandated that only a white camp official could trigger the shutter. The Pictures-generation Conceptual artist Sherrie Levine made a series *After Edward Weston,* rephotographing Weston's published images. And contemporary Latinx photographer Tarrah Krajnak's *Master Rituals II: Weston's Nudes* feature the photographer posing in attitudes once embodied by Weston's models, shutter release in hand. Interrogating Kodak's (in)famous 1888 promotional slogan, "You press the button, we do the rest," I argue that this gesture carries far greater importance than the brand's advertising campaign flippantly forsakes. This chapter traces the politics of exposure over several decades by circling out from an avowedly apolitical photographer, Weston. In so doing, I effectively (dis)place Weston into a new network, and hope to offer a more rigorous way of thinking about the politics of shutter release—beyond the totalizing lens of "the imperial shutter" or the superficial popular assimilation of "the decisive moment"—within the photographic event.

Toyo Miyatake

Two months after the Japanese bombing of Pearl Harbor, President Franklin D. Roosevelt signed Executive Order 9066, ordering all Japanese Americans to evacuate from the West Coast. In one of the grossest violations of civil liberties in US history, the order's execution resulted in the relocation of approximately 120,000 people, many of whom were US citizens, to one of ten internment camps located across the American West and Arkansas. The last of these closed in March 1946. Within these camps, special privileges were granted to *Nisei* (US-born, and thus often younger) citizens, while older, immigrant *Issei* were disadvantaged, upsetting traditional social norms. Among these differential opportunities was the right to make photographs and, more specifically, to release the shutter. At Manzanar for a time, even this privilege was further limited to "appointed personnel:" white guards. (Alinder, 2009: 81)

The established portrait photographer Toyo Miyatake, an Issei man, was among those interned at Manzanar, perhaps the most notorious Japanese internment camp in California. He smuggled in a lens and film plate holder, so that he might construct a camera inside. He eventually fashioned "a camera body from scraps of wood with the help of a carpenter." (Alinder, 2009: 80) Prior to this, he had run a successful portrait studio in the Little Tokyo neighborhood of Los Angeles since 1923. It was there that he met Weston, who frequented the bohemian enclave and held great affection for Japanese visual aesthetics during his Pictorialist phase. Miyatake was a commercial artist but also submitted art photographs to national and international salons, including the prestigious London Salon of Photography, where Weston showed. (Reed, 1986: 33, 35) It was Miyatake who first encouraged Weston to exhibit his photographs in Little Tokyo, in 1923, where he even sold a few prints—something Weston remarked upon in letters to his lover, Tina Modotti. (Alinder, 2009: 78, 81)

Owning a camera inside Manzanar was illegal. But camp administrators, especially its director, Ralph Merritt, recognized that effective "camp management" entailed certain concessions. Following a December 1942 strike that resulted in several deaths, the *Manzanar Free Press* announced the "approval for an establishment of a photo studio." (Alinder, 2009: 81) Japanese internees ran the studio, but only "appointed personnel" could snap the shutter—only they could "take" the picture—and, furthermore, they removed and secured the camera lens at the end of the workday. They thereby ensured that no unauthorized photographs would be taken of the camp and its inhabitants. In fact, the studio, operated by Miyatake, closed for a period of several months that spring when the staff could not identify a white person authorized to click the shutter. Merritt's response was to hire camp administrators' wives as chaperones, positioning their authority over photographic expertise. As art historian Jasmine Alinder writes, "the site of power resided not in the positioning of the subject but in the release of the shutter. The government's regulation over who clicked the shutter was an exertion of power over the right to self-representation." (Alinder, 2009: 81) Here is something at play more powerful and literal even than "the imperial shutter" that Azoulay theorizes.

The studio opened inside the camp on April 12, 1943. Its offerings were conventional: wedding, funeral, graduation, club, school group, and family photos, as well as individual portraits. (fig. 2) These were photographs made under conditions of unfreedom, yet that do not depict the deprivations that internees endured. They were, nonetheless, of unfathomable importance to their subjects. They marked important rites of passage, and visages before they changed, cementing familial bonds and those of friendship. Indeed, scholars of family photography note that a photograph commemorating such rituals is as much the event as the putative source of the gathering itself. (Zuromskis, 2003; Rose, 2010) Surely, preparing for and participating in the making of the photograph must have been of equal importance to the resulting photographic image: How ugly it was that for that event, characterized by an easy rapport established between the subjects and the photographer, Miyatake, whom they trusted, to be suspended or ruptured by the white chaperone at a key moment in the sitters' lives.

Eventually, Miyatake was granted permission to photograph without supervision, and outside of the studio. He could move "freely" about the camp and make photographs using his large view camera and tripod, with fresh supplies brought in from the outside. He believed that Weston intervened on his behalf with the camp's director, Merritt, an old friend of

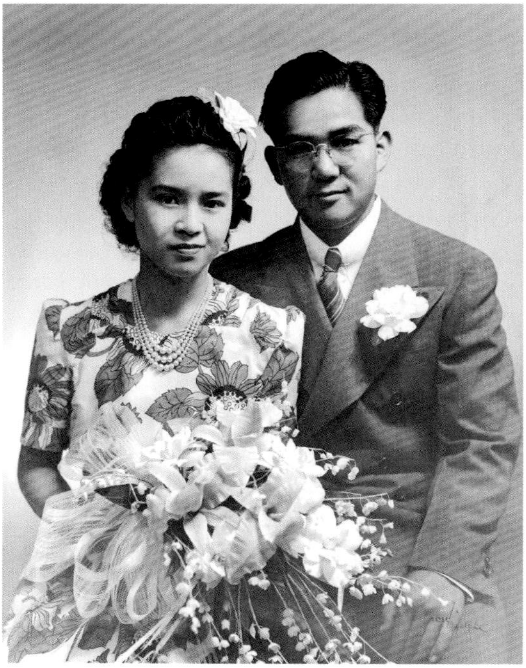

Fig. 2.
Toyo Miyatake, *Kenji and Hideko Yamamoto,* 1944. Courtesy of Toyo Miyatake Studio.
© Toyo Miyatake Studio.

Weston's, to gain Miyatake the privilege. Or at least, this is the conviction he retrospectively recounted to the contemporary artist Eikoh Hosoe. (Alinder, 2009: 86) Perversely, within the political economy of Northern California, it seems to have been Miyatake's pre-existing relationship with a particular white man and thus his own social currency that permitted him to regain full access to his medium. Weston did not save Miyatake's life, but perhaps he made his internment more bearable, and paved the way for Miyatake's production of new photographs: ones in which he pressed the button.

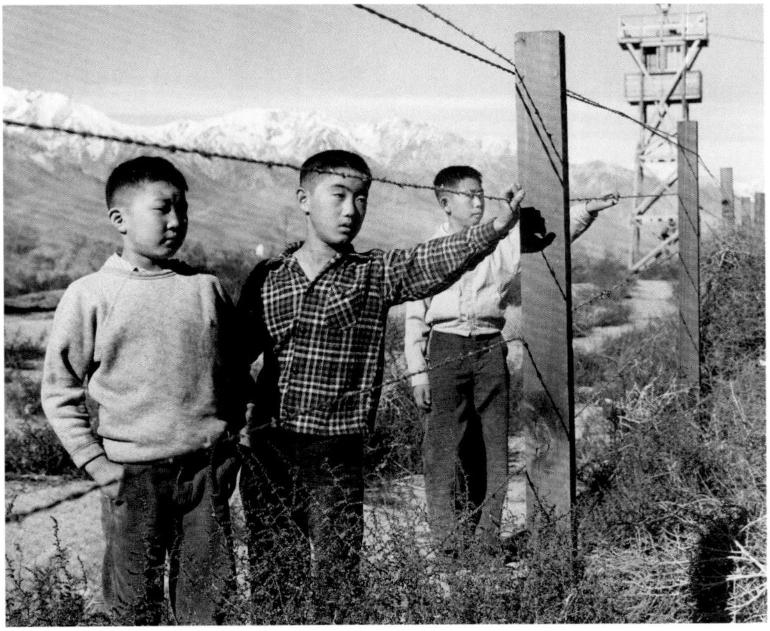

Fig. 3.
Toyo Miyatake, *Three Boys Behind Barbed Wire,* 1944. Courtesy of Toyo Miyatake Studio.

Alinder points out that one of these photographs, *Three Boys Behind Barbed Wire* (1943), has likely become the most reproduced photograph of Japanese internment, coming to stand in for the injustice as a whole. (fig. 3) Three boys stand wistfully, arrayed in a line along the fence; two of them even grasp the barbed wire awkwardly. They are enclosed by the humanmade border before them and the snowy peaks of the Sierra Nevada at their backs. Miyatake may have stood just outside of the fence to make the photograph. His proximity to the watch tower and the large size of his equipment would have made him easily recognizable to the guards, however, and not considered a flight risk. (Alinder, 2009: 92)

Through the dissemination of *Three Boys Behind Barbed Wire,* the photograph is sometimes misattributed or presented for variably aestheticizing and documentary purposes. These efforts sometimes seem

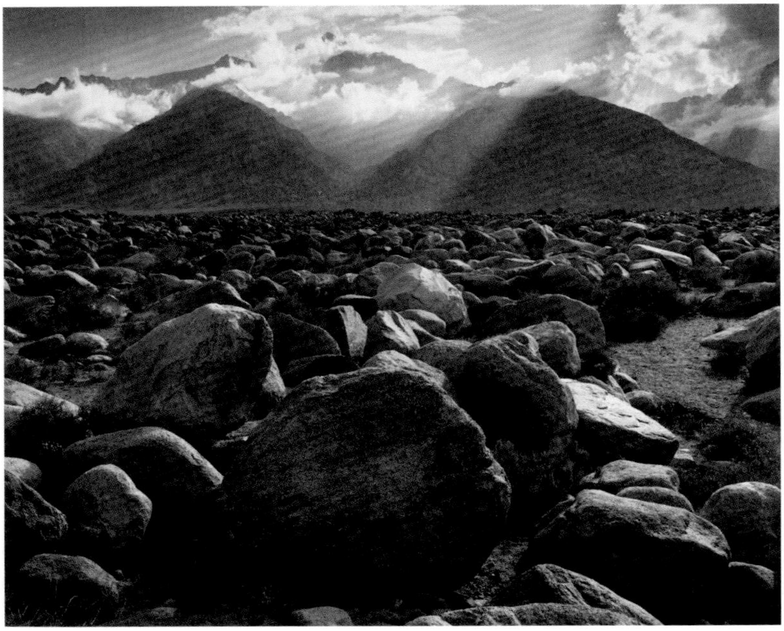

Fig. 4.

Ansel Adams, *Mount Williamson, Sierra Nevada, from Manzanar, California,*1944. Gelatin silver print, 15 1/8 × 18 ¾ in (38.4 × 47.6 cm). New York, Museum of Modern Art. © The Ansel Adams Publishing Rights Trust.

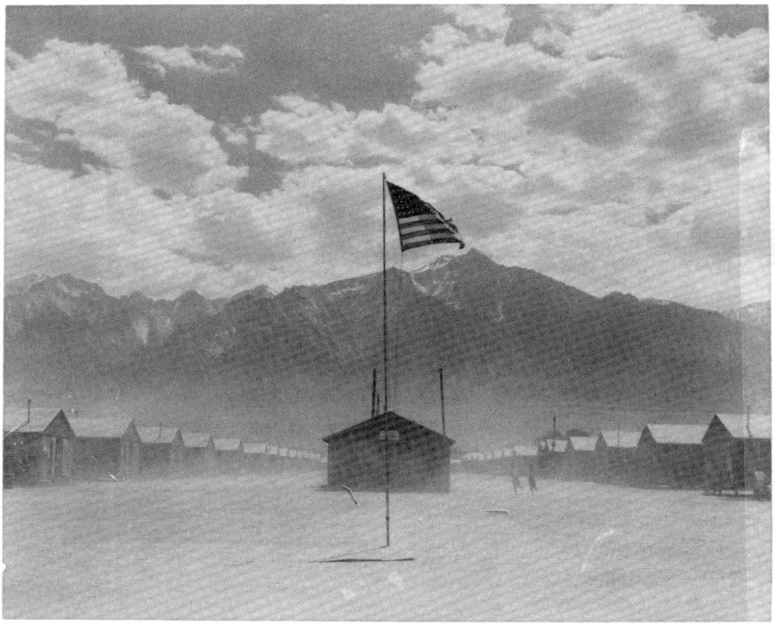

Fig. 5.

Dorothea Lange, *Dust Storm at Manzanar War Relocation Authority Center,* July 3, 1942. Washington, DC, US National Archives and Records Administration.

designed to make Miyatake's photographs conform to those of two better-known white photographers' depictions of Manzanar, often presented in terms of this binary: Ansel Adams's apparently apolitical landscapes published in his 1944 photobook *Born Free and Equal,* sometimes coupled with "patriotic" captions, and Dorothea Lange's bleak, government-sponsored documents. (Alinder, 2009: 88–91) (figs 4 and 5) Such caricatures do not justifiably characterize any of their holistic practices in Manzanar, which collectively picture an arid American West largely inhospitable to people.[1] But I would argue that the crucial difference in Miyatake's photographs is not just that he was an imprisoned inhabitant of Manzanar, and not just that he was in proximity to the events that transpired there. More fundamentally, unlike Adams and Lange, he had seen his access to and experience of photography impugned.

Such are the multiple overlapping ecologies—a term whose etymological roots refer to the study of home—that Miyatake and his photographs inhabited at Manzanar. He was one among several camera operators and, indeed, specialist photographers at Manzanar. Miyatake and Merritt were bound together not only as prisoner to warden, but as mutual friends of Weston. And Miyatake's lived and professional experience transcended the period of his internment. So, too, do his photographs. But the period of his incarceration, during part of which he was literally not free to represent his own community, mark a rupture between before and after, like the release of a shutter: perhaps more potent, more occlusive, but also revelatory in its own right.

Sherrie Levine

Miyatake was Weston's contemporary. Forty years later, Modernist photography was on the wane. In the Postmodern condition, with originality seemingly exhausted as a strategy within fine art practice, members of the Pictures generation, including Sherrie Levine, used a tactic of rephotography. The term typically refers to a return to the site where a photograph was first made, especially a given nineteenth-century landscape, and especially in the American West, as popularized by the practices of the Rephotographic Survey Project of the late 1970s. (Klett et al., 1984) More jaded in their efforts, artists of the Pictures generation, then all New York-based, took quite different sites—especially sites of publication and mediated memory—and much more recent photographic practices as their quarry. Richard Prince rephotographed Marlboro cigarette advertisements from magazines with strategic omissions, Cindy Sherman produced *Untitled Film Stills* (1977–1980) that refer to female archetypes, and Levine went straight to the source: published reproductions of Modernist photographic "masterpieces."

At once critiquing the recent commodification of photography against the backdrop of the 1970s market photo boom—especially in appropriating Walker Evans's now-valuable Depression-Era photographs of extreme poverty—and an institutional fetishization of her photographic ancestors, Levine completely evacuated the notion that her creative labor resided in the camera or the hand. (fig. 6) In her "after" series, the release of the shutter is not so much depoliticized as demonstrated, in its physicality, to be a near-impotent gesture, non-conforming to the models of analysis offered by Azoulay and Cartier-Bresson. Rather, like that of other Conceptual artists, Levine's work is presented as a purely intellectual exercise: of knowing the photographic canon and its history, making a meaningful selection, and then presenting the rephotographed image as her own.

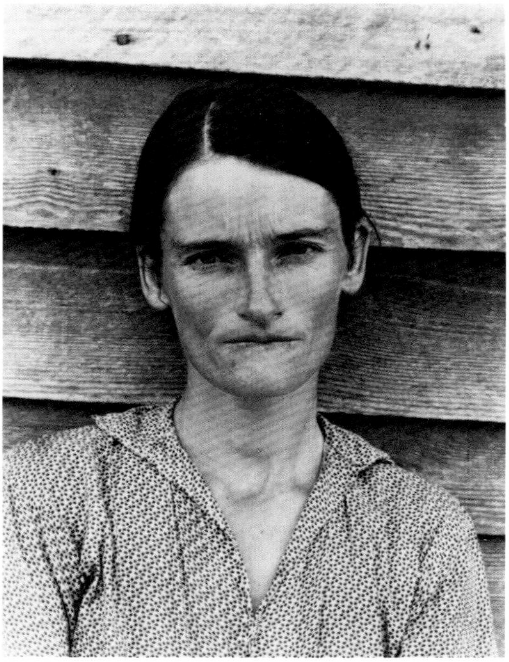

Fig. 6.
Sherrie Levine, *After Walker Evans: 4,* 1981. Gelatin silver print, 5 1/16 × 3 7/8 in (12.8 × 9.8 cm). The Metropolitan Museum of Art, New York. Gift of the Artist, 1995. © Walker Evans Archive, Metropolitan Museum of Art.

Levine's appropriation of Weston's *Torso of Neil* (1925) is especially intriguing as a meditation on creation and reproduction. (fig. 7) Weston was particularly proud of the small series he made of his second-to-youngest son's body, his offspring twice over. These photographs stand apart from the female nudes and landscapes for which Weston is better known. The model's poses are gracious; the body is delicately modeled, offset by the black background. Some of this series Weston printed in fine palladium to further draw out the tonal range of the image. Even before he printed the negative, he wrote in his daybook—a sort of diary he kept with an eye to posterity—that Neil "afforded me a visual beauty which I recorded in a series of Graflex negatives of considerable value." (Weston, 1962: 118–119) In that word "value" I hear simultaneously a reference to quality and to the photograph's potential translation into currency. In a later entry he remarked: "from the new negatives, the print on white stock which most pleases me, is a nude, a torso of Neil. It is a simple, artless conception and one of my best." (Weston, 1962: 143) For Weston and for its viewers, *Torso of Neil* resonated with the marble sculptures of Greco-Roman antiquity, aligning the photographer with great artists of the past—invoking the same logic of substitution that Frampton identified in Weston's practice. For Levine and her contemporaries in the 1970s, the photograph resonated anew during the photo boom, with its attendant desire for a quintessentially American photographic canon.

Aperture published the first volume of Weston's *Daybooks*—where these comments were made—in 1961, and Levine surely knew it. Aware of its associations, Levine photographed not an original print of Weston's "masterpiece," but its reproduced image in an exhibition catalog. Her decision to trigger the shutter before a distanced copy of the photograph extends the chain of reproduction, even as it snubs the original's value of what cultural theorist Walter Benjamin called aura, "a strange tissue of space and time." (Benjamin, 2008: 23) Yet, in a queer way, Levine has hewed to Weston's technique of pre-visualization in producing her series *After Weston.* She knew exactly what she would get from her printing session, once her exposure was made.

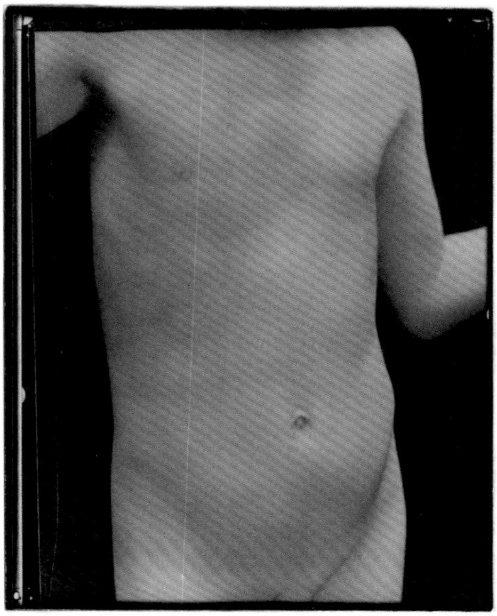

Fig. 7.
Edward Weston, *Torso of Neil,* 1925. Palladium print, 9 3/16×5 9/16 in (23.3×14.2 cm). Museum of Modern Art, New York. © Center for Creative Photography, Arizona Board of Regents / Artists Rights Society (ARS).

Levine's work raises, in a profound way, the question of labor's relationship to authorship. The same era that gave rise to Conceptual art practice gave us post-structuralism as well, including theorist Roland Barthes's "The Death of the Author." (Barthes, 1968) Yet the assignment and understanding of creative labor is important in itself—and not just as a means to investigation of authorship. In Levine's practice, it is the naming of her series, *After Evans, After Weston, After Rochenko,* etc., that designates *her* original production, not the release of a shutter, which could occur at any arbitrary moment in time *after* the artist made the source photograph. The titling further refers to an art-historical convention for naming less-skilled, otherwise anonymous artists who were followers of acknowledged masters (after Titian, after Botticelli, etc.). Levine knowingly claimed that positionality for herself, thereby short-circuiting any potential critic's journey to the punchline.

Tarrah Krajnak

If one way to understand the history of art photography is as a long unfolding, of one artist responding to another with their practice, then for a contemporary generation of artists, Weston has become a fruitful foil. In her series titled *Master Rituals II: Weston's Nudes,* Lima, Peru-born and now Los Angeles–based artist Tarrah Krajnak overtly responds to Weston's nudes with her own body. Striking poses once embodied by his well-documented models—including dancers and Weston's lovers—she

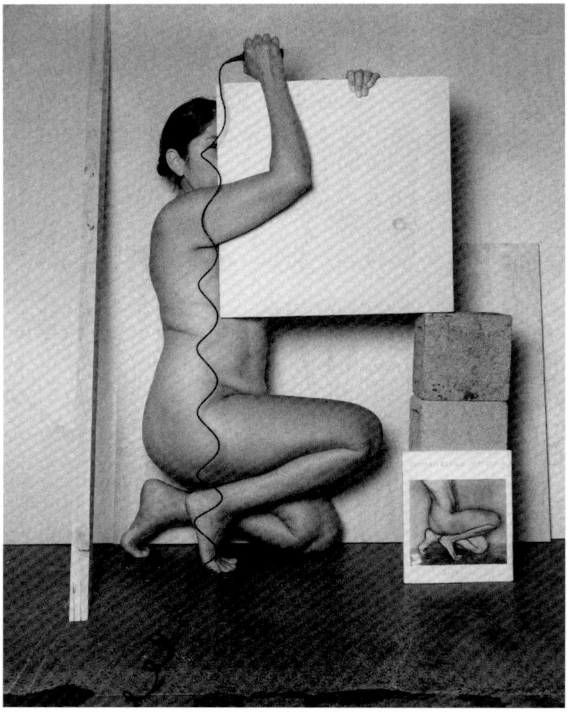

Fig. 8.

Tarrah Krajnak, *Self-Portrait as Weston/as Bertha Wardell, 1927/2020*, 2020. From the series *Master Rituals II: Weston's Nudes*.
© Tarrah Krajnak, courtesy of Zander Galerie.

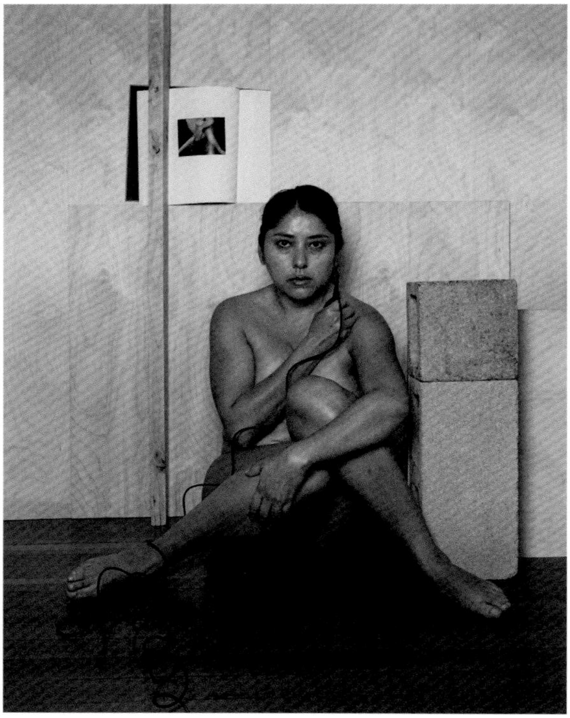

Fig. 9.

Tarrah Krajnak, *Self-Portrait as Weston/as Sonya Noskowiak (Holding Knee), 1934/2020*, 2020. From the series *Master Rituals II: Weston's Nudes*.
© Tarrah Krajnak, courtesy of Zander Galerie.

becomes both object and author of the photograph, gaining exposure through the association. Krajnak's face is often deliberately hidden by a wooden board, a mask, or even her hand holding the shutter release, at once emphasizing her control and referencing Weston's apparent objectification of the nude female body.

Most works in the series show the page of the source image in one of Weston's catalogs—in this example, *Edward Weston's Nudes* book cover—so that her homage is made explicit. (fig. 8) Here, the fist holding the shutter release is powerful and assertive, even as her body contorts awkwardly to match the dancer Bertha Wardell's. Krajnak does nothing to disguise the constructedness of the pose, the studio space, or the shutter release itself: Indeed, here it is thematized. She exposes Weston's (again reproduced!) seemingly effortless prints to be fictions of composure; Krajnak's emphasis is as much on the performance as on the finished print. Unlike Weston, and unlike Levine, she is deliberately "showing her work."

The title of the series, *Master Rituals II: Weston's Nudes* (which followed *Master Rituals I: Ansel Adams,* using quite different techniques of appropriation), confers the honorific of "master" upon Weston. "Rituals" refers to the intensely physical component of Krajnak's labor, as well as to repetitive action. Together, the phrase "master rituals" acknowledges the obeisance that a long lineage of photographers has paid to Weston, much as Frampton states. Yet Krajnak's experience is individualized, unlike Levine's, which is deliberately depersonalized. She tests her Indigenous body against those of the white models Weston posed and loved. And she made the photographs in a California studio, similar to the circumstances under which Weston worked. The studio props of seamed-cement or laminate flooring, water-damaged wooden boards, and concrete blocks invite haptic textual imagination, but also reflection on the haphazard. These are the circumstances available to an emerging artist without the means of renting a deluxe space or paying a studio assistant to make the exposure.

In the titling of *Self-Portrait as Weston/as Sonya Noskowiak (Holding Knee), 1934/2020* (2020), Krajnak references her dual positioning as both Weston and his model/partner (and photographer) Noskowiak, as both photographer and photographed, as well as the eighty-six years that have elapsed between the new photograph and its reference image's making. (fig. 9) Krajnak's pose is not sexy; it is mannered. Her face, which would have been omitted by Weston's camera, is a study in concentration—even confrontation—with the viewer.

Another contemporary photographer, Kelli Connell, recently posed her former partner, Betsy Odom, in the attitudes of one of Weston's favorite models, his second wife, Charis Wilson. The series, *Pictures for Charis*—part travelogue through the American West, part historical fiction—attempts to further displace Weston from the center of photographic history by paying homage to Charis rather than the "master." (Connell, 2024) *Doorway II* (2015), for example, captures Odom in a pose reminiscent of Weston's heralded *Nude* of 1936. (fig. 10) Whereas Wilson's pose is dancerly, Odom's is almost gangly, perhaps defeated. She is dressed in a tank top and shorts and is seated atop a low step with a doormat below her feet—the overall effect less timeless than Weston's *Nude.*

Pictures for Charis frequently evokes domestic space, whereas *Master Rituals II: Weston's Nudes* insistently foregrounds the studio. They thus register distinct ecologies of home-life and workspace, spaces that, for Weston, were often intertwined. A further, more consequential difference in Krajnak's work to Connell's is in the politics of shutter release. In the series *Pictures for Charis,* Connell and Odom are collaborators, but Connell ultimately controls the moment of exposure. In *Master Rituals II: Weston's Nudes,* Krajnak wields the implement for and against herself. It sometimes appears to the viewer almost like a grenade—its detonation a powerful weapon in the politics of image making.

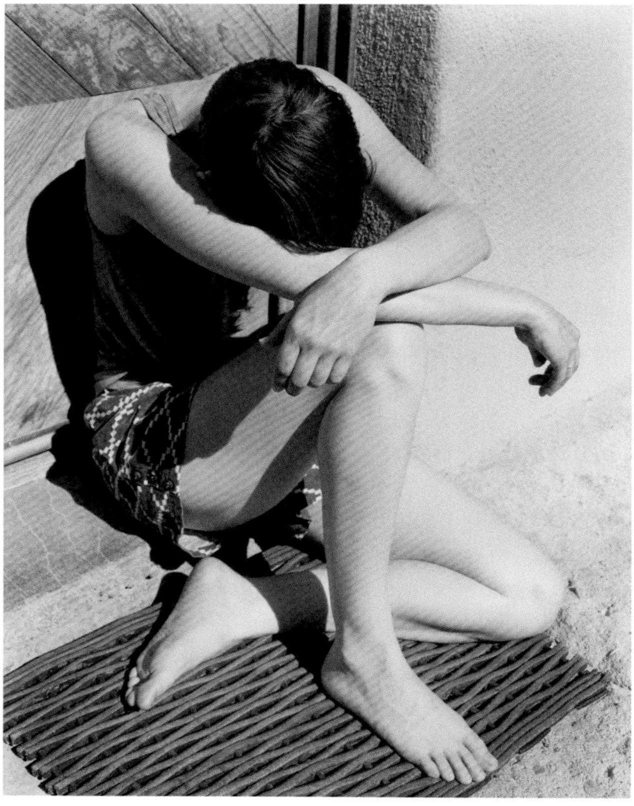

Fig. 10.
Kelli Connell, *Doorway II,* 2015. Gelatin silver print, 20 × 25 in (50.8 × 63.5 cm). From the series *Pictures for Charis.*
© Kelli Connell.

Conclusion

Perhaps I was ungenerous to claim that Frampton was uninterested in Weston. After all, the photographs sent him running to read Weston in his own words, to make meaning of what he had seen. The man, the photographs, really, may have literally saved his *life*. For better or worse, though the "taking" of photographs endangered Miyatake, made Levine's career, and may have injured Krajnak in her contortions, Weston and his photographs gave them life, too.

Frampton stated of Weston's photographs: "it may be that Weston's refusal to emancipate his images from the patriarchal house of his own perceptions amounts to nothing more than the simple declaration of a territorial claim." (Frampton, 1978: 58) The photographers discussed here, especially Levine and Krajnak, who brought their own anti-patriarchal critiques to their practices, effectively emancipate Weston's photographs from "his own perceptions." Their subjectivities and identities challenge the logic of substitution that Frampton believed inhered in Weston's photographs. They claim for his work something beyond his imaginings. In so doing, they displace Weston and his photographs into a distinct register and network of associations, decentering his place in the canon.

Since Cartier-Bresson—or even before that, since Kodak coined its famous slogan—it has become habitual to regard the triggering of the camera's shutter as the *ur*-photographic authorial gesture. The practices examined here challenge this assumption: Miyatake created the studio photographs at Manzanar even when disallowed to operate the shutter. Levine need hardly have operated the camera to reproduce the photographic "masterpieces" that became her subject. Only Krajnak's theatrical display of the remote shutter-release mechanism is fundamental to the meaning of the work. What is clear is that the unique combinations of their subject positions, sociopolitical contexts, and creative choices—in sum, their respective ecologies—results in differential outcomes for their practices to Weston's, with which they are nevertheless inextricably bound. Together, my analysis of these bodies of work demonstrates the plurality of politics exposed by shutter release, never simply the push of a button, nor a decisive moment, nor an imperial act.

Notes

1. An important corrective to this argument is offered by Phu, 2011: 54–83.

Bibliography

Jasmine Alinder, "The Right to Represent: Toyo Miyatake's Photographs of Manzanar," in *Moving Images: Photography and the Japanese American Incarceration* (Champaign: University of Illinois Press, 2009), 75–102.

Ariella Azoulay, *The Civil Contract of Photography,* trans. Rela Mazali and Ruvik Danieli (New York: Zone Books, 2008).

Ariella Aïsha Azoulay, *Potential History: Unlearning Imperialism* (London and New York: Verso, 2019).

Roland Barthes, "The Death of the Author" (1968), in *Image Music Text,* ed and trans. Stephen Heath (London: Fontana Press, 1977), 142–148.

Walter Benjamin, "The Work of Art in the Age of Its Technological Reproducibility" (1936, second version), trans. Edmund Jephcott and Harry Zohn, in *Walter Benjamin: Selected Writings,* eds Michael Jennings, Brigid Doherty, and Thomas Y. Levin (Cambridge, MA: Harvard University Press, 2008), 274–298.

Monica Bravo, *Greater American Camera: Making Modernism in Mexico* (New Haven, CT: Yale University Press, 2021).

Kelli Connell, *Pictures for Charis* (New York: Aperture, 2024).

Hollis Frampton, "Impromptus on Edward Weston: Everything in Its Place," *October* 5 (Summer 1978): 48–69.

Mark Klett, Ellen Manchester, and JoAnn Verburg, *Second View: The Rephotographic Survey Project* (Albuquerque: University of New Mexico Press, 1984).

Ellen Macfarlane, *Politics Unseen: Group f.64 Photography and the Problem of Purity* (Oakland, CA: University of California Press, 2025).

Thy Phu, *Picturing Model Citizens: Civility in Asian American Visual Culture* (Philadelphia: Temple University Press, 2011).

Dennis Reed, *Japanese Photography in America, 1920–1940* (Los Angeles: Doizaki Gallery, 1986).

Gillian Rose, *Doing Family Photography: The Domestic, the Public and the Politics of Sentiment* (Farnham, UK: Routledge, 2010).

Bryan R. Swopes, "16 December 1960," *This Day in Aviation: Important Dates in Aviation History*, 2018, https://www.thisdayinaviation.com/16-december-1960/ (accessed May 11, 2025).

Edward Weston, *The Daybooks of Edward Weston,* ed. Nancy Newhall, vol. 1, *Mexico* (Rochester, NY: George Eastman House, 1962).

Catherine Zuromskis, *Snapshot Photography: The Lives of Images* (Cambridge, MA: MIT Press, 2003).

PART 2
Natural Resource Extraction

CHAPTER 4
Missing the Forest for the Trees: Pulp, Pollution, and Landscape Photography in the Pacific Northwest

Katherine Mintie

In May of 1990, journalist Catherine Caufield published an essay in *The New Yorker* titled “The Ancient Forest.” The article examined the impacts of logging in the Pacific Forest, which runs along the West Coast of North America from Northern California up to Vancouver Island, in British Columbia. While sprawling across international borders, the Pacific Forest has long been identified with the lush natural beauty of the American West and has been photographed by Carleton Watkins, Ansel Adams, and many others who have shaped the contours of this diverse region in the popular imagination. Among other features, these artists were drawn to the fact that the Pacific Forest is the largest coniferous forest in the world and home to trees more than two thousand years old. (Caufield, 1990: 46) Despite the aesthetic appreciation of the Pacific Forest, the last of its ancient trees were in danger of being felled at the time of Caufield’s reporting due to demand for expendable consumer goods. Caufield cautioned her readers:

> Right now, most old-growth wood is used wastefully. Centuries-old Douglas firs are cut into two-by-fours and used to frame our houses; rare cedar planks are nailed into molds for poured concrete and then thrown away; ancient hemlocks are pulped and converted into rayon or cellophane, wrapping paper or disposable diapers. (Caufield, 1990: 79)

To her list of potentially wasteful uses of old-growth hemlock trees, Caufield could have added photographic paper.

Until recently, photography was practiced primarily as a paper-based medium. Bright white, smooth, and often glossy, photographic papers, with their radiant blankness, provided and continue to provide the foundation for billions of images. While these papers project an aura of unblemished purity, the production of photographic papers is resource intensive, chemically driven, and highly pollutive.

The history of the Rayonier Mill in Port Angeles, Washington, exemplifies the toxic legacies of the photographic paper industry and the paper industry more broadly in the Pacific Northwest. Rayonier, which operated in the Pacific Forest from the 1930s to the 1990s, produced huge amounts of pulp for the Eastman Kodak Company, a leading purveyor of photographic papers in the United States in the twentieth century, among other companies. (Funderburk, 2006: 19, 28–29) To bring to light some of the harms caused by the mass manufacture of photographic paper by Kodak, this essay will trace the processes that transformed sturdy hemlock trees into bleached paper pulp at Rayonier and then into polished pieces of photographic paper at Kodak's mill in Rochester, New York.

Beyond charting the effects of photographic pulp production in the Pacific Northwest, this essay will also consider how knowledge of the material history of photography can inform analyses of photographic prints. Specifically, I will examine prints from the series "Turning Back," produced by celebrated American photographer Robert Adams in the late 1990s and early 2000s. This body of work, which circulates as paper-based silver gelatin prints and in the pages of fine art books, focuses on the practice of clearcutting. A controversial form of logging perpetuated by the pulp industry among others, clearcutting involves felling whole sections of a natural forest and, in many cases, replacing them with genetically modified saplings that can be harvested when mature. (Adams, 2005: 34–35) While clearcutting is practiced in many regions, Adams focuses on clearcuts in the section of the Pacific Forest that runs through Oregon, where he has lived since the 1990s. The brutal images of clearcuts in "Turning Back," including *A Second Growth Stump on Top of a First Growth Stump, Coos County, Oregon* (1999; fig. 1) and *On Humbug Mountain, Clatsop County, Oregon* (1999; fig. 2), show the leveled stumps of old-growth trees and swaths of desolated, often burned land where a forest once grew thick. Adams produced these works to oppose deforestation in the region, and they have been read in this context by scholars. (see, for example, Schmelzer, 2022) However, the message of the series becomes more ambiguous when considered alongside the photography industry's long entanglement with the logging and paper industries. By joining an analysis of photographic imagery with photographic materiality, this essay points to unresolved tensions in the long-standing use of photography of a tool of environmental activism.

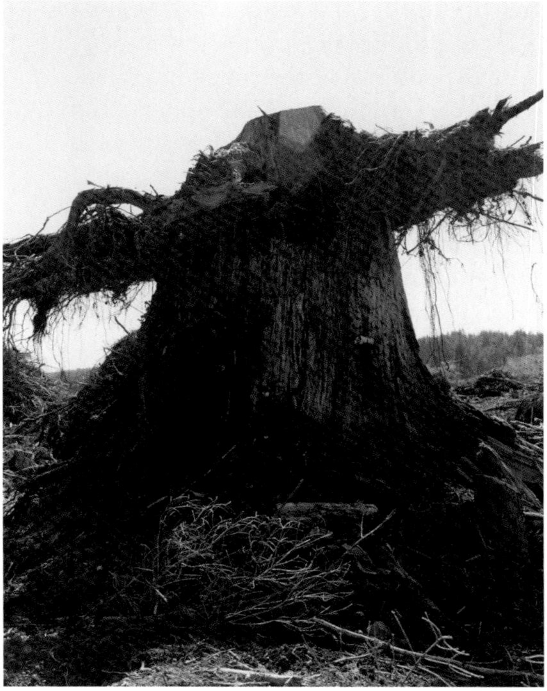

Fig 1.

Robert Adams, *A Second Growth Stump on Top of a First Growth Stump, Coos County, Oregon,* 1999. Gelatin silver print, 11×14 inches. © Robert Adams, courtesy of Fraenkel Gallery, San Francisco.

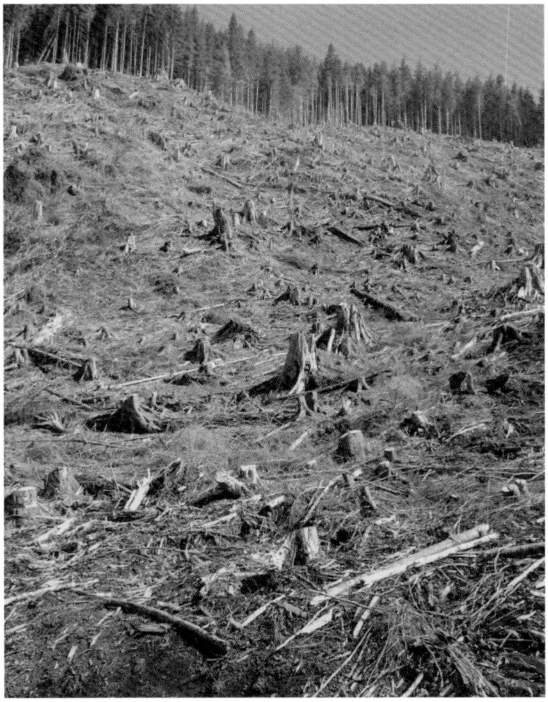

Fig 2.

Robert Adams, *On Humbug Mountain, Clatsop County, Oregon*, 1999. Gelatin silver print, 11×14 inches. © Robert Adams, courtesy of Fraenkel Gallery, San Francisco.

In thinking through the relationship between photographic images and their material underpinnings, this essay contributes to an emerging body of scholarship that engages with the business history of photography. (see, for example, Jenkins, 1987; Lovejoy, 2019; Davis, 2022; Angus, 2024; and Bravo; 2025) As photography historian Steve Edwards has argued, the renewed attention to the workings and impacts of the photography industry has allowed scholars to address not only the role of the medium in "the representation of ecological disaster," but also "the environmental impact of photographic production." (Edwards, 2020: 5) This approach to analyzing photographs, which attends to both photographic manufacture and imagery, offers a potential model for developing more holistic ecocritical histories of photography.

***

To situate the Rayonier Mill within the Eastman Kodak Company's broader manufacturing practices, it is useful to offer a brief overview of where Kodak sourced its photographic paper and pulp. One of the best sources on this topic is *History of the Paper Mills at Kodak Park* by Kit Funderburk, the former Senior Technical Manager at Kodak's Paper Mills in Rochester. Bringing together personal recollections and archival research, Funderburk offers an invaluable record of one of the most impressive papermaking operations of the twentieth century. (Funderburk, 2006: iv-vi)

As Funderburk relates, Kodak initially purchased its base papers from the Belgium-based General Paper Company, which was known since the nineteenth century for producing specialized paper for photographic purposes. (Jenkins, 1987: 195–199; Funderburk, 2006: 1) These early base papers were made from carefully processed cotton and linen rags sourced from European rag pickers and were coated with light-sensitive emulsions once they reached Rochester. (Funderburk, 2006: 4; Mintie, 2020) Kodak imported these base papers from the 1880s until the onset of World War I, when it was cut off from its European supplier and forced to experiment with making papers in-house. (Funderburk, 2006: 3–4) After a series of failures, Kodak ultimately crafted a stable base paper made from a combination of rags and wood pulp in the early 1920s. The initial wood pulp, made from northeastern spruce and balsam firs, was produced by the Brown Paper Company in Berlin, New Hampshire. (Funderburk, 2006: 4; Mintie, 2024: 36–51) After extensive testing, Kodak moved to the exclusive use of photographic paper made from wood pulp in the late 1920s because it was easier to source and more economical. (Funderburk, 2006: 9–10) The Brown Paper Company, Kodak's main supplier of pulp, began to face financial difficulties in the 1950s, which led Kodak to rely more heavily on pulp makers in the Pacific Northwest, including the Rayonier Mill. (Funderburk, 2006: 28–29) As noted earlier, the pulp produced at Rayonier was made primarily from western hemlock trees, which grew in abundance in the local area. Kodak shifted its sourcing of pulp once again in the 1980s, this time to the Aracruz mill in Brazil, which offered more affordable pulp made from non-native, genetically modified eucalyptus trees. (Funderburk, 2006: 37) This arrangement lasted until 2005, when Kodak permanently closed its paper mill at Rochester due to declining demand for photographic paper as digital photography surged. (Funderburk, 2006: 53)

As this brief history suggests, Rayonier was one of many pulp and paper mills that supplied Kodak over its long history. While this essay focuses on the practices at Rayonier, I want to emphasize that the mill was not unique in terms of the environmental and human damage it perpetuated. Indeed, pulp and papermaking mills ranked as top polluters in the United States for much of the twentieth century—in some years outranking the petroleum and chemical industries in the Environmental Protection Agency's Toxic Release Inventory (see, for example, United States Environmental Protection Agency [USEPA], 1989)—and these industries continue to be detrimental to the ecosystems and people of nations, such as Brazil, that have emerged as new leaders in the pulp industry. (Maxwell and Miller, 2012: 48–52; Smith, 2018: 24)

With this context in mind, let us now turn to Rayonier and its pulp-making practices. The mill was founded in 1930 in the town of Port Angeles, which is located in the northwest corner of the state of Washington, on the Strait of Juan de Fuca. (fig. 3) The site of the mill was historically the territory of the Lower Elwha Klallam Tribe, whose land base is now just west of Port Angeles. Though the mill initially focused on making pulp to produce rayon, hence its name, Rayonier quickly diversified and began to supply Kodak and others with paper pulp. (Erickson, 1963: 14–15)

The process of making pulp for Kodak began in the Pacific Forest. To source the hemlocks for its pulp production, Rayonier became one of the largest landowners in Washington state. In addition to acquiring the land of competitors in the pulp and timber industries, Rayonier also purchased significant tracts of forest from the Quinault Indian Nation. (Ficken, 1983: 12–14) At its peak, in the 1950s, the company possessed nearly 400,000 acres of land in the Pacific Forest—an area about the size of Rhode Island—which was logged extensively to produce the wood needed at Rayonier's pulp mill. (Erickson, 1936: 20) The effects of clearing the Pacific Forest at this large scale were well summarized by Caufield in *The New Yorker:* "This forest is home to a greater mass of life than even the tropical rain forests. It is crucial to the stability of the region's and the world's climate: it causes up to a third of local precipitation, and it stores more carbon than any other terrestrial ecosystem." (Caufield, 1990: 46) As Caufield emphasized, the impacts of cutting the Pacific Forest were and remain not only local but global.

Once the hemlock trees were felled by loggers, they were trucked to Rayonier's plant in Port Angeles. There the trees were run through a large woodchipper, and the chips were then placed in large steel urns called digesters to be broken down through the sulfite process. These steps in the pulp-making process are documented in a group of publicity photographs (fig. 4) circulated in a book published in the late 1930s by the Weyerhaeuser Timber Company, another regional supplier of pulp to Kodak, to mark the opening of its Sulfite Mill in Everett, Washington. (Funderburk, 2006: 28) Intended primarily for paper industry insiders, the book provides a detailed look at the operations of pulp making in the Pacific Northwest in the early part of the twentieth century and offers insight into the machinery and procedures employed by Rayonier. The photograph at the top left of the composition shows the exterior of the mill, with its heaving smokestacks. Moving counterclockwise, the image at bottom left shows wood chips moving on a conveyor belt toward the riveted digesters, which appear in the adjacent photograph. Finally, at upper right, a white workman appears to be adjusting the steam pressure on a digester as part of the process of "cooking" the chips.

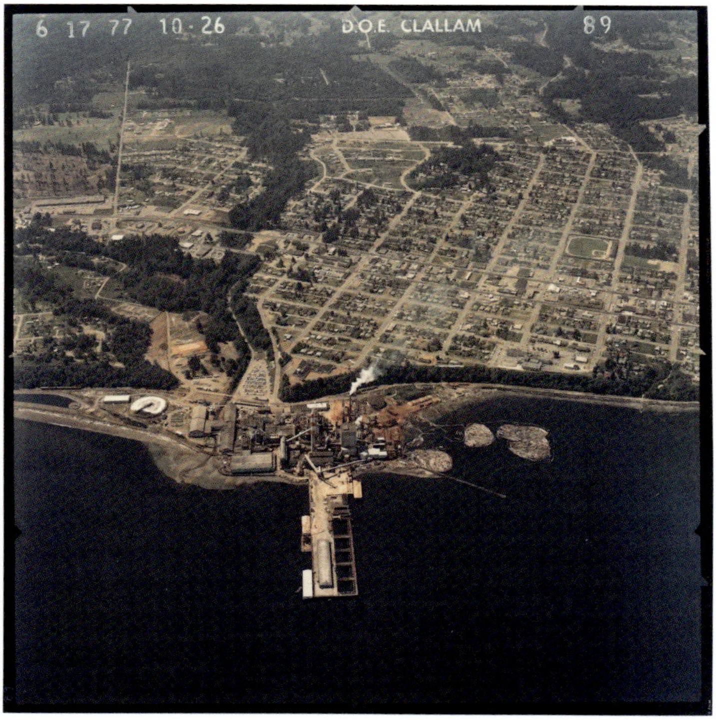

Fig. 3.

Clallam County Department of Ecology, *Aerial View of the Rayonier Mill,* 1977. Color transparency 4×5 inches. In the public domain.

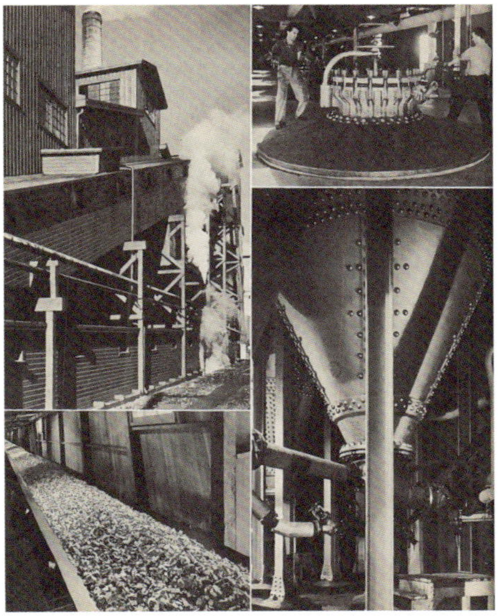

Fig. 4.

Unidentified photographer, *Untitled,* from *West Coast Hemlock Pulp: A Production of American Pulp Mills* (Weyerhaeuser Timber Company, 1938), 84. Halftones, 9.5×11.75 inches. © Weyerhaeuser Timber Company.

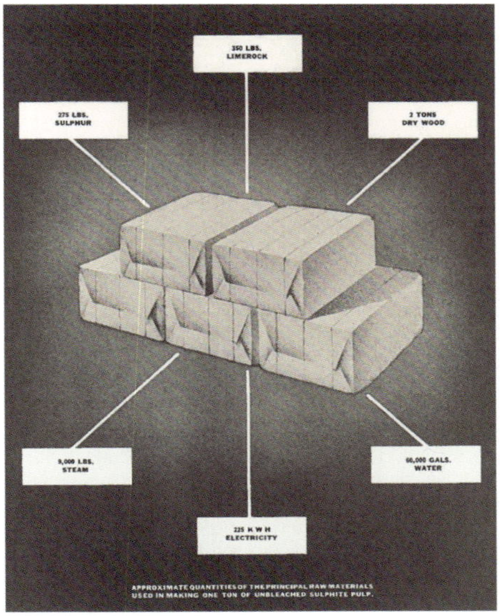

Fig. 5.

Unidentified designer, *Approximate Quantities of the Principal Raw Materials Used in Making One Ton of Unbleached Sulphite Pulp,* from *West Coast Hemlock Pulp: A Production of American Pulp Mills* (Weyerhaeuser Timber Company, 1938), 73. Halftone, 9.5×11.75 inches. © Weyerhaeuser Timber Company.

Once the wood chips entered the digester, they were subjected to a hot chemical bath composed of calcium carbonate and excess sulfur dioxide, which was produced by burning sulfur. (Sutermeister, 1954: 73) As shown in a diagram published by Weyerhaeuser (fig. 5), the process required substantial quantities of sulfur: about 275 pounds for each ton of pulp produced. The burning of sulfur at this level released significant amounts of sulfur dioxide into the local environment. Sulfur dioxide emissions, which the USEPA now regulates, would have negatively affected the respiratory systems of humans and animals in the region, decreased plant growth, and contributed to acid rain. (USEPA, 2024a) Indeed, in 1997 the Port Angeles community petitioned the Agency for Toxic Substances and Disease Registry (ATSDR) for a Public Health Assessment of the Rayonier site due to high reported levels of "respiratory problems (e.g., chronic cough, bronchitis, allergies and asthma), and other ailments such as headaches, miscarriages, childhood sickness, neural tube defects, and cancer." (ATSDR, 2004: 3) While the ATSDR could not conclusively link these severe health issues to air pollution from Rayonier, many members of the Port Angeles community, including health professionals at Olympic Memorial Hospital, held the mill responsible. (ATSDR, 2004: 3)

Once the chips were mixed with the chemicals in the digesters, steam was introduced to increase the pressure to "cook" the chips. Cooking is the chemical process in which the cellulose fibers in the wood chips, the central ingredient in pulp, are separated from the lignin, which binds the cellulose fibers together. Lignin contains acids that deteriorate paper over time, so pulp makers attempt to remove as much of it as possible from the cellulose. In the case of photographic paper, the desired paper was "acid-free" to ensure the long-term preservation of the images. Following the cooking process, the cellulose fibers are retained while the lignin is removed with the cooking acid. (Sutermeister, 1954: 83–86)

The mixture of cooking acid and lignin, known in the industry as waste liquor, was dumped by Rayonier directly into the Port Angeles Harbor, the large body of water adjacent to the Rayonier plant, as shown in a 1970s aerial view. (fig. 3) According to a 1974 USEPA report, Rayonier released about 50,000 pounds of noxious effluent into the harbor each day. (USEPA Office of Enforcement, 1974: 1) The Washington State Department of Ecology later documented that the high volumes of waste liquor disposed of in the harbor contributed to the death of local marine life and negatively impacted the local fishing industry. (Ecology and Environment, 2008: 2–2)

After being separated from the waste liquor, the remaining cellulose fibers were bleached to give the resulting photographic papers a luminous white tone. These bright whites came at a high cost, for the chlorine bleaching process employed by the mill led to the release of dioxins, which are dangerous carcinogens. (USEPA, 2024b) Rayonier and other twentieth-century pulp mills that employed chlorine-based bleaching emitted dioxins into the environment in three main ways. First, Rayonier discharged dioxins into the local air supply through the burning of chlorine. Second, Rayonier dumped effluent containing dioxins from the bleaching process directly into the Port Angeles Harbor, just like it did with the waste liquor from the digesters. Finally, Rayonier sent dioxins into the consumer market through products manufactured from its chlorine-bleached pulp, including photographic paper. (Boyd, 2002: 360–362; Department of Ecology, State of Washington, 2011: 1–3) As legal scholar William Boyd has described, the United States introduced regulations to limit bleaching processes that produced dioxins in 1997 after discovering the serious harms of these chemical compounds. (Boyd, 2002: 392–396) However, existing toxins cannot be removed from the environment, and all people within North America now have some level of dioxins within our body tissue. (Boyd, 2002: 348)

After bleaching, the pulp was pressed into thin sheets, dried, and shipped to the Eastman Kodak Company in Rochester. The process of transforming the whitened pulp into photographic papers involved several steps. First, the pulp would be mixed with sizing and dyes to improve the strength of the paper and determine its tone. (Funderburk, 2007: 23–24, 41–43) Next, a baryta layer would coat the papers to give them even surfaces and, in some instance, to increase whiteness and reflectance. (Funderburk, 2007: 25–31) Finally, a light-sensitive emulsion would be applied. Through variations in these procedures, Kodak produced the wide range of photographic papers that it offered throughout the twentieth century. While it is beyond the scope of this essay to address, it is important to note that the practices at the Kodak paper mills contributed to the devastating amounts of pollution emitted by Kodak into Rochester over the course of its operations and that the effects of these persistent contaminants continue to be felt by the community today. (Povinelli, 2017; Feser, 2023)

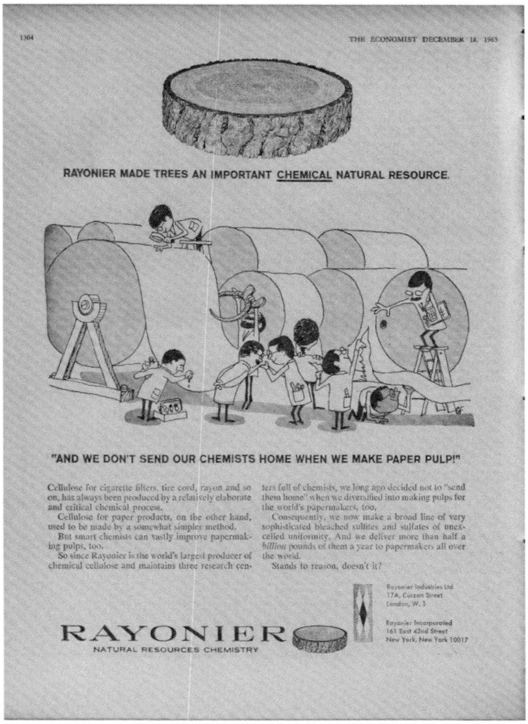

1304 THE ECONOMIST DECEMBER 18, 1965

RAYONIER MADE TREES AN IMPORTANT CHEMICAL NATURAL RESOURCE.

"AND WE DON'T SEND OUR CHEMISTS HOME WHEN WE MAKE PAPER PULP!"

Cellulose for cigarette filters, tire cord, rayon and so on, has always been produced by a relatively elaborate and critical chemical process.

Cellulose for paper products, on the other hand, used to be made by a somewhat simpler method.

But smart chemists can vastly improve papermaking pulps, too.

So since Rayonier is the world's largest producer of chemical cellulose and maintains three research centers full of chemists, we long ago decided not to "send them home" when we diversified into making pulps for the world's papermakers, too.

Consequently, we now make a broad line of very sophisticated bleached sulfites and sulfates of unexcelled uniformity. And we deliver more than half a *billion* pounds of them a year to papermakers all over the world.

Stands to reason, doesn't it?

RAYONIER
NATURAL RESOURCES CHEMISTRY

Rayonier Industries Ltd
17A, Curzon Street
London, W. 1

Rayonier Incorporated
161 East 42nd Street
New York, New York 10017

Fig. 6.
Unidentified designer, "Rayonier Made Trees an Important Chemical Natural Resource." *The Economist,* December 18, 1965, 1304. Halftone, 7.5×10 inches.
© Rayonier ITT.

As this overview of the manufacturing practices of Rayonier and Kodak suggests, the process for making photographic pulp and paper was driven by chemistry and chemicals. This key aspect of the pulp-making process is highlighted in an advertisement (fig. 6) that Rayonier ran in *The Economist* in December 1965. Set between simple line drawings of an abstracted, lifeless tree stump and a group of bespeckled men in white lab coats energetically examining large rolls of white paper, a bold line of text declares: "Rayonier made trees an important chemical natural resource." Though the phrasing is clunky, the slogan and illustrations clearly convey that Rayonier saw the ancient trees of the Pacific Forest as valuable only when combined with the humanmade innovations of chemistry, which transformed them into cellulose commodities. It was its dependence on toxic chemicals, however, that led to the downfall of Rayonier's Port Angeles operation.

After the establishment of the EPA in 1970 and the implementation of legislation aimed at curbing pollution, Rayonier attempted to remedy its operations, but it remained a major polluter. In 1993, the mill in fact ranked as Washington's top polluter and was emitting more than 3.6 million tons of discharge into the air and water annually. (USEPA, 1993) Between the costly improvements necessary to bring down toxic releases and the rising cost of wood in the region, Rayonier ultimately closed the Port Angeles plant in 1997 (Ecology and Environment, 2008: 3–1)

Though no longer actively polluting the area, the mill left behind a wide range of harmful contaminants, some of which are listed in a chart (fig. 7) published by the Washington State Department of Ecology, the organization in charge of the cleaning up of the site. The process of remediating the former Rayonier Mill has been ongoing since 1997, and 30,000 tons of hazardous materials have been removed to date. (Department of Ecology, Washington State, 2023: 2) Though we are nearly thirty years into the decontamination process, much remains to be done, and there is no end date in sight.

Contaminant Type	Soil?	Groundwater	Surface Water?	Air?	Sediment?	Bedrock?
Metals - Metals Priority Pollutants?	C?	C?				
Metals - Metals - Other?		C?			C?	
Halogenated Organics - Polychlorinated biPhenyls (PCB)?	C?				C?	
Non-Halogenated Organics - Petroleum Products-Unspecified?	C?	C?			C?	
Non-Halogenated Organics - Phenolic Compounds?		C?			C?	
Halogenated Organics - Dioxins/Furans?	C?	C?			C?	
Non-Halogenated Organics - Polycyclic Aromatic Hydrocarbons?	C?	C?			C?	
Metals - Arsenic?	C?				C?	
Other Contaminant - Other Deleterious Substances?					C?	
Metals - Lead?	C?					
Metals - Mercury?					C?	
Other Contaminant - Bioassay Failure?					C?	
Non-Halogenated Organics - Petroleum-Diesel?	C?					
Pesticides - Pesticides-Unspecified?		C?				

S	Suspected	**RA**	Remediated-Above
C	Confirmed Above Cleanup Levels	**RB**	Remediated-Below
B	Below Cleanup Levels	**R**	Remediated

Fig. 7.
Washington State Department of Ecology, *Contaminant at the Port Angeles Rayonier Mill Site,* ca. 2010. Digital chart. In the public domain.

***

Having traced the pulp-making process at Rayonier and its various impacts, let us now return to Robert Adams's series "Turning Back." As discussed earlier, Adams produced this body of work in response to his horror at the practice of clearcutting in the Pacific Forest, specifically in areas near his home in northwestern Oregon. Like Washington, Oregon has long been and continues to be a center for the timber and paper trade, and there are several pulp mills that continue to operate there. (Adams, 1951: 21–23)

The destructiveness of the contemporary logging industry to the forest landscape is readily visible in Adams's *A Second Growth Stump on Top of a First Growth Stump, Coos County, Oregon.* (1999; fig. 1) This silver gelatin print is dominated by the presence of a wide, dark tree stump atop which a thinner, shaggy tree has fallen horizontally. The wrecked trees stand alone amid the scattered branches and roots of their felled neighbors. Together the two trees create the shape of a cross against the blank sky and conjure a ragged human form. The visual suggestion of a body is not accidental, as Adams has described industrial clearcutting as "de-limb[ing]" trees. (Adams, 2005: 34–35) By anthropomorphizing the fallen trees in his photographs and descriptions of them, Adams forces his viewer to reckon with the practice of clearcutting as the decimation of fellow living entities. More difficult for Adams to convey is the long-term damage to the larger forest ecosystem, a sphere that largely evades the superficial gaze of the human eye and photography. (Nixon, 2011: 2)

Where *A Second Growth Stump on Top of a First Growth Stump, Coos County, Oregon* brings its audience close to the trees felled in a clearcut, *On Humbug Mountain, Clatsop County, Oregon* (1999; fig. 2) offers a broader view of deforestation in the region. Almost the entire composition shows the defoliated slope of a clearcut hill that was once densely covered with trees but is now populated only by desolate stumps and scattered branches. Given the complete devastation of the land, it is not surprising that critics have suggested that this and other photographs in "Turning Back" resemble the grim aftermath of a battlefield. (Greenough, 2021: 59) Just beyond the ruins, at the very top of the composition, stands a thin section of the remaining forest. While sturdy and thick, the surviving trees appear fragile in the face of the catastrophic effects of the clearcut and the foreboding that they will be felled next. The scale of the harm to the ancient trees of Pacific Forest has troubled Adams deeply. As he told curator Sarah Greenough, clearcutting is not just "a matter of exhaustion of resources [...] there is involved an exhaustion of spirit." (Greenough, 2021: 57)

While the visual strategies employed by Adams in "Turning Back" make a compelling and urgent case against clearcutting, the material nature of the photographs and the ways in which they circulate are a thornier issue. In making his protest, Adams, like many artists and activists, relies on paper-based media, particularly paper photography and photographic books. As we have seen, the creation of this paper, even if not produced at Rayonier, requires the felling of many trees as well as the release of significant pollution into the local environment. The tension between the message of Adams's work and its physical manifestation poses broader questions about the role of photography within environmental activism and ecocritical approaches to art history. Indeed, in a parallel manner to Adams, I have relied here on numerous paper photographs and the printed pages of books to raise awareness of the deforestation and pollution caused by Rayonier. Given the fraught duality of photography as both a product of and weapon against environmental degradation, what are artists and scholars to do? One answer comes from anthropologist and historian of photography Christopher Pinney, who points out that the competing "potentialities and propensities latent in photography" can be difficult to navigate but also open new ways of looking at and thinking about the medium. (Pinney, 2012: 155)

Adams himself has acknowledged the simultaneous limits and capacities of photography regarding environmental activism. Writing about "Turning Back," he asks "Of what significance is minority evidence? Photography is inherently fragmentary, and I find I base my faith on perfect moments." (Adams, 2005: 163–64) I find Adams's emphasis on the fragmentary useful in thinking through the tensions that arise when considering photography's relationship to the environmental movement and the development of ecocritical histories of photography. Beyond pointing to the fact that photographs necessarily represent only a selected, partial view of any subject, the notion of the fragmentary also conjures the ways in which scholars conventionally view and analyze photographs: how we tend to focus on what is represented and to overlook the material histories that are also embedded within photographs.

While Adams has reconciled himself to photography's fragmentary nature, I would encourage historians of the medium to move beyond our splintered approach to analyzing photographs and to try to understand them in a more holistic manner. This work will require studying photographs not only visually, but also materially and ecologically, an approach I have attempted to model here. When photographs are read in this more expansive manner, we might begin to see the forests for the trees.

Bibliography

W. Claude Adams, "The History of Papermaking in the Pacific Northwest: I," *Oregon Historical Quarterly* 52, no. 1 (March 1951): 21–27.

Robert Adams, *Turning Back: A Photographic Journal of Re-Exploration* (San Francisco: Fraenkel Gallery; New York: Matthew Marks Gallery, 2005).

Agency for Toxic Substances and Disease Registry (ATSDR), *County, Washington* (Washington, DC: Agency for Toxic Substances and Disease Registry, 2004). https://www.atsdr.cdc.gov/hac/pha/RayonierInc_Final%20PHA_05-13-2004/RayonierInc_Final%20PHA_05-13-2004.pdf (accessed November 20, 2025).

Siobhan Angus, *Camera Geologica: An Elemental History of Photography* (Durham, NC: Duke University Press, 2024).

William Boyd, "Controlling Toxic Harm: The Struggle Over Dioxin Contamination in the Pulp and Paper Industry," *Stanford Environmental Law Journal* 21 (2002): 345–419.

Monica Bravo, "Mercury is a Messenger: Photography's Dependence on Quicksilver and Mining Labor," *American Art* 39, no. 1 (Spring 2025): 16–25.

Catherine Caufield, "A Reporter at Large: The Ancient Forest," *The New Yorker* (May 14, 1990): 46–84.

Heather Davis, *Plastic Matter* (Durham, NC: Duke University Press, 2022).

Department of Ecology, State of Washington, "Rayonier Mill Off-Property Soil Dioxin Study: Draft Report Available for Public Review and Comment" (July 2011), https://apps.ecology.wa.gov/cleanupsearch/site/2270#site-documents (accessed June 5, 2024).

Department of Ecology, State of Washington, "Status Update: Port Angeles Cleanup Sites and Natural Resource Damage Assessments," (March 2023), https://apps.ecology.wa.gov/cleanupsearch/site/2270#site-documents (accessed June 5, 2024).

Ecology and Environment, for Department of Ecology, Department of Ecology, State of Washington, "Rayonier Mill Off-Property Soil Dioxin Study: Draft Report Available for Public Review and Comment" (July 2011), https://apps.ecology.wa.gov/cleanupsearch/site/2270#site-documents (accessed June 5, 2024).

Stephen Edwards, "Why Pictures?: From Art History to Business History and Back Again," *History of Photography* 44, no. 1 (2020): 3–15.

Russell F. Erickson, *The Story of Rayonier Incorporated* (Princeton, NJ: Princeton University Press, 1963).

Ali Feser, "Was It Worth It?" *American Anthropologist* 125 (2023): 662–668.

Robert E. Ficken, "Pulp and Timber: Rayonier's Timber Acquisition Program on the Olympic Peninsula, 1937–1952," *Journal of Forest History* 27, no. 1 (January 1983), 4–14.

Kit Funderburk, *A Guide to Surface Characteristics: Kodak Fiber Based Black and White Papers* (Rochester, NY: self-pub., 2007).

Kit Funderburk, *History of the Paper Mills at Kodak Park* (Rochester, NY: self-pub., 2006).

Sarah Greenough, *American Silence: The Photographs of Robert Adams* (Washington: National Gallery of Art; New York: Aperture, 2021).

Reese V. Jenkins, *Images and Enterprise: Technology and the American Photographic Industry 1839–1935* (Baltimore: John Hopkins University Press, 1987).

Alice Lovejoy, "Celluloid Geopolitics: Film Stock and the War Economy, 1939–47," *Screen* 60, no. 2 (Summer 2019): 224–241.

Richard Maxwell and Toby Miller, *Greening the Media* (Oxford and New York: Oxford University Press, 2012).

Katherine Mintie, "Material Matters: The Transatlantic Trade in Photographic Materials during the Nineteenth Century," *Panorama* 6, no. 2 (Fall 2020), https://journalpanorama.org/article/re-reading-american-photographs/material-matters/ (accessed April 28, 2026).

Katherine Mintie, "Pâte à Papier, Pollution, et Déclin: Pour un Prise en Compte des Externalités Négatives de l'Industrie du Papier Photographique au XXe Siècle," *Photographica* 8 (April 2024): 36–51.

Rob Nixon, *Slow Violence and the Environmentalism of the Poor* (Cambridge, MA: Harvard University Press, 2011).

Christopher Pinney, "Seven Theses on Photography," *Thesis Eleven* 113, no. 1 (2012): 141–156.

Elizabeth Povinelli, "Fires, Fogs, Winds," *Cultural Anthropology* 32, no. 4 (2017): 504–513.

Paul Schmelzer, "If Robert Adams Wants His Eco-Conscious Photography to Change Anything, It's Us," *National Gallery of Art Blog*, August 12, 2022, https://www.nga.gov/blog/robert-adams-eco-concious-photography.html (accessed June 13, 2024).

Maureen Smith, *The U.S. Paper Industry and Sustainable Production: An Argument for Restructuring* (Cambridge, MA: MIT Press, 2018).

Edwin Sutermeister, *The Story of Papermaking* (Boston: S. D. Warren Company, 1954).

United States Environmental Protection Agency (USEPA), "TRI On-site and Off-site Reported Disposed of or Otherwise Released (in Pounds), for All chemicals, by Industry, U.S., 1989," *Toxic Release (TRI) Explorer*, https://tinyurl.com/54bcar38 (accessed November 7, 2025).

United States Environmental Protection Agency (USEPA), "1993 Dataset for Washington," *Toxic Release Inventory (TRI) Explorer*, https://tinyurl.com/vz7bvfu3 (accessed June 4, 2024).

United States Environmental Protection Agency (USEPA), "Sulfur Dioxide Basics," https://www.epa.gov/so2-pollution/sulfur-dioxide-basics (accessed June 5, 2024a).

United States Environmental Protection Agency (USEPA), "Learn About Dioxin," https://www.epa.gov/dioxin/learn-about-dioxin (accessed June 5, 2024b).

United States Environmental Protection Agency Office of Enforcement, *Evaluation of ITT Rayonier, Inc. Outfall Port Angeles Harbor, Washington* (Denver: National Field Investigation Center, 1974). https://nepis.epa.gov/Exe/ZyPURL.cgi?Dockey=9100LL4P.txt (accessed June 10, 2024).

CHAPTER 5
Looking at Cows, Reanimating Gelatin: The Biopolitics of Photographic Emulsion

Alan C. Braddock

I begin with two photographs of Yosemite taken nearly half a century apart. One of them, a picture from the 1870s or 1880s by Carleton Watkins (1829–1916) and now in the Huntington Library collection, shows a "view down the valley from Union Point." (fig. 1) The other, a "view of the valley" by George Eastman (1854–1932), founder of the Eastman Kodak camera and film company, was taken during a trip to California in 1913 and today resides at the Eastman Museum in Rochester, New York, near the firm's headquarters. (fig. 2) Both photographs depict an iconic park in the American West using well-established conventions of landscape representation, with vertical elements in nature framing a panoramic view and invoking a nationalized romantic sublime. As representations of a famous "natural" place, these pictures suggest thematic and compositional continuity, prompting the usual interpretive considerations regarding the history of national parks, tourism, environmental conservation, colonial settlement, Indigenous dispossession, and the like.[1]

Yet, an exclusively iconographic, or representational, approach to understanding these apparently similar images elides substantial differences between them in materials and processes—differences that embody significant changes in photographic production and related supply chains over time. A key distinction concerns their different photographic emulsions. The Watkins picture is an albumen print, made by exposing an emulsion of egg white and chloride on a heavy glass plate that needed to be coated with a light-sensitive silver nitrate solution just before use. Eastman's photograph was made with an emulsion of silver halide crystals suspended in dry gelatin on film, requiring no liquid sensitization prior to exposure. In other words, whereas Watkins had to

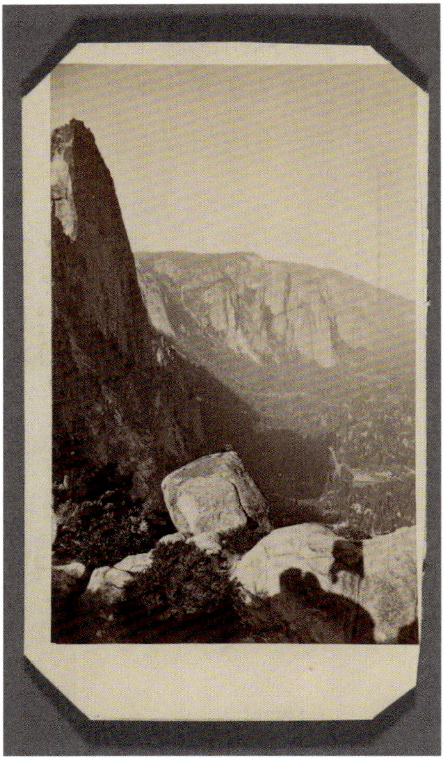

Fig. 1.

Carleton Watkins, *View down the Valley from Union Point, Yosemite*, 1870–1888. Albumen print, 16 × 12 cm. San Marino, Calif., Huntington Library.

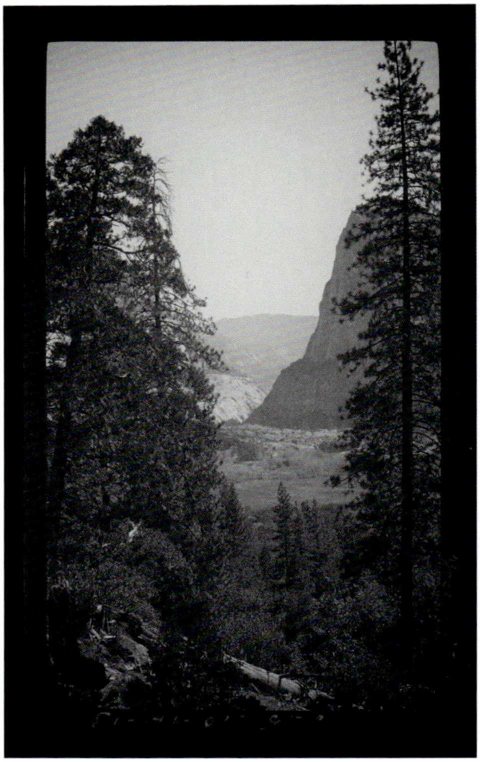

Fig. 2.

George Eastman, *View of the Valley, Yosemite Trip*, 1913. Negative, gelatin on nitrocellulose roll film, 5 ½ × 3 ¼ in. Rochester, International Museum of Photography at George Eastman House.

carry cumbersome equipment and apply wet chemicals when taking his picture, Eastman could simply point and shoot with a portable Kodak camera, then develop the image later at his leisure. The implications of this difference are profound and far-reaching, for they involve not only esoteric details about changes in photographic technology, but also broader impacts on the environment, commerce, and visuality.[2]

In this essay, I explore those impacts by focusing on gelatin and its emergence as the dominant emulsion material in photography during the late nineteenth century, when it largely replaced albumen and other substances, thanks in no small part to the industrial innovations of George Eastman. Throughout the nineteenth century, gelatin was made of both plant and animal matter for various purposes, including uses in food and pharmaceuticals, but Eastman and his scientists at Kodak were leaders in crafting photographic gelatin from specially purified collagen that originated in animal bones, hides, and other materials—primarily

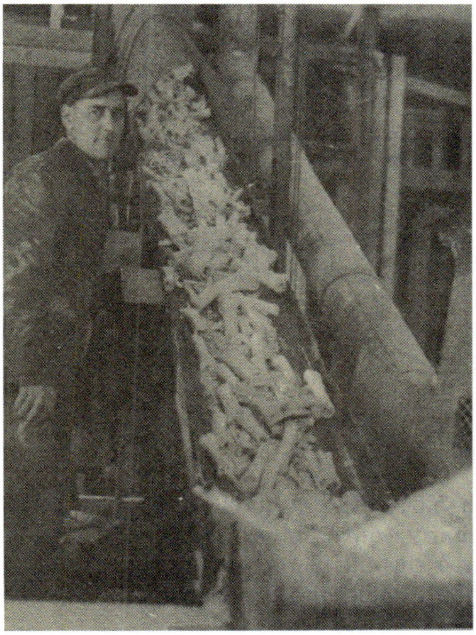

Fig. 3.

Unidentified photographer, bones on conveyor belt for gelatin production at Eastman Kodak, in "This is Eastman Gel," *Kodakery*, February 11, 1960, 1.

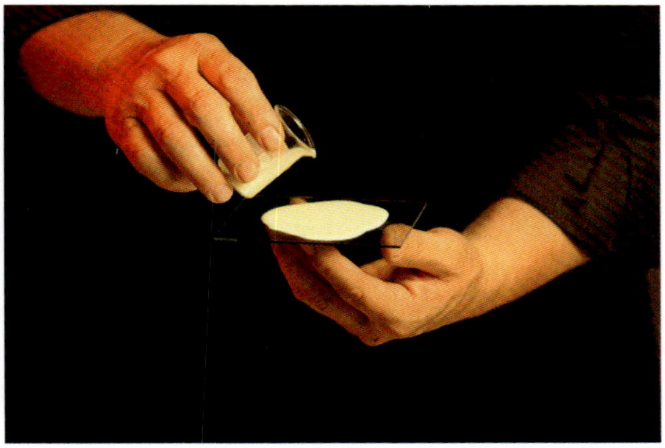

Fig. 4.

Gelatin emulsion poured on a glass plate at George Eastman House. Eastman Museum website.

https://www.eastman.org/event/talks-workshops/online-gelatin-dry-plates-george-eastman-demonstration-history

of cows and calves—obtained from industrial tanneries and slaughterhouses as far away as Chicago, where cattle from the American West were processed by the millions for a global market. Photography thus leveraged western ingredients for an international supply chain. Illustrations from a twentieth-century Kodak magazine show scenes of the company's photographic gelatin production facility in Rochester, including one documenting a conveyor belt with a never-ending stream of cow bones to be crushed and treated with steam, acid, and other chemical processes to extract and purify the substance. (fig. 3) The finished product appears as a milky fluid poured from a beaker in an Eastman Museum demonstration video about gelatin dry-plate photography. (fig. 4) After the addition of silver halides, the liquid gelatin emulsion was coated onto a photographic plate or film and dried transparently, attaching to the surface without losing sensitivity to light and ready to be taken almost anywhere. Indeed, as Eastman's scientists eventually discovered, their purified form of gelatin enhanced exposure sensitivity while creating ease of access and use.[3]

Animal-based gelatin provided a stable, portable, and commercially abundant emulsion material, the convenience of which cannot be overstated. By making it possible to expose photographs on dry, portable glass plates or film without the need for wet chemicals, gelatin facilitated a veritable revolution in consumer access to the medium, epitomized in Kodak's famous advertising slogan, "You press the button, we do the rest," referring to how Eastman's company ingeniously de-coupled the taking of snapshots from the messy front- and back-end processes of preparing coated films and later developing them. By the late nineteenth century, consumers could simply shoot pictures and send their Kodak camera back to the company for printing, after which the device was returned newly loaded with film. Gelatin enabled this with its stability and portability, but it became almost invisible to photographers under the newly compartmentalized, industrial regime that Eastman perfected. Nevertheless, its advantageous properties immanently transformed the medium by facilitating greater logistic mobility and exposure sensitivity, both of which ensured that gelatin silver printing and related processes would dominate photography for more than a century, into the digital era. The resulting expansion of gelatin production had substantial impacts, however, not only on the medium and its end users, but also on the animals and people that provided the necessary material and labor for this historic development.[4]

An ecocritical perspective on gelatin invites us to think about materiality, labor, and animality together by looking beyond familiar aesthetic and anthropocentric considerations about artistic style, representation, and humanistic social context. Instead of regarding gelatin as a neutral substance of marginal interest, the present essay seeks to center and politicize this essential material by interpreting its production, ubiquity, and seeming invisibility as epitomizing reification—capitalism's tendency to transform labor and living beings into exchangeable commodities while concealing knowledge and historical memory of them. To counteract this reifying tendency and make visible a crucial component of photography that is hard to see in pictures, I use ecocritical strategies for reanimating gelatin and tracing aspects of its commodity chain to recover what I call the biopolitics of photographic emulsion.[5]

Ecocriticism as a mode of inquiry has long attended to problems of invisibility, because unseen agents have often produced substantial environmental and cultural impacts that demand scrutiny—especially since the advent of modern industry. For example, concerns about insidious, subliminal, and subterranean toxins permeate Rachel Carson's *Silent Spring* (1962), a landmark environmental text describing DDT and other industrial chemicals as "A grim specter [that] had crept upon us almost unnoticed," "flowing unseen through the earth" such that they "defy detection." (Carson, 2011: 2) More recently, in another widely influential book (citing Carson as an important inspiration), Rob Nixon defined environmental "slow violence" as "a violence that occurs gradually and out of sight, a violence of delayed destruction that is dispersed across time and space, an attritional violence that is typically not viewed as violence at all," since it fails to be "immediate in time, explosive and spectacular in space, and [...] erupting into instant sensational visibility." Nixon (2011: 2) Dialectics of visibility and invisibility also inform an extraordinary study by Timothy Pachirat titled *Every Twelve Seconds:*

Industrialized Slaughter and the Politics of Sight (2011), which critically examines the modern abattoir as an environment deliberately hidden from public view to maintain strategic fictions about its supposed systematic order, civilized rationality, and cleanliness. Basing his book on firsthand experience working in an industrial slaughterhouse, Pachirat intended it "to provoke reflection on how distance and concealment operate as mechanisms of power in modern society," where the abattoir functions as a "zone of confinement" at once "invisible" and "on the whole inaccessible to ordinary members of society," thereby making the mass killing of animals "tolerable." Although Pachirat identifies as a political scientist and not as a scholar of the environmental humanities, his attention to visuality has attracted attention in ecocritical art history. (Pachirat, 2011: 3, 4).[6]

Of course, the medium of photography enjoys a long-standing reputation for enhancing visibility by augmenting the powers of human perception, as in the famous nineteenth-century chronophotographic and "animal locomotion" experiments conducted by Étienne-Jules Marey (1830–1904) and Eadweard Muybridge (1830–1904), who used cameras to reveal phases of movement in horses, humans, and other creatures that were invisible to the naked eye. And yet, I argue in this essay, photography has concealed much about its material constituents and processes to maintain fictions akin to those described by Pachirat in the context of industrialized slaughter, while also occluding environmental violence—both slow and fast—of the sort analyzed by Carson and Nixon. To that extent, my project has something in common with that of Siobhan Angus, whose 2024 book *Camera Geologica* examines the extractivist underpinnings and environmental effects of photography in great detail, noting, "[o]f all mediums, it is the most effective at concealing its materiality." (Angus, 2024: 6) The focus of Angus's investigation, however, is the industrial mine, not the slaughterhouse, as a key site of photographic extraction, and this is where our studies differ. Except for very brief introductory remarks about animal material, Angus concentrates on subterranean substances—bitumen, silver, platinum, iron, uranium, and rare earth elements. Although certain materials discussed by Angus—such as bitumen and other fossil fuels—originate in prehistoric animal and plant matter, I contend that their extraction entails rather different ethical and environmental stakes from those relating to modern industrial slaughter. Thus, despite its many strengths, her book instantiates a familiar tendency in much ecocriticism of marginalizing non-human life by showing deference to what I would call terrestrial materiality or minerality as opposed to the sentient vitality that figures prominently in critical animal studies. A similar tendency informs the recent collection titled *Capitalism and the Camera*, edited by Kevin Coleman and Daniel James (2021) which contains little discussion of gelatin and no reference to slaughterhouses. As the product of industrial slaughter, gelatin invites analysis from an ecocritical perspective informed by considerations of animality to counteract the reification of sentient life as a central component in photographic production. By connecting the dots between photography's immanent and seemingly invisible material infrastructure (the emulsion) and its consumer commodity (the image) ecocriticism here forges an interesting rapprochement with not only critical animal studies but also aspects of Marxist analysis.[7]

Accordingly, I take a cue from the Marxist literature scholar Nicole Shukin, whose 2009 book *Animal Capital* contains this brief but provocative statement about photographic gelatin:

> It is here, in the material convolutions of film stock, that a transfer of life from animal body to technological media passes virtually without notice. [...] One needs to tease out the animal ingredients of film stock via a material history of photographic gelatin. [...] Gelatin can be excavated as one site where the production of capitalist culture can be seen to always also involve the rendering of nature. (Shukin, 2009: 104)

As a scholar of literature, Shukin discusses no specific photographs in this context, nor does she dig deeply into the historical archives of Eastman Kodak or other entities along the gelatin commodity chain, but her penetrating observation prompted me to spend several weeks with those archives in Rochester, New York. What has become clear from my research is that gelatin merits ecocritical consideration in its own right, as a material that connects photography with a far-reaching network of industrial production and consumption premised on reification—the transformation of living beings by the millions into an abstract thing at once invisible and essential to photography. The idea of reification has long reverberated in Marxist theory, from Lukács to Althusser and beyond, but here is the key passage from Karl Marx in the first volume of *Capital* (1867):

> A commodity is therefore a mysterious thing, simply because in it the social character of men's labour appears to them as an objective character stamped upon the product of that labour; because the relation of the producers to the sum total of their own labour is presented to them as a social relation existing not between themselves, but between the products of their labour. This is the reason the products of labour become commodities, social things whose qualities are at the same time perceptible and imperceptible by the senses [....] There it is a definite social relation between men, that assumes, in their eyes, the fantastic form of a relation between things. (Marx, 1906: 83)

Marx's reference to a "mysterious thing" at once "perceptible and imperceptible" aptly describes gelatin as an emulsion substance made of formerly living beings rendered largely invisible through chemical purification, yet ubiquitous and crucial to the production of photographs. However, an ecocritical perspective on photographic gelatin arguably expands Marxist analysis by viewing the "social character" of labor and "social relations" in more-than-human terms. After all, gelatin and photography are not merely the products of what Marx calls "men's labour" or a "social relation between men," since countless cows have contributed their lives to its "fantastic form." In gelatin silver print photography, we are literally looking at cows—not just as subject matter but as the material surface and substrate of the visual artifact. As proof of how easy it is to overlook or forget this fact, I note that even Shukin—who analyzes other pictures—does not mention gelatin's implication in the cover image of her book, which depicts living cows (which she never identifies or discusses). Ecocritical interpretation that is attentive to non-human life therefore helps reanimate photographic gelatin by reminding us of its animality while exposing a gap between traditional Marxist political economy and a more expansive biopolitics.[8]

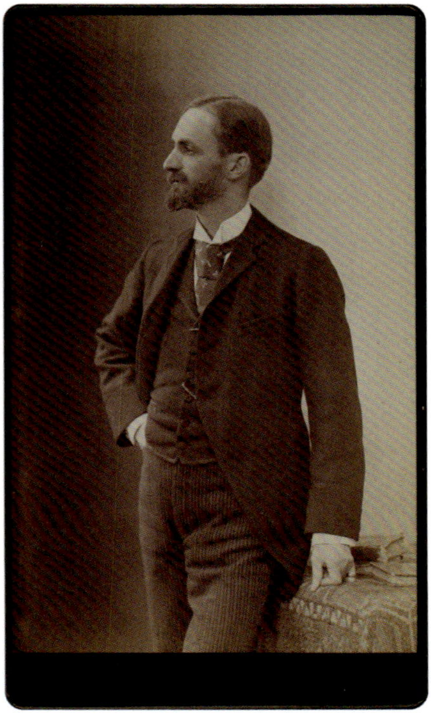

Fig. 5.
Nadar (Gaspard-Félix Tournachon), *Portrait of George Eastman*, 1890. Albumen silver print, 19.6 × 12.5 cm. Rochester, International Museum of Photography at George Eastman House.

To further illuminate the biopolitics of gelatin emulsion and to reanimate its non-human constituents, I want to look more closely at George Eastman's signature innovations in photography beginning around the turn of the twentieth century. Eastman appears in an 1890 photograph by Nadar, the famous French portraitist, during one of the American entrepreneur's trips to Europe. (fig. 5) This elegant picture was taken as Eastman vigorously promoted the engines of his commercial success: gelatin-coated dry plates (which had already begun to displace the albumen process used by Nadar) and the 1888 Kodak camera, a revolutionary consumer model loaded with gelatin-coated film. As the Nadar portrait suggests, within a decade of founding his company in 1880, Eastman was regularly traveling—both abroad and around the United States—cultivating various business ventures, including deals for the provision of gelatin in mass quantities to ensure reliable sourcing of the material for his rapidly growing enterprise.[9]

In a letter written March 31, 1911, to one of his representatives in Surrey, England (S. V. Haus), under the heading "Manufacture of

Gelatine," Eastman discussed the installation of a new gelatin production plant in Rochester to be modeled on one in Bad Sobernheim, Germany. (George Eastman Correspondence) After ordering his operative to visit and gather information about the German gelatin factory, Eastman pondered how to source his proposed new gelatin plant with what he calls "the clippings of calf skins that we need" by using middlemen in the meat and leather industries, whom he describes as "the parties who collect them from the slaughter houses and before they have gone to the tannery." "These clippings," he said, "consist almost entirely of head skins, including the ears and nose, and also the clippings from the feet," which "aggregate about 2–1/2 to 3 lbs. from each animal." A subsequent letter to the same operative on June 1, 1911, referred to "Armour & Co." as a source of animal material. When Eastman founded Kodak in 1881, Armour & Company was already one of the biggest meatpackers in the world, operating large facilities in Chicago and Omaha, Nebraska, with a network of cattle producers that extended as far west as California by the early twentieth century. In a postscript to his letter of March 31, 1911, Eastman concluded by saying "Our man says it takes 5 lbs. of clippings to make 1 lb. of gelatine"—a calculation that hints at the magnitude of animal material needed for such production, not to mention the breathtaking scale of violence and reification taking place in adjacent industries upon which the commercialization of photography increasingly relied.[10]

One peculiarly violent episode in the emergence of gelatin silver print photography underscores another aspect of gelatin's irresistible convenience as an emulsion material. In 1881, US military engineers at Willet's Point, New York, conducted an experiment to demonstrate the rapid photographic exposure time afforded by gelatin emulsion. Attaching dynamite to the head of an aging mule, the engineers used an electric current to simultaneously trigger an explosion and a camera loaded with a gelatin dry plate to capture that explosion. The result was a gelatin silver print photograph, showing the mule beheaded but still standing at the instant of the blast. (fig. 6) The picture offers another clear indication of prevailing biopolitical realities, in which non-human life was relentlessly sacrificed on the altar of technological progress and commerce. Those realities become clearer still in a subsequent publication about the experiment in an issue of the periodical *Scientific American*, where an article titled "Instantaneous Photography" displays before-and-after illustrations of the exploded mule with a text touting "the remarkable sensitiveness of photo-gelatine plates." (fig. 7) In pursuit of speed and convenience in photography, the sensitivity of the gelatin plates clearly mattered more than ethical sensitivity toward the non-human animal. And yet, ironically, as a growing body of scientific research would demonstrate in the coming years, the properties

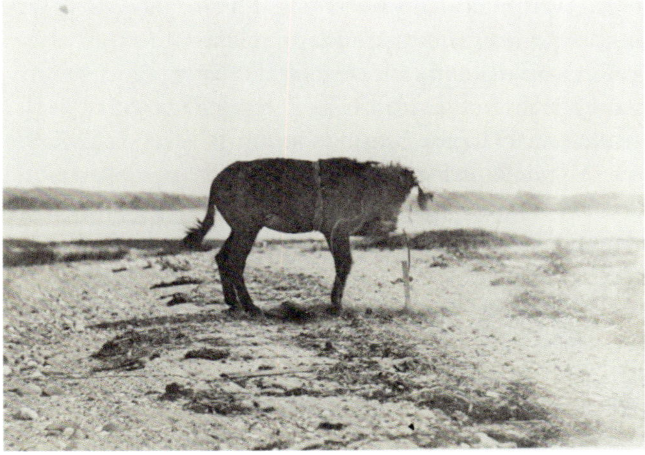

Fig. 6.

Unidentified photographer, *Mule Decapitated by Dynamite Charge*, 1881. Half-tone print on paper, 20×25 cm., after original gelatin silver print taken at the U.S. Engineer School, Willet's Point, New York. Boston, National Archives.

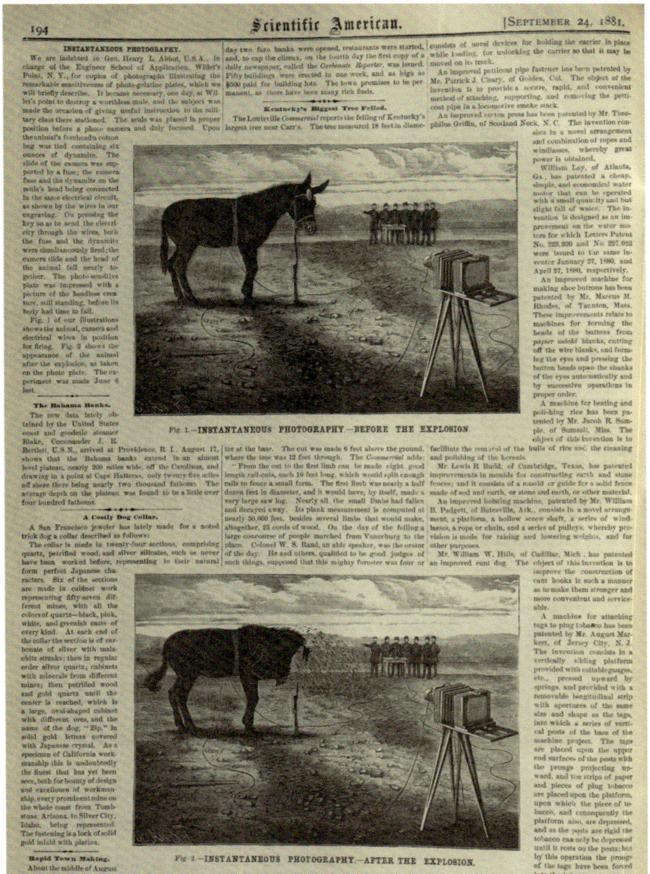

194 Scientific American. [September 24, 1881.

INSTANTANEOUS PHOTOGRAPHY.

We are indebted to Gen. Henry L. Abbot, U.S.A., in charge of the Engineer School of Application, Willet's Point, N. Y., for copies of photographs illustrating the remarkable sensitiveness of photo-gelatine plates, which we will briefly describe. It became necessary, one day, at Willet's point to destroy a worthless mule, and the subject was made the occasion of giving useful instruction to the military class there stationed. The mule was placed in proper position before a photo camera and duly focused. Upon the animal's forehead a cotton bag was tied containing six ounces of dynamite. The slide of the camera was supported by a fuse; the camera fuse and the dynamite on the mule's head being connected in the same electrical circuit, as shown by the wires in our engraving. On pressing the key so as to send the electricity through the wires, both the fuse and the dynamite were simultaneously fired; the camera slide and the head of the animal fell nearly together. The photo-sensitive plate was impressed with a picture of the headless creature, still standing, before its body had time to fall.

Fig. 1 of our illustrations shows the animal, camera and electrical wires in position for firing. Fig. 2 shows the appearance of the animal after the explosion, as taken on the photo plate. The experiment was made June 6 last.

Fig. 1.—INSTANTANEOUS PHOTOGRAPHY.—BEFORE THE EXPLOSION.

The Bahama Banks.

The new data lately obtained by the United States coast and geodetic steamer Blake, Commander J. R. Bartlett, U.S.N., arrived at Providence, R. I., August 17, shows that the Bahama banks extend in an almost level plateau, nearly 200 miles wide, off the Carolinas, and drawing to a point at Cape Hatteras, only twenty-five miles off shore there being nearly two thousand fathoms. The average depth on the plateau was found to be a little over four hundred fathoms.

A Costly Dog Collar.

A San Francisco jeweler has lately made for a noted trick dog a collar described as follows:

The collar is made in twenty-four sections, comprising quartz, petrified wood, and silver silicates, such as never have been worked before, representing in their natural form perfect Japanese characters. Six of the sections are made in cabinet work representing fifty-seven different mines, with all the colors of quartz—black, pink, white, and greenish casts of every kind. At each end of the collar the section is of carbonate of silver with malachite streaks; then in regular order silver quartz; cabinets with minerals from different mines; then petrified wood and gold quartz until the center is reached, which is a large, oval-shaped cabinet with different ores, and the name of the dog, "Zip," in solid gold letters covered with Japanese crystal. As a specimen of California workmanship this is undoubtedly the finest that has yet been seen, both for beauty of design and excellence of workmanship, every prominent mine on the whole coast from Tombstone, Arizona, to Silver City, Idaho, being represented. The fastening is a lock of solid gold inlaid with platina.

Fig 2.—INSTANTANEOUS PHOTOGRAPHY.—AFTER THE EXPLOSION.

Rapid Town Making.

About the middle of August a stampede took place from Deadwood, Central City, and Lead City, Dakota, to a new and wonderfully promising silver region about ten miles from Deadwood. A town was laid out, lots drawn for by all present, rules of government agreed to, and the place named "West Virginia City." In forty-eight hours the town contained nearly one thousand inhabitants, and nine saloons were in operation. On the third day two faro banks were opened, restaurants were started, and, to cap the climax, on the fourth day the first copy of a daily newspaper, called the *Carbonate Reporter*, was issued. Fifty buildings were erected in one week, and as high as $500 paid for building lots. The town promises to be permanent, as there have been many rich finds.

Kentucky's Biggest Tree Felled.

The Louisville *Commercial* reports the felling of Kentucky's largest tree near Carr's. The tree measured 18 feet in diameter at the base. The cut was made 6 feet above the ground, where the tree was 12 feet through. The *Commercial* adds:

"From the cut to the first limb can be made eight good length rail-cuts, each 10 feet long, which would split enough rails to fence a small farm. The first limb was nearly a half dozen feet in diameter, and it would have, by itself, made a very large saw log. Nearly all the small limbs had fallen and decayed away. Its plank measurement is computed at nearly 30,000 feet, besides several limbs that would make, altogether, 25 cords of wood. On the day of the felling a large concourse of people marched from Vanceburg to the place. Colonel W. S. Rand, an able speaker, was the orator of the day. He and others, qualified to be good judges of such things, supposed that this mighty forester was four or five centuries old. There were to be distinctly seen over 300 rings, and it is not known how long since it completed its growth, as it has been dead for many years.

MECHANICAL INVENTIONS.

An improved elevator and carrier has been patented by Mr. Daniel Dockstader, of Fonda, N. Y. The invention consists of novel devices for holding the carrier in place while loading, for unlocking the carrier so that it may be moved on its track.

An improved petticoat pipe fastener has been patented by Mr. Patrick J. Cleary, of Golden, Col. The object of the invention is to provide a secure, rapid, and convenient method of attaching, supporting, and removing the petticoat pipe in a locomotive smoke stack.

An improved cotton press has been patented by Mr. Theophilus Griffin, of Scotland Neck, N. C. The invention consists in a novel arrangement and combination of ropes and windlasses, whereby great power is obtained.

William Lay, of Atlanta, Ga., has patented a cheap, simple, and economical water motor that can be operated with a small quantity and but slight fall of water. The invention is designed as an improvement on the water motors for which Letters Patent No. 223,930 and No. 227,023 were issued to the same inventor January 27, 1880, and April 27, 1880, respectively.

An improved machine for making shoe buttons has been patented by Mr. Marcus M. Rhodes, of Taunton, Mass. These improvements relate to machines for forming the heads of the buttons from *papier mâché* blanks, cutting off the wire blanks, and forming the eyes and pressing the button heads upon the shanks of the eyes automatically and by successive operations in proper order.

A machine for beating and polishing rice has been patented by Mr. Jacob R. Sample, of Summit, Miss. The object of this invention is to facilitate the removal of the hulls of rice and the cleaning and polishing of the kernels.

Mr. Lewis R. Budd, of Cambridge, Texas, has patented improvements in moulds for constructing earth and stone fences; and it consists of a mould or guide for a solid fence made of sod and earth, or stone and earth, or other material.

An improved hoisting machine, patented by Mr. William B. Padgett, of Batesville, Ark., consists in a novel arrangement, a platform, a hollow screw shaft, a series of windlasses, a rope or chain, and a series of pulleys, whereby provision is made for raising and lowering weights, and for other purposes.

Mr. William W. Hills, of Cadillac, Mich., has patented an improved cant dog. The object of this invention is to improve the construction of cant hooks in such a manner as to make them stronger and more convenient and serviceable.

A machine for attaching tags to plug tobacco has been patented by Mr. August Markert, of Jersey City, N. J. The invention consists in a vertically sliding platform provided with suitable gauges, etc., pressed upward by springs, and provided with a removable longitudinal strip with apertures of the same size and shape as the tags, into which a series of vertical posts of the base of the machine project. The tags are placed upon the upper end surfaces of the posts with the prongs projecting upward, and the strips of paper and pieces of plug tobacco are placed upon the platform, upon which the piece of tobacco, and consequently the platform also, are depressed, and as the posts are rigid the tobacco can only be depressed until it rests on the posts; but by this operation the prongs of the tags have been forced into the tobacco.

An improved sash cord fastener has been patented by Messrs. Thomas P. Dunne and Paul Bath, of New York city. The object of this invention is to facilitate the attaching of the sash cord from which the balancing weight is suspended to the sash in such a manner that neither the stop beads, parting strips, nor the sashes need be removed for attaching the cord or chain.

Fig. 7.

Unidentified artist, illustrations in "Instantaneous Photography," *Scientific American*, September 24, 1881.

of gelatin itself contributed to—and even enhanced—the sensitivity of photographic emulsion, revealing a kind of lingering agency or animation in the animal material long after its slaughter, extraction, and reification. ("Instantaneous Photography," 1881).

Though not a trained scientist himself, George Eastman seems to have understood something about this ineffable animal agency through his many commercial investments in photographic gelatin and countless personal experiments with emulsion formulas. In addition to designing leather bindings for his cameras and promoting the "instantaneous" exposure of his gelatin plates and films, Eastman hired legions of scientists and technicians to examine photographic gelatin in meticulous detail in order to refine and exploit its properties. One important example of such scientific inquiry is Samuel Edward Sheppard's book *Gelatin in Photography*, published by Eastman Kodak in 1923. Sheppard was a British chemist hired by Eastman in 1913 to run experiments on physical and colloid chemistry at the Kodak Research Laboratory in Rochester. Based on more than a decade of study, Sheppard's treatise contains these declarative statements about the significance of gelatin in photography:

> Since all modern photographic processes depend upon the use of gelatin as the medium in which the silver salts are suspended, the properties and behavior of gelatin are of the utmost importance in photographic theory. [...] As is commonly known, gelatin and its humbler relative, glue, are products of animal origin, the result of the action of hot water or steam upon certain tissues and structures of the body. Primarily, it is obtained from the connective tissues by hydrolytic separation from the skeletal, fleshy, and fatty substances which are suspended in it in the organism. Though by manufacture somewhat of an artificial substance of varied character, it has originally a definite physiological and biochemical unity which may be diminished, but is neither destroyed nor created, in the process of preparation. (Sheppard, 1923: vii, 25)

The latter statement by Sheppard is particularly interesting, for it acknowledges that even after extensive processing, gelatin retains characteristics of the "organism" from which it was extracted. Sheppard proceeds to explain and illustrate the gelatin "process of preparation" in exacting detail, including "Liming," "Washing," "Neutralizing," "Cooking," "Clarifying or Filtering," "Concentrating," and "Bleaching." A series of remarkable illustrations shows blown-up elevation views of a cow hide and epidermis at different stages in the gelatin-making process, beginning with a "Normal calfskin." (fig. 8) followed by "Calfskin after 66 hours in lime water," "after 120 hours in lime water," and finally "after bathing." (fig. 9) As these illustrations suggest, the transformation

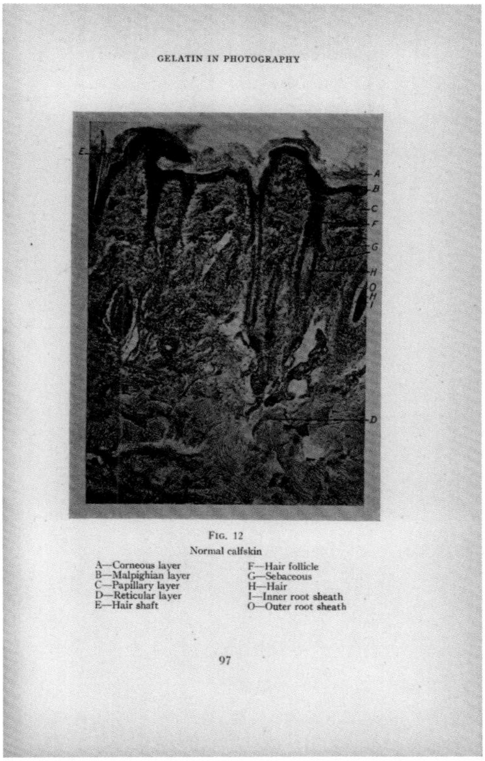

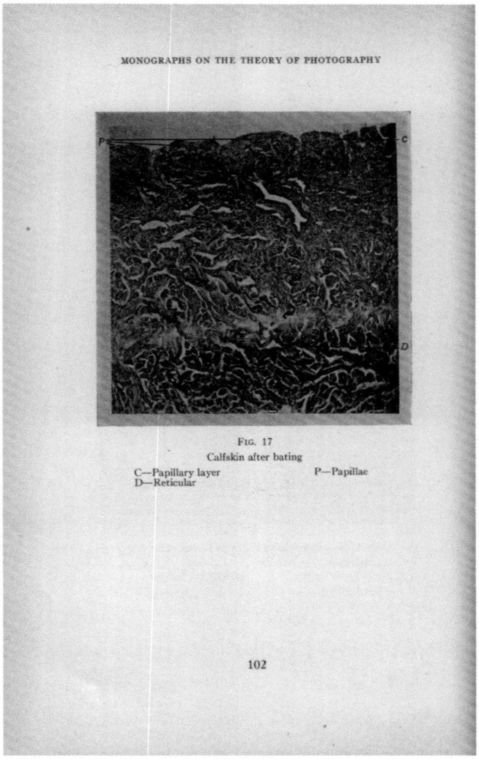

Fig. 8.

"Normal Calfskin," in S. E. Sheppard, *Gelatin in Photography* (Van Nostrand/Eastman Kodak, 1923), 97.

Fig. 9.

"Calfskin after Bat[h]ing" in S. E. Sheppard, *Gelatin in Photography* (Van Nostrand/Eastman Kodak, 1923), 102.

of living flesh into gelatin entails a gradual process of homogenization and abstraction leading eventually to liquefaction. The elevation views provided in these images also underscore the relevance of gelatin production as a form of extractivism analogous to industrial mining of subterranean materials. Both industrial processes—gelatin production and mining—entail the vertical penetration of layers, or strata, to remove and refine matter essential to photography. If we are to regard mining for that medium as "geological" (per Angus, 2024), then I would designate gelatin production as *biopolitical.* (ibid., 26–32, 97–102)

Gelatin exemplifies the almost miraculous power of capitalism to extract resources and produce surplus value from the unlikeliest of places and materials—in this case from the detritus of slaughterhouses and tanneries. Even an outspoken socialist critic such as Upton Sinclair conceded the astonishing efficiency of the meatpacking industry in *The Jungle*, his famous 1906 muckraking novel about the Union Stock Yards in Chicago, saying:

> No tiniest particle of organic matter was wasted. [...] Out of the horns of the cattle they made combs, buttons, hairpins, and imitation ivory; out of the shinbones and other big bones they cut knife and toothbrush handles, and mouthpieces for pipes; out of the hoofs they cut hairpins and buttons, before they made the rest into glue. From such things as feet, knuckles, hide clippings, and sinews came such strange and unlikely products as gelatin, isinglass, and phosphorus, bone black, shoe blacking, and bone oil. (Sinclair, 1906: 46–47)

Sinclair did not mention the mysterious thing known as gelatin emulsion in photography, but *The Jungle* illuminated its origins in an industry that helped establish modern standards of systematic slaughter, mass production, and the reification of life and labor.[11]

It is fitting in this context to cite a 1926 article in the *Daily Worker*, a socialist newspaper, comparing labor conditions in the Eastman Kodak Gelatin Plant at Rochester to those in Chicago described by Sinclair in *The Jungle* twenty years before. As the anonymous author explained,

> If Upton Sinclair ever visits this 'Flower City' [a nickname for Rochester] [...] I shall petition the officials of the Eastman Kodak Co. to show him thru the gelatine department of the Kodak Park plant, commonly referred to by the workers as the "gut factory." I have concluded, after reading his description of the filthy conditions under which the slaves of the packing industry toll [...] in his 'The Jungle,' that no other person could describe the plight of these gelatine workers in the same realistic manner as could Sinclair. ("Filth in Kodak Plant is Hard Even to Relate," 1926: 5)

The article proceeded to describe "Rotten Hides [...] white with maggots," "The Vat Gang" of laborers "plastered with lime from head to foot," a stench that made Sinclair's "Stockyards Smell Mild," and such "Low Pay" for "working under these conditions" that employees ought to "organize themselves into a union." Such conditions indicate that the industrial production of photographic gelatin at Eastman Kodak entailed the biopolitical extraction of life and labor across species. Although George Eastman proudly touted his welfare capitalism in the form of various worker benefit programs (including profit sharing, life insurance, disability benefits, retirement annuity accounts, and more), such policies were motivated by his economic desire to deter unions, strikes, and associated costs. He also implemented a biased hiring policy that privileged White, Anglo-Saxon workers while excluding African Americans, Catholic immigrants, and Jews. These strategies largely succeeded for Eastman during his lifetime (he committed suicide in 1932 at age 77 while suffering from a painful degenerative spinal disease).

Worker loyalty remained high at Eastman Kodak through the 1950s, but labor issues began to arise in the 1960s, notably in 1966, when the Rochester-based civil rights organization FIGHT (which stands for freedom, integration, God, honor, today) protested the company's discriminatory hiring practices and succeeded in negotiating an agreement to add 600 African American employees.[12]

In 1930, a few years after the publication of the *Daily Worker* article, Eastman Kodak purchased a century-old factory in Peabody, Massachusetts, called the American Glue Company, renaming it Eastman Gelatine Corporation. Peabody was the historic center of the Massachusetts leather tanning industry, so it had well-established networks for obtaining animal material. Kodak needed the facility to meet growing demand for photographic gelatin. Eastman Gelatine Corporation supplied two-thirds of Kodak gelatin during the twentieth century, while the company plant in Rochester provided the rest. Demand for gelatin became so great during the twentieth century, however, that Eastman Kodak had to supplement American supplies of animal material with cow bones imported from South America. According to a 1957 article in the *Boston Herald*, Eastman Gelatine Corporation produced 2,500 tons of gelatin per year, along with 4,500 tons of fertilizer and animal food products; 2,500 tons of gelatin translates into 5,000,000 pounds, which in turn required roughly 25,000,000 pounds of raw animal material in the form of hides and bones, based on the calculation I quoted earlier from Eastman's correspondence. In the early twenty-first century, Kodak sold Eastman Gelatine Corporation to a European conglomerate amid the rise of digital photography and growing legal problems associated with air and water pollution generated by the facility. (Eastman Gelatine Corporation, 1945; "Eastman Kodak Subsidiary Pulls Weight in Production of Gelatine," 1957: A75; Dumanoski, 1983; Miller, 1984; Massachusetts Department of Public Health, 1986; Castellucio, 2011)[13]

Digital photography hardly eliminated demand for gelatin in photography, though. To the contrary, gelatin has remained central to photographic printing, and demand has continued to grow, as confirmed by recent market reports, like one stating that "analysis of the photographic gelatin market reveals a positive outlook for the industry," marked by "increasing demand for traditional photography products" along with "growing popularity of analog photography and the nostalgia associated with it." The article also lists "major market players" in photographic gelatin, including corporations in India, China, Europe, and the United States. Another recent market analysis predicts more than 100 percent growth in overall animal gelatin production in the United States during the coming decade, citing among other factors how "the rising demand for gelatin photography will fuel the growth of the industry." So, just as Angus (2024) has demonstrated the ongoing extractivism of photography vis-à-vis rare earth materials, there is ample evidence confirming the medium's persistent use of animal material in the form of gelatin even in the digital regime. (Precedence Research, 2022; Market Edge Solutions, 2023)[14]

As a way of concluding, I illustrate one of the many photographs George Eastman took of prize cows and calves that he owned and raised on the grounds of his palatial estate in Rochester and at his rural Oak Lodge retreat in Halifax County, North Carolina. (fig. 10) In terms of genre, such animal photographs constitute an agricultural corollary to his picturesque snapshots of western scenery at places such as Yosemite, an example of which I presented at the beginning of this essay. Often given individual names and tended by hired human wranglers, the animals cultivated and photographed by Eastman provided him with milk, butter, cheese, and meat for his private consumption. In voluminous personal correspondence, he wrote obsessively to his staff and to agricultural experts about how to optimize his herd for food production. At once pastoral hobby, scientific experiment, and photography project, Eastman's cows reveal his complex personality, which combined impulses of gentleman farmer with corporate industrialist, animal husbandman with mass producer of photographs and photographic gelatin. Not surprisingly, Eastman used the word "stock" to describe quantities of both film and cows. His pictures of the latter, as gelatin prints, embody and represent cows. No explicit verbal acknowledgment of this intersection of iconography and materiality appears in Eastman's correspondence, perhaps because it was so obvious to him as to not require comment. For many of us today, the connection between photographic gelatin and cows may not be so obvious, but ecocritical understanding helps us see and comprehend their inextricable

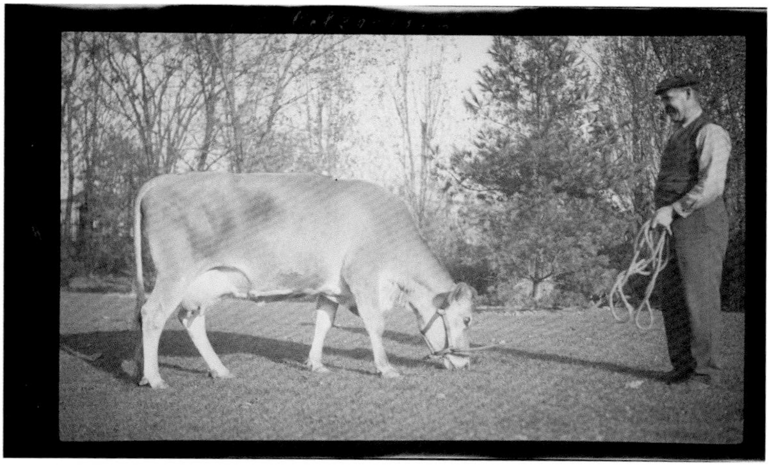

Fig. 10.

George Eastman, *Cow "Sylvia" on Eastman's Property*, 1916. Gelatin on nitrocellulose roll film, 3 ¼ × 5 ½ in. Rochester, George Eastman Collection, International Museum of Photography at George Eastman House.

interrelationship. Such an understanding requires that we look beyond the form and content of the photographic image to consider its material substrates, infrastructures, and commodity chains, which tend to remain invisible to analysis focused on aesthetics and iconography. Gelatin epitomizes this elusive yet essential dimension of the medium. Ecocriticism—especially when attentive to non-human life—thus enlarges photography by revealing its implication within a broader field of historical and material agents.[15]

Notes

I thank Erin Fisher at the University of Rochester Library for assistance with my research in the Kodak Historical Collection, as well as Ken Fox and John Elsbree at the Eastman Museum for helping me navigate the Menschel Library and the Eastman Legacy Collection.

1. On Yosemite, see Scott, 2006, with references to Watkins (11, 56, 60–65, 84, 116, 131, 173, 203, 208–209), Eastman (116), and Native Americans (3, 6, 68, 91–113, 120, 161, 187, 199, 201). For more on Watkins, see Green, 2020; Naef, 2009. On Eastman, see Brayer, 2006. See also Spence, 1999.

2. For a helpful discussion and comparison of these and other photographic processes, see http://www.graphicsatlas.org/compareprocesses/ (accessed April 28, 2026).

3. On slaughterhouses, meatpacking, and western supply chains shaping "the great bovine city of the world," see Cronon, 1991: 207. The Kodak magazine illustrations appear in Lawrence, 1960, unpaginated. On the discovery by Sheppard and Kodak scientist C. E. Kenneth Mees about how cows that ate mustard seeds improved gelatin sensitivity, see James, 1990: 129–134. For the Eastman Museum demonstration image, see https://www.eastman.org/event/talks-workshops/online-gelatin-dry-plates-george-eastman-demonstration-history. On gelatin in photography, see Jenkins, 1975: 65–95. For a general history of gelatin, see Mikhailov, 2023.

4. On gelatin dry-plate photography, consumer access, and "You press the button, we do the rest" as an Eastman Kodak advertising slogan, see Collins, 2021: 1–22, 61. See also Jenkins, 1975: 65–95.

5. For the classic Marxist definition of *reification*, see Lukács, 1971: 83–109.

6. For one example of Pachirat's influence, see Braddock, 2023.

7. Regarding critical animal studies, see, for example, Matsuoka and Sorenson, 2018; Taylor and Twine, 2014. On ecocriticism and animal studies, see Braddock, 2015, 447–467, esp. 457–462. On Marey and Muybridge, see Braun, 1992, and Solnit, 2004.

8. Recent scholarship has complicated dominant assumptions about Marx's anthropocentrism by recovering neglected ecological dimensions of his writings. See, for example, Saito, 2017.

9. On Eastman's early promotional travels in Europe, including his encounter with Nadar and the efforts of Nadar's son, Paul, as Eastman's agent in France, see Brayer, 2006: 73–93, 121–136.

10. George Eastman to S. V. Haus, Surrey, England, March 31 and June 1, 1911. On Armour & Company as one of the "Big Six" meatpackers of the late nineteenth century in America, see Pacyga, 2015, xvi-xvii, 106–124. See also Cronon, 1991: 234–235, 242–243, 254, 340, 375. In 1907, Armour & Company built a meat smoking and packing facility at 1050 Battery Street in San Francisco, now on the US National Register of Historic Places. https://noehill.com/sf/landmarks/nat2009001117.asp. For a representation of Armour & Company's reach, showing cattle farms in California and throughout the Great Plains, see Armour's Food Source Map, 1922.

11. For an illuminating study of Sinclair's novel and its historical context from the perspective of critical animal studies, see Lundblad, 2013. On various dimensions of the meatpacking industry in Chicago, see also Weems, 2016: 106–143;

Pacyga, 2015; Cronon, 1991; Giedion, 1948: 209–246.

12. On Eastman's welfare capitalism and exclusionary hiring practices, see Jacoby, 1997: 57–70. Regarding FIGHT, see Wadhwani, 1996: 59–75.

13. The first two items can be found in the Kodak Historical Collection, Box 97, Folder 2. On the former Eastman Gelatine Corporation factory in Peabody as an Environmental Protection Administration Superfund site, see https://cumulis.epa.gov/supercpad/CurSites/csitinfo.cfm?id=0100461 (accessed April 28, 2026).

14. On rare earths, see Angus, 2024: 196–221.

15. The author thanks John Elsbree at the Eastman Museum, Rochester, for help in navigating the Eastman Legacy Collection, which contains copies of George Eastman's original letters, including many written during the 1910s describing his private "stock" of cows and calves. See, for example, George Eastman to Almon R. Eastman, July 9, 1916 (George Eastman Correspondence).

Archival Collections

George Eastman Correspondence, Eastman Legacy Collection, Eastman Museum, Rochester.

Bibliography

Siobhan Angus, *Camera Geologica: An Elemental History of Photography* (Durham, NC: Duke University Press, 2024).

"Armour's Food Source Map: The Greatness of the United States is Founded on Agriculture" (Chicago: Armour & Company, 1922/2024a), https://commons.wikimedia.org/wiki/File:Armour%27s_Food_Source_Map_.jpg (accessed April 28, 2026).

Alan C. Braddock, "From Nature to Ecology: The Emergence of Ecocritical Art History," in *A Companion to American Art*, eds John Davis, Jennifer A. Greenhill, and Jason D. LaFountain (Malden, MA: Wiley Blackwell, 2015).

Alan C. Braddock, *Implication: An Ecocritical Dictionary for Art History* (New Haven, CT: Yale University Press, 2023), unpaginated ebook.

Marta Braun, *Picturing Time: The Work of Etienne-Jules Marey (1830–1904)* (Chicago: University of Chicago Press, 1992).

Elizabeth Brayer, *George Eastman: A Biography* (Rochester: University of Rochester Press, 2006).

Rachel Carson, *Silent Spring* (Boston: Houghton Mifflin, 1962).

John Castelluccio, "Kodak Enters Agreement to Sell Eastman Gelatine Factory," *Patch* (December 23, 2011), https://patch.com/massachusetts/peabody/kodak-enters-agreement-to-sell-eastman-gelatine (accessed April 28, 2026).

Kevin Coleman and Daniel James (eds), *Capitalism and the Camera: Essays on Photography and Extraction* (London: Verso, 2021).

Douglas Collins, *The Story of Kodak* (New York: Harry N. Abrams, 1990).

William Cronon, *Nature's Metropolis: Chicago and the Great West* (New York: W.W. Norton, 1991).

Eastman Gelatine Corporation, *Handbook for the Men and Women of Eastman Gelatine Corporation* (Peabody, MA: Eastman Gelatine Corporation, 1945).

"Eastman Kodak Subsidiary Pulls Weight in Production of Gelatine," *Boston Sunday Herald*, October 27, 1957: A75.

"Filth in Kodak Plant is Hard Even to Relate," *Daily Worker*, December 10, 1926: 5.

Siegfried Giedion, *Mechanization Takes Command: A Contribution to Anonymous History* (New York: Oxford University Press, 1948).

Tyler Green, *Carleton Watkins: Making the West American* (Oakland, CA: University of California Press, 2020).

"Instantaneous Photography," *Scientific American*, September 24, 1881.

Sanford M. Jacoby, *Modern Manors: Welfare Capitalism since the New Deal* (Princeton, NJ: Princeton University Press, 1997).

Thomas Howard James, *A Biography-Autobiography of Charles Edward Kenneth Mees: Pioneer of Industrial Research* (Rochester, NY: Photographic Research Laboratories, Eastman Kodak Company, 1990).

Reese Jenkins, *Images and Enterprise: Technology and the American Photographic Industry, 1839 to 1925* (Johns Hopkins University Press, 1975).

Bob Lawrence, "This is Eastman Gel," *Kodakery* (February 11, 1960), unpaginated.

György Lukács, The Phenomenon of Reification," in *History and Class Consciousness: Studies in Marxist Dialectics*, trans. Rodney Livingstone (1923; Cambridge, MA: MIT Press, 1971), 83–109.

Michael Lundblad, *The Birth of a Jungle: Animality in Progressive-Era U.S. Literature and Culture* (New York: Oxford University Press, 2013).

Market Edge Solutions, "Photographic Gelatin Market Size, Growth and Forecast, 2023–2030," *LinkedIn*, December 11, 2023, https://www.linkedin.com/pulse/photographic-gelatin-market-size-growth-fore-cast-nmu6f/ (accessed April 28, 2026).

Karl Marx, *Capital: A Critique of Political Economy. Volume I: The Process of Capitalist Production*, trans. Samuel Moore and Edward Aveling (1867; Chicago: Charles H. Kerr, 1906).

Massachusetts Department of Public Health, "An Investigation into Pancreatic Cancer Mortality in Peabody, Massachusetts," September 1986, https://archive.org/stream/investigationintoomass_0/investi-gationintoomass_0_djvu.txt (accessed April 28, 2026).

Atsuko Matsuoka and John Sorenson, *Critical Animal Studies: Trans-species Social Justice* (Lanham, MD: Rowman & Littlefield, 2018).

Oleg V. Mikhailov, "Gelatin as It is: History and Modernity," *International Journal of Molecular Sciences* 24, no. 3583 (2023), https://doi.org/10.3390/ijms24043583 (accessed April 28, 2026).

Mark Miller, "Cleanup Still Leaves Loose Ends: Questions on Health Remain in Peabody," *Boston Globe*, July 27, 1984.

Weston Naef, *Carleton Watkins in Yosemite* (Santa Monica: J. Paul Getty Museum, 2009).

Rob Nixon, *Slow Violence and the Environmentalism of the Poor* (Cambridge, MA: Harvard University Press, 2011).

Timothy Pachirat, *Every Twelve Seconds: Industrialized Slaughter and the Politics of Sight* (New Haven, CT: Yale University Press, 2011).

Dominic A. Pacyga, *Slaughterhouse: Chicago's Union Stockyard and the World It Made* (Chicago: University of Chicago Press, 2015).

Precedence Research, "Gelatin Market Size, 2023–20232 (USD Billion)," *Precedence Research*, 2022, https://www.precedenceresearch.com/gelatin-market (accessed April 28, 2026).

Kohei Saito, *Karl Marx's Ecosocialism: Capital, Nature, and the Unfinished Critique of Political Economy* (New York: Monthly Review Press, 2017).

Amy Scott (ed.), *Yosemite: Art of an American Icon* (Los Angeles: Autry National Center in association with University of California Press, 2006).

S[amuel] E. Sheppard, *Gelatin in Photography* (Van Nostrand/Eastman Kodak, 1923).

Nicole Shukin, *Animal Capital: Rendering Life in Biopolitical Times* (Minneapolis: University of Minnesota Press, 2009).

Upton Sinclair, *The Jungle* (New York: Doubleday, 1906).

Rebecca Solnit, *River of Shadows: Eadweard Muybridge and the Technological Wild West* (New York: Penguin, 2004).

Mark David Spence, *Dispossessing the Wilderness: Indian Removal and the Making of the National Parks* (New York: Oxford University Press, 1999).

Nik Taylor and Richard Twine (eds), *The Rise of Critical Animal Studies: From the Margin to the Centre* (New York: Routledge, 2014).

R. D. G. Wadhwani, "Kodak, FIGHT, and the Definition of Civil Rights in Rochester, New York, 1966–1967," *Historian* 60, no. 1 (September 1996): 59–75.

Jason Weems, "Scale, A Slaughterhouse View: Industry, Corporeality, and Being in Turn-of-the-Century Chicago," in *Scale*, ed. Jennifer L. Roberts (Chicago: Terra Foundation for American Art, 2016).

CHAPTER 6
The Death of 7000 Songbirds: Photography's Toxic Affinities

Alice Cazenave

This short provocation examines photography's history by thinking about the metals and salts that make up photographs. Silver grounds the magic of analog photography; it is the essential light-sensitive coating of photographic negatives. In the early 1920s, Kodak manufactured 147,000 miles of motion picture film a year—enough to circle the earth six times. (Fenton, 1921) This piece extends important inquiries into the elemental base of photography by tracing the chemical worlds that surround silver extractivism in the American West. (Angus, 2024; Bravo, 2025) It theorizes photography as a series of molecular relations that make marks on the world in delayed and dispersed ways. (Schuppli, 2020; Tucker, 2022) It thinks about these cloudy relations as the ecologies of photography: an intricate metabolism of chemicals, industries, earthly matter, and bodies that mobilize in and through each other. They are cyanide, songbird, mercury, and roots that coalesce into new bodies on their journeys through the mine. In tracing these chemical assemblages, this piece ask: What are some of the distinct conditions that have enabled the development of photography, and how have these participated in the construction and destruction of more-than-human worlds?

Landscapes of the American West: Silver, Willow, and Settler Colonialism

This work draws on recent fieldwork in regions known today as the Comstock, Nevada. This is the homeland of Nüümü (Northern Paiute), Newe (Shoshone), and Wašišiw (Washoe) peoples, and in 1859 the region was discovered to hold the densest deposits of silver in the United States. Silver booms in the region coincided with early developments

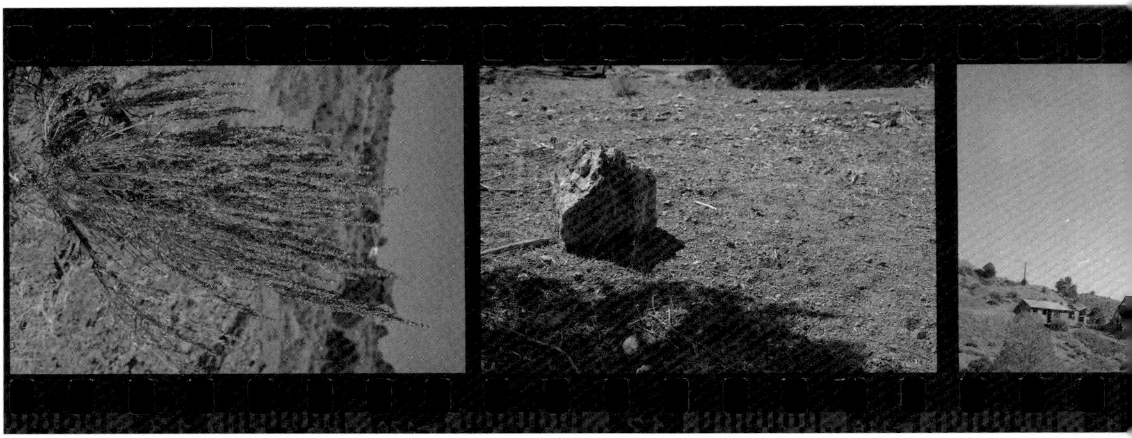

in photography, and across three decades, between 1860–1995, nearly seven million tons of silver were extracted from the Comstock. This extraction produced nearly half a billion dollars' worth of capital in today's currency. (Vischer, n.d.) In 1880, Kodak was founded in Rochester, New York, and at its peak, Kodak consumed 13 tons of silver a week to transform into analog film. (Eastman Kodak Company, 1949: 13)

Extractive industries violently uprooted Native American peoples, as well as the ecologies that sustained, and were sustained by, them. Coinciding with this in the 1870s was the development of landscape photography that depicted the American West as wild, empty, and a resource for the taking. (Hore, 2022) As visual technologies that moved through cultural, governmental, and military spaces, landscape photographs of the American West constructed new ways of thinking about territory and nationhood. (ibid.) They shaped settler-colonial ideas of progress and industrial growth, and these ideas played out materially in the form of precious-metal extraction. The legacies of this extraction were violent in more ways than one—ask the Pinyon forests, the Pinyon jays, or the rivers that absorbed the mercury. (Cazenave, 2025) (fig. 1)

Conflicts between settler-colonists extracting silver and Native American peoples defending their lands inspired early Westerns films. (Dickson, 1894; Ford, 1926) The year is 1894. Hollywood is being born, and the film era is still silent. Nitrate film crackles and pops as it speeds through the projector. Flashes of smoke explode on the screen to wide-eyed audiences. They clap and cheer the motion picture. It looks like cowboys and gunfights, but really it is something else: This is the story of film's materiality unraveling on the silver screen.

After a century of violent silver extraction from Native American lands, mining corporations are required to ask for permission. The year is 1954. The prospecting requests are now situated in archives, but

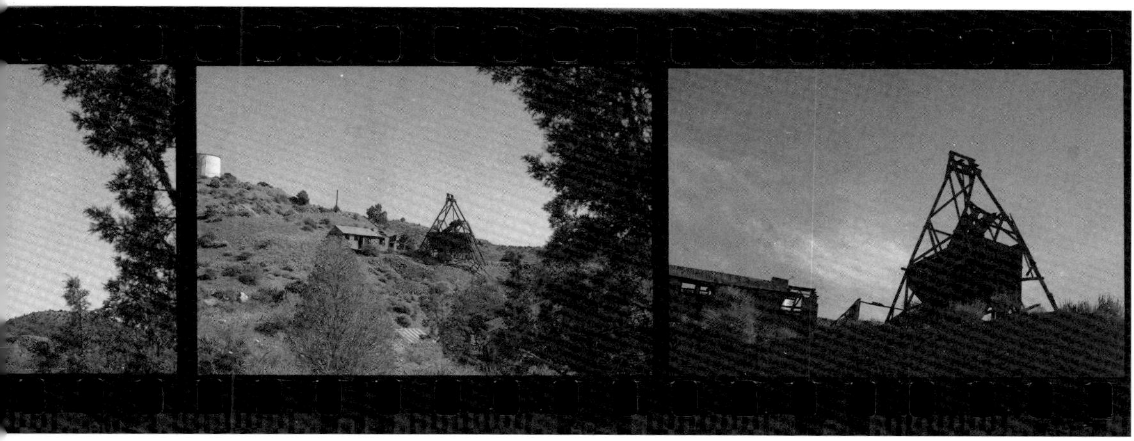

they once passed the hands of the Pyramid Lake Paiute Tribal Council (University of Reno, Special Collections). The letters ask for leases to mine—they assure the Council that the Council has "nothing to lose, and everything to gain." (letter by Daniel W. Foster, 1940) The Council asks where the corporations will source their water. The question is handwritten in the margins of the request letter to be raised at the next Council meeting. (letter by Raymond Patrick, 1954) Sourcing water was a critical issue because water was sacred, but to mining industries it powered the machines that pounded extracted ore. This was the work of stamp mills and their giant battering rams.

Fig. 1.
Alice Cazenave, *Untitled (Silver City Mines)*, 2022.
© Alice Cazenave, 2025.

As many as two hundred stamp mills lined the Carson River (Toinahukwa in Northern Paiute and Watahshemu in Washoe). Their stamps towered twenty feet tall and moved in unison to crush extracted ore. This created a fine, silvered powder that was then flushed with mercury, which caught the silver and adhered to it. This mercury-silver amalgam was then heated to evaporate off the mercury and leave the isolated silver. As a result of just thirty years of nineteenth-century operations, fourteen million pounds of mercury entered the Carson River Basin. (USEPA, 2022) While the river is now dry, willow trees line where the water once flowed, and they make an excellent ingredient for plant-based photo chemistry. (Neilsen et. al, 2017) (fig.2)

My photographic practice involves hand-mixing low-toxicity photo chemistries with plants. I forage plants from landscapes that have been touched by silver extraction and for analog film industries. I mix them into chemistry to create experimental photographs that tell stories about the ecologies, geologies, and lives that form part of photography's history. (Cazenave, 2025) The chemistry that I make is known in photographic worlds as the developer chemistry. This is the first chemistry that analog photographers use to develop their initial images

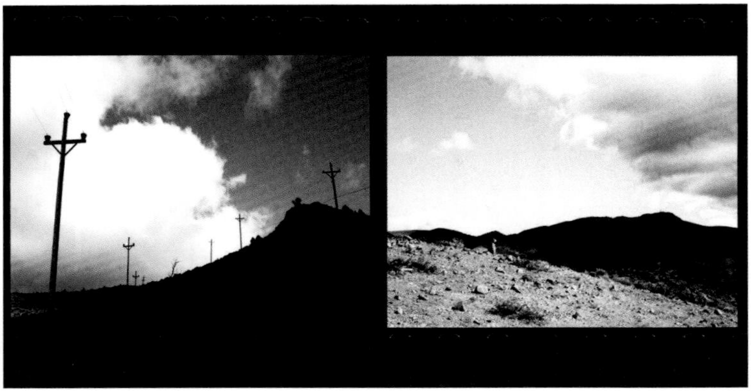

Fig. 2.

Alice Cazenave, *Untitled (Silver City Mines)*, 2022. © Alice Cazenave, 2025.

on film or paper. The formation of the image occurs when the developer chemistry interacts with the silver in the film's emulsion. Upon mixing, the silver emerges in the trace of a photographed scene. This reaction is then stopped with a water wash and fixed with sodium thiosulfate. This last step stops the photograph from wandering off and continuing its exposure to light. The result is a miniature world rendered visible by silver. It is this alchemy of the darkroom and the materiality of its processes that grabs some photographers by the scruff of their necks.

Willow leaves are long and crisp and elegant in shape. I have a habit of crunching them between my fingers when I pick them—a legacy from childhood. I grab about two handfuls and boil them. I then leave the brewed solution outside to cool, and a measure each of washing soda and vitamin C join the mix. I whirl the bowl by its edges to dissolve the ingredients that mix with the motion. A silver miner I met practiced the same motion when he separated silver from ground-up ore. The silver pooled to the side of the tub as he whirled it with fast, elegant motions—a beautiful moment when the body thinks and the brain rests.

The base of a plant-developer chemistry is washing soda and vitamin C. These ingredients create sodium ascorbate, which powers the photo chemistry. The plants add what I call messiness; they can increase contrast, soften tones, or speed up development times of the photo chemistry. (fig. 3)

Working with plant photo chemistry is slow but enriching. Definitely humbling. Recently I have found it addictive. It is never certain how the chemistry will work—it is a summoning toward something that feels worthy or curious but has no decipherable form. (Manning, 2015)

To develop a film, a handheld developing tank is loaded with film—this takes place in the total darkness of a light-tight changing bag. The home-made chemistry washes through the tank and covers the film in warm, amber liquid. Once the chemistry enters, you can no longer see it. It falls beneath a level that protects the film from daylight. The tank must now be agitated to allow the chemistry to rinse over it evenly. There are strict rules about how long to do this for. Film companies spend hundreds of hours researching and developing precise timings for developing film. Too long, and the film over-develops—blowing out tonal range. Too short, and the film under-develops—creating images that are only partly visible. I find that plant photo chemistry is much more forgiving and develops well without strict time-keeping regimes. I agitate the tank at loose intervals when I remember, for roughly ten minutes. (fig. 4)

There is an international movement of analog photographers that is moving away from industrially produced photographic chemistries because of their environmental impacts. (Carr & Fletcher, 2024) They join those in academic circles who are researching the effects of increasingly toxic worlds (Nading, 2020) and the colonial infrastructures that frame patterns of pollution and their uneven global distribution. (Murphy, 2017; Liboiron, 2021) Other scholars explore how people living in toxic habitats make sense of these spaces (Li, 2015), and the roles of corporate powers in masking the impacts of their chemical operations. (Renfrew and Pearson, 2021) Silver itself is not toxic to humans, but in ionic forms, it can wreak havoc on marine and microbial worlds. (Choi et al., 2008) It is for this reason that some photographers remove silver from their waste fixer chemistry before disposing of it. (Carr & Fletcher, 2024) Historically, this was done for economic reasons to

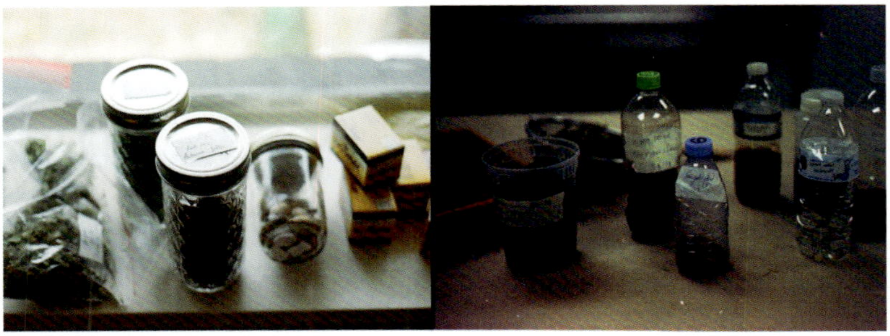

Fig. 3.

Alice Cazenave, *Untitled (Foraged Plants and Photo-Chemistry)*, 2024. © Alice Cazenave, 2025.

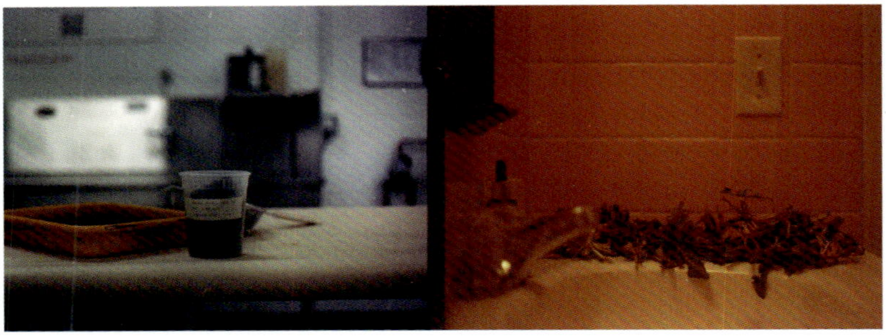

Fig. 4.

Alice Cazenave, *Untitled (Plant Extractions and Foraged Roots)*, 2024. © Alice Cazenave, 2025.

recover the silver for resale. All of the black-and-white photographs that accompany this piece were developed with willow photo chemistry. The willow was foraged from the now-contaminated Carson River Basin, where the stamp mills once stamped silver from its ore. My focus is to examine silver's toxicity through the lens of mercury and cyanide, which enable its extraction, and it is with these two lethal substances that I start my journey.

Geologies of Photography: Mercury, Cyanide, and Photographic Silver

In Nevada, people took me on driving tours around abandoned mines in the desert while I was visiting as an artist-in-residence at Silver City. People read the landscape through the changing colors of its rock. They showed me mountains of pale yellow rubble that posed as hillsides but were in fact piles of tailings. Tailings is a mining term used to describe earth that has been cast away as waste rock. It is earth that has been displaced by tunneling, and it often contains toxic metals, such as mercury or arsenic, that have been discarded during extraction. Sometimes, mercury emits as a vapor from the rubble piles. (Lechler, 1992) In the Comstock, the tailings piles go on for miles and miles. They are easy to spot—they are pale and yellow from being spat out of the mining process. (fig. 5)

We passed mountainsides fractured by open-pit mining operations. These operations used explosives to extract minerals closer to the earth's surface. This style of mining is different from the historical nineteenth-century operations, which tunneled below ground. Rock faces exposed by mine blasts can uncover sulfide minerals that would otherwise be buried within the rock. (USEPA, 1994) Metal mines, such as silver, copper, and gold mines, are often rich in sulfides, and these can leach out as sulfuric acid when it rains. This sulfuric acid can further dissolve the rock and, depending on its chemical composition, release other harmful metals, such as arsenic and cadmium. (ibid.) Sulfuric acid leaching from fractured rock is extremely corrosive. Creeks receiving sulfuric acid run-off have a pH so low that it matches that of battery acid. I imagine it vividly: battery acid creeks powering mountains—charged and churning to the steady beat of mine blasts. Again, and again, and again, with positive rocks and negative rocks grinding the mountain-machine perpetually forward. The creek glistens; its mercury masked as progress.

We follow the creek down to the Carson River Basin, where the plains become a huge, flat expanse. Today, the Carson River is categorized as a National Priorities Superfund site. These are toxic sites allocated extra funds by the United States Environmental Protection Agency for

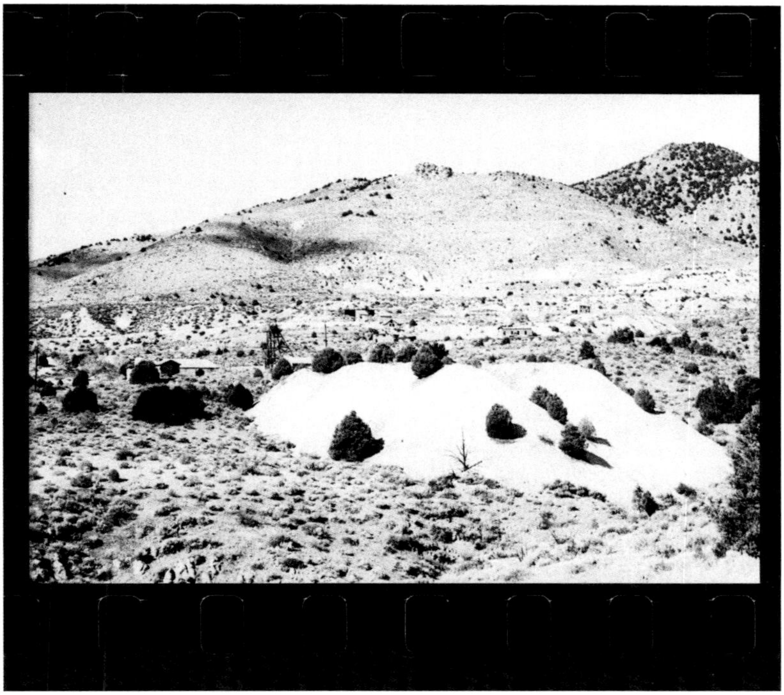

Fig. 5.
Alice Cazenave, *Untitled (Silver City Tailings)*, 2022.
© Alice Cazenave, 2025.

urgent clean up. (USEPA, 2022) The people I met in the Comstock played in the Carson River as children. They hauled rocks to dam the creek and create swimming holes. I think about the mercury and wince at the thought. Tiny ankles in the glistening water. In the summer, when the water was low, the children hunted for mercury that gathered beneath river boulders. Because it was heavy, it lived under the boulders. It could be retrieved by shoving a silver spoon, an actual silver spoon, into the hole. The mercury stuck to the silver and was scraped off into a jar with a plastic poker card.

Like mercury, cyanide also has an affinity for silver—it finds silver when it is near it. Eventually in the United States, mercury was phased out of mining operations and replaced with cyanide. Cyanide was introduced for several reasons; one was that it was considered less persistent in environments: It did not stick around like mercury does, which bioaccumulates in the fat tissues of fish and works its way up the food chain into bears and humans that eat the contaminated fish.

Cyanide works by catching silver from piles of extracted ore called heap leach piles. To work, the cyanide must be oxygenated. It is then sprayed over the heap leach piles to permeate and collect the silver. They say cyanide smells like almonds. I look over to a pile as we drive through the desert and picture it vividly—vats of cyanide wearing oxygen masks, and huge, rubbly lungs rising and falling with the sweet vapor of the mix.

The cyanide solution percolates through the rock-lungs and along pipes that network under the heap leach pile. The cyanide-silver is collected into specialized ponds near the pile. The mining term for this silver-rich solution is *pregnant solution*. I wince at the association. Pregnancy feels like a gift. This has a distinct quality of theft.

Toxic Assemblages: Cyanide, Songbirds, and White-Tailed Sea Eagles

Cyanide ponds holding extracted precious metals can measure hundreds of hectares in surface area, depending on the size of the mining operation. Birds drink from these ponds and die from their effects, but the pools remain, like some kind of cyanide-siren song. Physical barriers, such as gauze, are installed to cordon off unsuspecting creatures, but despite this, between 1980 and 1989, around 7000 dead songbirds were recovered from leach ponds in the western states of the United States. (Allen, 1990; Eisler and Wiemeyer, 2004) Of all the recorded animal deaths, this felt especially plaintive. The silence. I learned that the voice box of a songbird is the size of a raindrop, and it stirred something in me—a strong sense of protection. This increased when I learned that it takes songbirds years to practice their songs. They inherit a template song from their parents and fill it out by learning from other birds. It is thought that they sing to defend their territories. This carries extra weight in this context—these are the territories colonized by photography, and the ecologies silenced with its extractivism.

Cyanide is lethal when ingested, but it is considered a low-probability risk in mining because it evaporates quickly. It is also thought to be degraded by bacteria and protozoa into less harmful compounds. (Eisler and Wiemeyer, 2004) For these reasons, cyanide is widely used in both gold and silver mining operations, which often occur together. However, when mining cyanide spills do occur, the impact is colossal. In 1995, the failure of a dam in Guyana, South America, released 3.3 × 109 liters of cyanide waste into the nation's primary waterway. (Da Rosa and Lyon, 1997) That is 3.3 billion liters—a quantity so large it feels unreal. The cyanide waste killed fish along an 80 km stretch of river and caused widespread contamination of irrigation and drinking water systems. (ibid.) Five years later in the Danube River Delta, heavy metals and cyanide plumes flowed into the Black Sea, 1000 km away from the original spill location. (Koenig, 2000) At least 200 tons of fish were killed in this spill, as well as endangered European otters and white-tailed sea eagles that consumed the contaminated fish.

Today in the United States, mining corporations are under strict regulation to manage the environmental risks associated with their cyanide use. Solutions of lye, an alkaline material, are used to control the pH of heap leach piles to prevent them from becoming acidic. (Eisler and Wiemeyer, 2004) This is required because in acidic conditions, the formation of lethal hydrogen cyanide gas occurs. But no one is there to prevent acidity once cyanide leaves the parameters of the mine. A bird's stomach for instance, like ours, is incredibly acidic. Some scientists suggest that even low concentrations of cyanide are lethal to birds because the cyanide is activated by the acidity of their stomachs. (Clark and Hothem, 1991) In the scientific world, this is known as additive toxicity—when toxicity gathers force as it metabolizes in new chemical environments.

Genealogies of Chemistry: Chemical Weapons and the Photographic

How, then, did we get here, when the toxicity of cyanide and its derivatives has been observed since ancient times? While the toxicity has been observed, the exact molecular composition and mechanism of cyanide was only discovered in recent history—in 1815, by the French chemist Joseph Louis Gay-Lussac. It feels strange to imagine a world without molecular science; it feels like chemicals have always been with us, and the knowledge of their whereabouts, a given. They are, after all, clouds of atoms that have been floating around since the beginning of time. Chemical evolutionists describe a world where all life arose from non-living matter. Over millions of years, atoms clustered and coalesced into simple molecules, suspended in the soup of pre-biologic life. Dark, murky, and full of potential. Molecular chains curled like the edges of a leaf as it dries. They folded into three-dimensional structures, forming early proteins and lipids, which became trapped by larger chains, enveloping them into spherical forms—the progenitors of all cellular life. It feels beautiful—the dance of carbon and hydrogen under storm-stricken skies, when the earth was young and still deciding on weather and water and time.

The formation of chemicals, then, feels natural, inevitable even. But in Europe and the United States, the nineteenth century was the chemical century, and it helped design and mass produce hazardous chemicals synthesized specifically for mass destruction. (Szinicz, 2005) This was a different kind of chemistry than that of nebulous earth. This was a powerful and calculated command of the molecular world, known in industrial worlds as synthetic chemistry.

Before they entered mining operations, cyanide derivates took center stage as military weapons. They were prioritized because of their high toxicity and fast action. During World War I, at least forty-one toxic compounds were specifically designed for war operations. (ibid.: 171) Synthetic chemicals were researched and developed to administer a range of deadly effects to humans. These included reworkings of cyanide, itself a naturally occurring substance, into a range of equally lethal derivatives, such as bromobenzyl cyanide (tear gas agent) and cyanogen chloride (blood agents). It is thought that the majority of deaths in World War I were caused by chemical weapons. (ibid.) Cyanide derivatives were later abandoned because they evaporated too quickly. They did not stick around long enough to gas enemy soldiers with the military's desired efficiency.

When cyanide does meet a human body, it poisons us by latching onto the iron nestled at the center of our blood proteins, because we are made of metals, too. These proteins are the body's honorable carriers of oxygen, and cyanide inhibits them by blocking the iron in their elaborate folds. The iron is critical because it receives and releases oxygen. When cyanide finds the iron, it blocks this function. The blood, unable to let go of its oxygen, turns bright red. As the blood chokes on excess, the body chokes on scarcity. Held captive in the blood cells that carry it, the oxygen is starved from the body that needs it, and bodily systems fail in minutes. (Kuyucak and Akcil, 2013) Cyanide poisoning in the blood teaches us that cyanide has an affinity for iron. An affinity so strong that it is lethal. The affinity between these two substances is at the center of an early photographic process, too, named cyanotype.

So far, I had experienced the silver mines of Nevada in strips of black and white and willow. With cyanotype, they jump to blue. (fig. 6) A deep blue, like sapphires. Where the chemistry pools on the paper, it is deeper still, like blueberries. Invented in the 1840s and still practiced today, cyanotype printing uses cyanide salts and iron salts to produce Prussian-blue photographs. These salts are present in the photo chemistry that is coated into paper to make the prints. The cyanide and the iron bind together to produce a rich cyan color; this is the mind-bending moment when chemical worlds produce color worlds. Unlike in bodily systems, the strong bond between cyanide and iron salts in cyanotypes renders them safe to make. (Bishop-Thorpe, 2016) The cyanide cannot liberate itself to cause damage; it is locked up with the iron.

In some mining regions, this same cyan color can emerge of its own accord in the ground. (Welman-Purchase and Hansen, 2023) This happens when spilled cyanide solutions bind to iron-rich sediments in the earth. The cyanide finds the iron and forms cyan compounds organically. I imagine it vividly: whole mountainsides deepening from brown to blue. A giant, geochemical photograph, exposing under the desert sun.

Historical and industrial uses of cyanide span military, mining, and photographic worlds. In World War I, cyanide was too transient—it evaporated too quickly to administer its desired lethality. This so-called

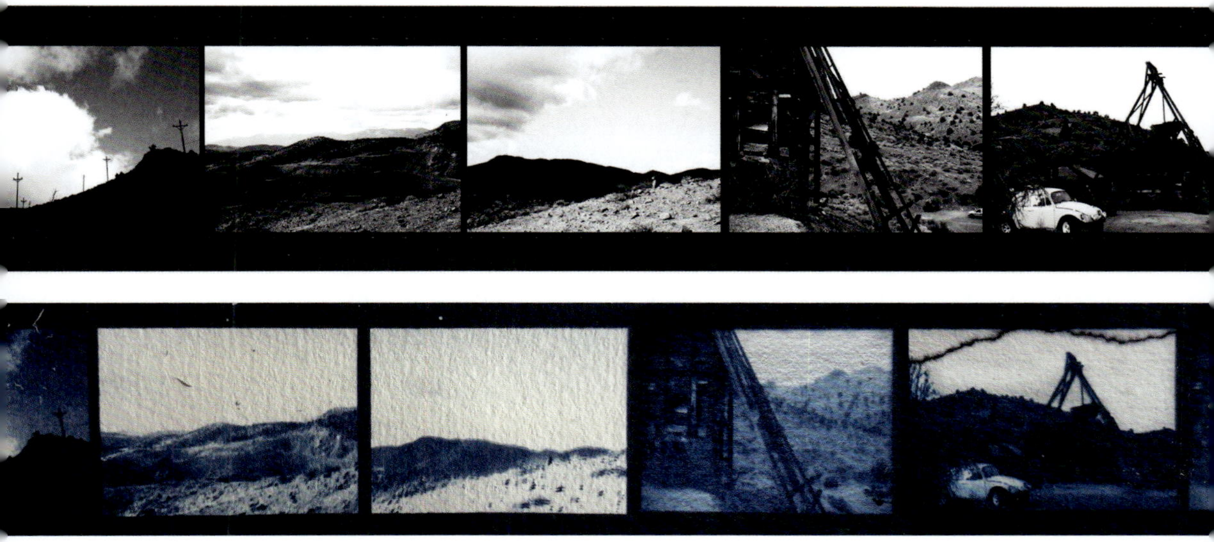

Fig. 6.
Alice Cazenave, *Untitled (Willow-Developed Film and Cyanotype Print)*, 2022. © Alice Cazenave, 2025.

ephemerality is used by corporations to argue for the safety of its use in mining operations. How, then, can we trace the toxicity of chemicals that evaporate into thin air? How can we talk about the toxicity of photography through a medium that is inherently toxic?

Plant Photo Chemistry: Tracing Toxic Affinities

I return to the messy work of plant photo chemistry. Willow trees are native to the American West; they are considered shade-intolerant species and prefer full sun exposure, which is generously provided in this region. Willows can be found along the banks of rivers, water systems that were exploited by mining operations to mill extracted ore. In the Carson basin, the willows grew alongside cottonwoods. The primary range of this woodland ecosystem is in the foothills of the Sierra Nevada, with smaller populations in Oregon and Nevada. Willow trees are significant to Native American populations for their medicinal properties and for weaving baskets. In floodplain ecosystems, such as the Carson River Basin, willows play a critical role in stabilizing soil, regulating water systems, and filtering pollutants.

I chose willow to tell the stories of photographic silver extraction because willow can remediate mercury- and cyanide-polluted landscapes. (Ebbs et al., 2003) Through a process known as phytoremediation, willow trees draw up these lethal ions through their roots. Once inside, the ions are metabolized by the tree and become assimilated into its proteins. (Larsen et al., 2004) While the cyanide is no longer visible, it is the stories of cyanide that willow holds in its botanical matter. For me, this is its most creative potential. It has a portal-like quality. Any willow tree could provide the compounds required to mix photo chemistry. Here, I am interested in the situatedness of the Carson River willow trees and what they materially witness (Schuppli, 2020). I am interested in how extractivism in the American West becomes archived in willow trees, and how this plant matter can be used experimentally to make visible this violence. In doing so, this approach asks questions about the provenance of photographic materialities and thinks critically about these in the broader context of photography and landscape photography, whose rich history is materially and culturally linked to the American West.

Mixing plant photo chemistry is a means to trace what is not always immediately visible—it is a way of starting with landscapes as they present themselves to you and working backward. As beings that connect aquatic, atmospheric, and metallic worlds together, plants carry stories of these assemblages that are made and encountered as they grow. In the Carson River, the willows carry stories of photography's toxic and colonial relations. They can be read in the mercury and cyanide assimilated through the willows' roots or in the silence of songbirds as they fall to the ground. In the words of Donna Haraway:

> It matters what matters we use to think other matters with; it matters what stories we tell to tell other stories with; it matters what knots knot knots, what thoughts think thoughts, what descriptions describe descriptions, what ties tie ties. It matters what stories make worlds, what worlds make stories. (Haraway, 2016: 12)

Through their situatedness, willow trees suspend these disparate but connected worlds into a single plane. This is not to say that these worlds become condensed or simplified, but, rather, that they are held together in a dense meshwork in the hidden chambers of the plant. (Ingold, 2016) Foraging and mixing plant-chemistry feels like reaching inside the belly of a photograph and pulling it into view from its guts. It is a literal getting your hands dirty with the stories held in the landscapes that form the work.

Throughout my PhD fieldwork, the practice of tracing became a significant research method. (fig. 7) It allowed the complex flows of chemicals, temporalities, and bodies that made up this research to be visualized on a single plane. This method of tracing is also an act of breaking down—of attending to what makes up entities and tracing their constituent parts. This act of breaking down is a practice lifted directly from my photographic practice, which centers on hand-mixing low-toxicity chemistries. This involves researching industrial chemistries and finding replacement, less harmful ingredients (Pardo, 2024, 2025). Phenolic

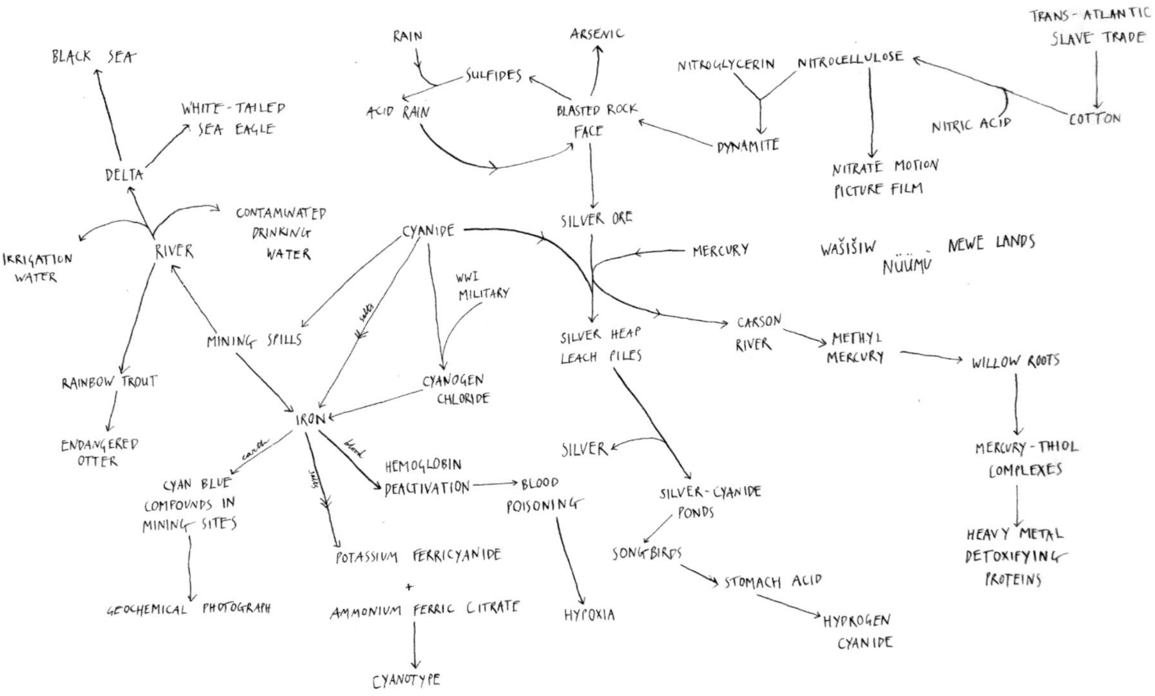

compounds are a good example. Traditionally sourced from fossil fuels, phenolic compounds can be sourced from the plant world at your fingertips rather than from its ancient remains beneath the earth's crust. This was the role of the willow—it provided phenols to the mix.

Fig. 7.
Alice Cazenave, *Metabolism of Film (Nevada)*, 2025. © Alice Cazenave, 2025.

Foraging for replacement photo ingredients has become a significant part of my research and visual practice. It is a way of experiencing field sites that provide your photo materials. The creative focus shifts away from the aesthetics of the final photograph—which is typically valued in photo practices—toward what grows where and why. Anthropologist Tim Ingold argues that the world opens up differently to us when we shift our attention to it. (Ingold, 2023) Foraging brings the plant world into your presence. It beckons to you through the act of looking and making. It is a moment when the eyes become alive. Using plants as experimental photo chemistry does not necessarily mean interpreting or understanding the plants, but, rather, going along with them. There are many unknowns and many experiments. Working this way feels like a long conversation between photographic practice and the ecologies that support it. Ingold describes this as mutual responsiveness, where the act of engaging shapes and is shaped by the person studying it. (ibid.) Something shifts. The culture of making changes.

A respect develops. You only pick what you need, and never more than a third of a plant. You feel more aligned and part of the same living space. You owe yourself to the plants, and they owe themselves to you. You breathe out the gases they need, and they expel the ones you need. They suck up toxicants from the earth, and you remove them from waste chemistry—one less job for them to do. You help each other out. You make each other's worlds.

My use of plants is not intended as a romanticization of the landscapes that I work with. Nor is it, to paraphrase Sean Cubitt, a colonial yearning for pristine landscapes destroyed by colonialism. (Cubitt, 2014) While it is difficult to escape the use of silver in film emulsions, using plant photo chemistry at least reduces the toxicity of developing images. Working with plants is a way of acknowledging photography's toxic history while working away from it. (Tuck, 2009) It is a method of uncovering narratives that are otherwise backgrounded in the stories we tell ourselves about photography and its impact. The use of willow here offers a reading of photography as a collision of many worlds. It is silver, mercury, cyanide, and songbird; violence, water, life, and death; displaced and dispersed. These are the ecologies of photography, the intricate metabolisms that transform, persist, and evaporate—at once present, past, and in the making.

Archival Collections

George Eastman Correspondence, Eastman Legacy Collection, Eastman Museum, Rochester.

Mining request letters, Box No. NC16; Pyramid Lake Paiute Tribe of the Pyramid Lake Reservation, University of Reno Nevada, Special Collections.

Bibliography

C. Allen, "Mitigating Impacts to wildlife at FMC Gold Company's Paradise Peak Mine," in *Proceedings of the Nevada Wildlife Mining Workshop, Reno, Nevada*, ed. R. McQuivey. March 27–29 (Reno: University of Reno, 1990).

Siobhan Angus, *Camera Geologica: An Elemental History of Photography* (Durham: Duke University Press, 2024).

Alex Bishop-Thorpe, "Cyanotype Safety," last modified 2016, https://www.analoguelab.com.au/cyanotype-safety/ (accessed April 20, 2025).

Monica Bravo, "Mercury is a Messenger: Photography's Dependence on Quicksilver and Mining Labor," *American Art* 39, no. 1 (2025): 16–25.

Alice Cazenave, "Parallel Archives: Photographic Silver & Its Landscapes," *Trajectoria: Anthropology, Museums, and Art* 6 (2025) (Special Issue).

Edd Carr and Hannah Fletcher. "The Sustainable Darkroom," last modified 2024, https://sustainable-darkroom.com/ (accessed March 18, 2025).

Okkyoung Choi, Kathy Kajun Deng, Nam-Jung Kim, Louis Ross Jr, Rao Y. Surampalli, and Zhiqiang Hu, "The Inhibitory Effects of Silver Nanoparticles, Silver Ions, and Silver Chloride Colloids on Microbial Growth," *Water Research* 42, no. 12 (2008): 3066–3074.

Donald R. Clark and Robert L. Hothem, "Mammal mortality in Arizona, California, and Nevada gold mines using cyanide extraction," *California Fish and Game* 77 (1991): 61–69.

Sean Cubitt, "Decolonising Ecomedia," *Cultural Politics*, 10, no. 3 (2014): 275–286.

Carlos D. Da Rosa and James S. Lyon (eds), *Golden Dreams, Poisoned Streams* (Washington D.C. Mineral Policy Center, 1997).

Eastman Kodak Company, *Kodak Park: Its Product and People* (Rochester N.Y, Eastman Kodak Company, 1949).

Stephen Ebbs, Jospeh Bushey, S. Poston, Dylan Kosma, M. Samiotakis, and David Dzombak, "Transport and Metabolism of Free Cyanide and Iron Cyanide Complexes by Willow," *Plant, Cell and Environment* 26 (2003): 1467–1478.

Ronald Eisler and Stanley N. Wiemeyer, "Cyanide Hazards to Plants and Animals from Gold Mining and Related Water Issues," *Reviews of Environmental Contamination and Toxicology* 183 (2004): 21–54.

Donna Haraway, *Staying with the Trouble: Making Kin in the Chthulucene* (Durham, NC: Duke University Press, 2016).

Jarrod Hore, *Visions of Nature: How Landscape Photography Shaped Settler Colonialism* (Oakland: University of California Press, 2022).

Tim Ingold, *Lines: A Brief History* (London and New York: Routledge, 2016).

Tim Ingold, "On Not Knowing and Paying Attention: How to Walk in a Possible World," *Irish Journal of Sociology* 31, no. 1 (2023): 20–36.

Robert Koenig, "Wildlife Deaths Are a Grim Wake-Up Call in Eastern Europe," *Science* 287 (2000): 1737–1738.

Nural Kuyucak and Ata Akcil, "Cyanide and Removal Options from Effluents in Gold Mining and Metallurgical Processes," *Minerals Engineering* 50–51 (2013): 13–29.

Morten Larsen, Stefan Trapp, and Angelo Pirandello, "Removal of Cyanide by Woody Plants," *Chemosphere* 54, no. 3 (2004): 325–333.

Paul Lechler, *Mercury Contamination of the Carson River* (Reno: Nevada Bureau of Mines and Geology, 1992).

Fabiana Li, *Unearthing Conflict: Corporate Mining, Activism, and Expertise in Peru* (Durham, NC: Duke University Press, 2015).

Max Liboiron, *Pollution is Colonialism* (Durham, NC: Duke University Press, 2021).

Erin Manning, "Against Method," in *Non-Representational Methodologies: Re-Envisioning Research*, ed. Phillip Vannini (Durham, NC: Duke University Press, 2015).

Michelle Murphy, "Alterlife and Decolonial Chemical Relations," *Cultural Anthropology* 32, no. 4 (2017).

Alex Nading, "Living in a Toxic World," *Annual Review of Anthropology* 49 (2020): 209–224.

Richard Nielsen, Lauren Bon, and Tristan Duke, "Silver and Water: An Interview with Metabolic Studio's Optics Division," last modified 2017, https://www.metabolicstudio.org/owens-valley/optics-division (accessed November 10, 2025).

Andrés Pardo, *Back to Basics I: A Guide to Ecological Photochemistry* (London: Folium, 2024).

Andrés Pardo, *Back to Basics II: Extended Recipes for Ecological Photochemistry* (London: Folium, 2025).

Daniel Renfrew and Thomas W. Pearson, "The Social Life of the "Forever Chemical": PFAS Pollution Legacies and Toxic Events," *Environment and Society: Advances in Research* 12 (2021): 146–163.

Susan Schuppli, *Material Witness: Media, Forensics, Evidence* (Boston, MA: MIT Press, 2020).

Ladislaus Szinicz, "History of chemical and biological warfare agents," *Toxicology* 214, no. 3 (2005): 167–181.

Eve Tuck, "Suspending Damage: A Letter to Communities," *Harvard Educational Review* 79, no. 3 (2009): 409–428.

Jennifer Tucker, "Chemical Affinities: Photography, Extraction, and Industrial Heritage in Nineteenth-Century Northern England," *Nineteenth-Century Contexts* 44, no. 5 (2022): 525–539.

United States Environmental Protection Agency (USEPA), "Abandoned Mine Drainage Prediction," last modified 1994, https://www.epa.gov/sites/default/files/2015-09/documents/amd.pdf (accessed February 2, 2025).

United States Environmental Protection Agency (USEPA), "Carson River Mercury Site: Dayton NV Clean Up Activities," last modified 2022, https://cumulis.epa.gov/supercpad/SiteProfiles/index.cfm?fuseaction=second.cleanup&id=0903020 (accessed December 4, 2025).

Edward Vischer, "The Comstock Lode," Stanford University, https://exhibits.stanford.edu/mining/feature/the-comstock-lode (accessed January 21, 2025).

Megan D. Welman-Purchase and Robert N. Hansen, "Cyanide within Gold Mine Waste of the Free State Goldfields: A geochemical Modelling Approach," *Environmental Pollution* 318 (2023): 1–9.

Filmography

William K. L. Dickson, *Buffalo Bill* (Edison Studios, 1894). 16 seconds.

Paul Fenton, *A Movie Trip Through Film Land* (1921). 23 min.

John Ford, *3 Bad Men* (Twentieth Century Studios, 1926). 93 min.

CHAPTER 7
The Nomadic Alchemist

Andrés Pardo Piccone

The instant-coffee developer turns 30. In 1995, Professor Scott Williams approached his Technical Photographic Chemistry class at the Rochester Institute of Technology with an assignment that ended in the creation of a film developer based on instant coffee, sodium bicarbonate, and potassium hydroxide. Years passed until the idea of reformulating this unnamed developer arose. This "new" approach to film processing ends in what we now know as Caffenol-C—a reformulation of the Rochester Institute of Technology class formula plus vitamin C. It works based on the ability of sodium ascorbate (a by-product of mixing vitamin C with a sodium salt) to reduce exposed silver and is enhanced or modified by the phenols in instant coffee. The chemistry behind Caffenol-C is not the same as the original one; it works and keeps the idea of building a developer with familiar and easy-to-get sources, but a fundamental distinction must be made. While the original formula based the ability of film processing only on the polyphenolic acids of instant coffee, in Caffenol-C the ability is based on sodium ascorbate, and the phenols modify the ascorbate behavior but do not do the bulk of the job. This difference defines the developers based on ascorbate plus something commonly known as plant-based developers and others.

Thirty years later, what happened in between tells another story of resistance and survival. As digital technology swept across the industry, film processing labs began to close one after another. What could have been the end of the medium instead became the beginning of something different. Out of the ruins of commercial production, a network of artist-run laboratories emerged—small collectives salvaging contact printers, Lomo tanks, and chemistry from abandoned facilities. In Europe, places such as Atelier MTK and L'Abominable became the heart of this rebirth. Later, Process Reversal in Colorado, Filmwerkplaats in Rotterdam, and Kinolab in Bogotá extended the map. What united them was not nostalgia, but the will to stay autonomous, to keep the knowledge of the lab alive by doing it themselves. As Genevieve Yue observed, "we are not in an economy but an ecology" (Yue, 2015), and as Kelly Egan reminds us, "film labs are perhaps the most important elements of cinema's infrastructure [...] the heart of filmmaking, film distribution, film exhibition, and film preservation." (Egan, 2018: 173) It was an act of repair, of learning again how to touch the medium, how to mix, test, and fail. These spaces filled the thirty-year gap between the Rochester Institute of Technology experiment and today's sustainable darkroom: a lineage of reappropriated tools; open sharing; and slow, collective chemistry. Much of this work extended into the use of plants and locally sourced materials, yet few of its practitioners attempted to define or categorize these developers. Establishing distinctions between them—by origin, function, or degree of transformation—is essential for understanding what truly constitutes a natural or sustainable process.

Plant-modified developers are a group of working solutions for the film lab, which use a plant extraction—ranging from plain water to a solvent—as the vehicle in which the chemicals are dissolved. It is essential to distinguish between the different types of developers to understand the chemical agent on which the development is based. Seven groups can be identified:

1. Plant-based or infusion developers work solely with naturally extracted chemical compounds as the developing agent.

2. Hybrid developers are solutions where the user combines a plant extraction rich in reducers (phenols, flavonoids, tannins, etc.) with a synthetic developer agent. The most common is ascorbic acid. In these developers, both the plant extraction and the synthetic agent work in conjunction to process silver halides into metallic silver, with the plant extraction playing an active role in the development process.

3. Ascorbate-plant developers are solutions where most of the silver reduction is performed by sodium ascorbate, while plant extraction plays a secondary role. Though the effect of the plant extraction on silver halides is negligible, it can act as a pH modifier, buffer, restrainer, preservative, stain reducer, or color illusion inducer.

4. Upcycled developers are working solutions for the film lab that use waste-derived materials as their base, repurposing discarded substances into functional chemistry. These can include spent coffee grounds, rusty metals, food scraps, plant residues, or other

byproducts that would otherwise be discarded. Some upcycled developers work solely with the extracted compounds as the developing agents, while others combine these recovered materials with additional chemicals—such as ascorbic acid—to enhance their performance. The key aspect of upcycled developers is their focus on circularity, reducing environmental impact by transforming waste into a valuable photographic resource.

5. Distilled developers are solutions where specific plant compounds are isolated and purified through advanced techniques—such as distillation or solvent extraction—to obtain more concentrated developing agents. These developers focus on specific chemicals, such as quercetin or catechins, and offer more control over the concentration and chemical interactions involved in silver halide conversion. They represent a more refined approach than infusions, offering precision in the development process while remaining plant-based.

6. Energy-driven developers are solutions that utilize electromechanical power reduction instead of relying on phenolic compounds to donate electrons for silver halide reduction during film processing. These developers apply energy inputs, such as Bach solarization, to enhance the chemical reactions in the development process. The reduction of silver halides is driven by electromechanical forces. A clear example of this can be observed in solar prints, where an excess of light energy, rather than chemicals, controls both exposure and development, highlighting how energy can directly impact the photographic process.

7. Metal-assisted plant developers are a group of developers where metal ions—either naturally occurring in plant extracts or intentionally added—enhance or catalyze the reduction of silver halides. For example, iron-rich botanicals or copper-infused plant extractions can act synergistically with plant phenols—and ascorbate, if added—to improve development efficiency. These developers operate similarly to traditional iron-based processes but remain rooted in plant-derived chemistry.

Note that there is a difference between a reducer and a developer: While a reducer is a chemical with the ability to turn the silver halides present in a silver emulsion into metallic silver, a developer holds a special property: It can differentiate between exposed and unexposed silver halides. That way it can render the image, working only where the exposed halides are.

Though the focus of the Rochester Institute of Technology class was primarily chemical experimentation rather than ethics, the use of instant coffee—often made from beans sourced from Brazil—quietly introduced questions about origin and responsibility into the darkroom (Winton, 2022). Formulating a truly local developer solution, one that is not based on international supply chains, requires rethinking the relationship between materials, place, and practice. This is where principles of ecological design and sustainable thinking are essential, ruling how we source, process, and dispose of materials in the lab.

Working from the Global South changes the meaning of every act in the lab. Here, nothing is neutral; even a teaspoon of soda or a handful of coffee carries the weight of history. Macarena Gómez-Barris calls Latin America an "extractive zone," and the description feels accurate, even if theory tends to orbit more than act. (Gómez-Barris 2017, xvi) What matters is what happens in the sink. Chemistry travels thousands of kilometers before touching local water. To think of a local developer is to move against that current, not to fix the world, but to stop pretending it is clean. Coffee is a good example: grown here, exported, then sold back as instant powder. Using spent grounds collected nearby reverses that route. Waste becomes source, residue becomes knowledge. It is not activism—it is attitude.

The Hannover Principles—a set of sustainable design guidelines developed for the 2000 World's Fair in Hannover, Germany—can be useful as a framework for this approach. The Hannover Principles general guidelines are: insisting on the rights of humanity and nature to coexist, recognizing interdependence, respecting relationships between spirit and matter, accepting responsibility for the consequences of design, creating safe objects of long-term value, eliminating the concept of waste, relying on natural energy flows, understanding the limitations of design, and seeking constant improvement by the sharing of knowledge (McDonough and Braungart, 1992). Their application in the darkroom and film lab feels natural despite them being originally intended for architecture. Photography, after all, is a design process, a set of choices about materials, energy, and waste. Vilém Flusser argues that cameras are designed with built-in programs that dictate the range of possible photographs they can produce. Each photograph is, therefore, a realization of one of these pre-existing possibilities within the camera's program—reducing freedom—but sustainable photography expands that freedom. (Flusser, 2000: 26)

The Hannover guidelines and the principles of eco-design emphasize that "solutions grow from place". Ecological design begins with an intimate knowledge of a particular place. Therefore, it is small-scale and direct, responsible for both local conditions and local people. If we are sensitive to the nuances of place, we can inhabit without destroying."

(Edwards, 2005: 103) Applied to photography, it means that materials and processes should, ideally, be derived from the local environment. However, this is hard to put into practice. Thinking just on the basic chemicals needed to build low-toxic developers—sodium carbonate and ascorbic acid—are produced in countries far away, especially for people living in the Global South.

Two of the Hannover Principles are essential for the photographer seeking a more sustainable approach to lab work: eliminating the concept of waste and relying on natural energy flows. Thinking about Caffenol-C, the idea of using instant coffee is that the powder holds many natural polyphenols from the plant that can convert exposed silver halides into metallic silver. A polyphenol is a natural compound found in plants, acting as one of their defense tools against stress, disease, and damage from ultra-violet light. For humans, they are what make green tea, berries, or red wine not just tasty but also beneficial, as they help combat free radicals and reduce inflammation. In the film lab, polyphenols can be incredibly useful, as they can act as developing agents or restrainers.

The difference between the 1995 Rochester Institute of Technology coffee developer and Caffenol-C is clear when we look at the seven groups of developers: While the former is a plant-based developer, the latter is a hybrid developer that combines the development ability of sodium ascorbate and phenols. It is essential to make this distinction in the search for a local developer agent, no matter where you are: Asia, South America, Europe, or the American West. To rely on a local agent and to apply eco-design, a source that can work on its own is essential. Because no matter what is mixed with ascorbic acid, the developer will work, but, a truly local developer, will connect to the place if the process is at least divided by the developer agents like a hybrid developer. Each plant-based developer is tied to its place. The mix changes with the soil, the kind of water, the air. Nothing is stable. A formula that works one way in Montevideo behaves differently in Rotterdam or Tokyo. That difference is not a flaw but part of the meaning; the image absorbs its surroundings.

Testing the plant alone is an essential experiment in this quest for a developer. The most important aspect of the Rochester Institute of Technology class is that it reintroduces the seed of sustainability in photography, shows that it is possible, and that the results depend upon experimentation, not just by following a set of pre-established rules.

Much unconfirmed information about Caffenol-C can be read, such as that during WWII coffee was used as a film developer, that it is based on caffeic acid as a developing agent, that decaf will not work, that it is a case of the cheaper the better... What is true is that it works; it can be compared to any commercial developer, but is mixed with easy-to-get chemicals, and the quality is not compromised. What is not true is that it is necessarily a sustainable alternative or eco-friendly, especially if you live in the Global North. Most instant coffee comes from global coffee powerhouses, such as Brazil and México—from where it is exported to many places, including Kenya. Given that Kenya is itself a coffee producer, the economic decisions made to move the goods are strange. (Winton, 2022)

It is clear that processing film with instant coffee in places far away from where coffee is grown, such as Europe, is not as sustainable as it might sound for being a "natural source." But repurposing spent grounds offers a chance to rethink our relationship with place and waste. Polyphenols from instant coffee can process film, and coffee comes from a plant, yet relying on waste, such as used grounds, instead of a non-waste product, especially one that can easily be replaced in the developer mix, may be a smarter move. Changing even one element will have an impact on the economy, the environment, and our connection with local people. Then more questions arise, for example: What is sustainable photography? Can a form of art attached directly to industrial products and capitalism be sustainable? When does photography stop being sustainable?

One intelligent move to relink photography with sustainability is to turn to coffee grounds waste, which can be gathered locally from coffee shops, offices, or our homes. It is the first step to avoiding market control and to start escaping the program, as described by Vilém Flusser, where the apparatus predetermines the range of possible images (Flusser, 2000).

The next step is to know how we can extract—at home, at school, or in the lab—those chemicals that can process film. Relying on natural energy flows is not easy in all parts of the world: While during summer, the hot sun in the Global South can enhance a fermentation process to build a supercharged polyphenols extraction, in cities such as London this process may be trickier, but it is possible. The key is patience and a self-imposed slowdown. Slowing down can reconnect contemporary practice to the origins of photography. It is well documented that back then, processes were slow, film speed was a snail, and people needed to pose for up to half an hour in front of the camera for the silver salts to be exposed.

Fox Talbot was a key figure in photographic history, and little can be added to his contributions to the process. But when we look deeply from a more sustainable point of view, some clues for our future arise from his process, notebooks, and letters. He uses fewer toxic compounds compared with his French competitor. He builds his emulsions and relies on a chemical extracted from a plant to process the images. Reading Talbot's letters, it is possible to identify two developer agents used by the father of modern photography. One is iron sulfate (protosulfate of iron is the chemical name he uses in his letters), and the other and more interesting for the present is gallic acid. Gallic acid was, and still can be, extracted—through fermentation, full evaporation, and purification of the resulting crystals—from oak galls, a material formed by a small wasp that lives on oak trees, but galls are not the only source. Oak leaves and bark, black walnuts, berries, and pomegranate leaves are some natural sources that bring to the table tannic acid, a complex polyphenolic acid that, after fermentation and degradation, ends in gallic acid.

In a letter dated August 12, 1842, Robert Hunt, a British poet, scientist, and pioneer photographer, asks Talbot, "Do you really find any great advantage in the *crystalised* gallic acid over the Infusion of galls?" (Talbot Correspondence Project) Hunt mentions not only infusion but also "tincture." A tincture is used to prepare an infusion, a diluted form of the initial fermented brew. It has been known since the early days of photography that plant extraction can process film effectively, so why did and do we rely on chemicals that are unsafe for the user; toxic to the environment; and come from fossil, nonrenewable sources? These materials are closely attached to extractivist practices, control, capitalism, and a programmatic final appearance of the image, in which photographic manuals state how the correct negative and print should look, which is far from a free form of expression. Sustainability at its root is freedom, justice, and not relying on mainstream products and trends. What happened is that the industrial and extractivist capitalistic photography apparatus deliberately hid the idea of building developers from natural sources for its benefit by an overwhelming bombardment of advertisements.

This historical view is more than just nostalgia—it shows an early sense of care for place that echoes in today's ideas of sustainability. Talbot's chemistry was locally rooted: His gallic acid came from oak galls. That link between matter and landscape tied his practice to a specific ecology. In Montevideo, where there are no oak wasps, that source does not exist, and the absence invites another way of thinking—how to replace gallic acid with what grows, falls, or breaks down nearby. What once came out of necessity now becomes a way to work sustainably.

But the Industrial Revolution and capitalism were not the only ones to blame, so, too, was a way of life where it was easy to get things pre-done rather than do them yourself. And this continues to happen, and not only in photography: Most people do not know how to prepare the soil, how to grow plants to feed their family, or how to heal different diseases using herbs. Photography is not the exception in the race for commodities. The answer to the question What happens in the middle? is easy: Capital greed and a population shaped to operate under a pre-established system rather than build its own. Friendly publications, such as the Kodak booklets, explaining how to shoot, process, and print using Kodak chemistry are examples of how far this programming goes. Even worse is Kodak's infamous "You press the button, we do the rest" advertising campaign of 1888 (Wikipedia contributors, 2025).

The point may be arguable, but I suggest Talbot shows how photography at its birth was a truly sustainable form of art: It complies with contemporary understandings of sustainability and is now being pursued as the new photographic praxis. Talbot and Hunt, among others, teach users how to use oak galls, patience, dedication, and experimentation, and not to give up. In *All Art is Ecological,* Timothy Morton might say that Talbot was designing for the future, because "the future emerges directly from the objects we design." (Morton, 2021: 62) Talbot was a visionary, a magician of light, and a teacher for the future. The "Three Dimensions of Liberty" by Billy Bragg (2019) stamps on a tiny book how liberty, equality, and accountability dimensions must coexist for freedom to be meaningful and sustainable, particularly in the face of rising authoritarianism and corporate influence. Sustainable photography is not just about processing using plants; it can be sparked from there, which is why it is so important for all the workshops, publications, and artists working this way across the world. It is also important to trace this practice into history to continue with unfinished research: Talbot's notebooks and letters are one of the first creative commons examples of knowledge that should be reintroduced.

Flusser (2000) asks: Who predominates in the photographic act? Is it the photographer or the apparatus? Does the camera determine the photographer, or vice versa? The camera offers limited options, but photography is freedom. You choose almost everything about the process: the camera, or no camera at all; the lens; the film speed and emulsion; the format; the angle and height of the camera; the focal point; the amount of light hitting the emulsion. So many choices, until you press the shutter. "You press the button, we do the rest" strips away that freedom. Traditional film processing is like running a computer program; you follow certain steps and get a predictable result. Stick to the manual, and you end up with a "well-developed" negative. But the freedom you had while shooting disappears as you strictly follow lab rules designed to produce a "correct" negative.

It is important to slow down and look back in time to reclaim freedom and to be as sustainable as you can be. Walking is a practice that precedes the photographic act, like the lunar society members needed to walk as a way of thinking, we should walk the fields looking for elements that will become part of the lab. Fallen nuts, bark, leaves, branches, or flowers will be gathered to start the working solutions. The idea is to create a tincture, but not exclusively from galls. Here is where "solutions come from the place" becomes relevant and can be linked to the American

West or to any other place. A guide can establish how to create a truly sustainable photographic method, but if the main element is not present everywhere, it should be possible to replace it by testing similar available goods; this is the focus here.

The democratization of photography that we were told would happen with the new technologies is not happening as it was promised; what is happening is slavery. Slowing down and degrowth in photography may mean going down to fewer than thirty-six exposures, even going to single images created with time and patience and then producing the needed solutions to render those images within a natural and human time scale. To become a sustainable photographer is to turn into a "nomadic alchemist"—a breed of image makers who move through urban and rural environments, gathering natural or discarded materials and transforming them into creative solutions for photographic processes. This practice blends adaptability, resourcefulness, and a deep connection to place, merging art and ecology through the craft.

Turning Alchemy into Matter

The following tests were performed using black walnut tinctures, an infusion, and spent coffee grounds to demonstrate nature's power. The *Silver Sunbeam*, a book by John Towler from 1864, clearly describes how to obtain gallic acid from oak galls, and this method also works with walnuts and coffee grounds.

For the black walnut extraction, mix fallen nuts with water and expose them to air for two or three months, topping up the water as it evaporates, to maintain volume. Occasionally break down the nuts' green skin—now turning brown as it degrades—by stomping with a broomstick or similar tool. After filtration, the resulting liquid resembles the tincture mentioned by Hunt. It can then be allowed to fully evaporate and purified to obtain phenolic acids in crystal form, as one does with gallic acid. Alternatively, it can be used as is or diluted for use as an infusion. Add ten cloves per liter of water, to act as a natural preservative and prevent bad odors.

The process is simpler using spent coffee grounds, as the grounds are already in powder form and the increased surface area makes it easier to extract phenolic compounds. Within a week, a potent solution will be ready to use. In both the walnut and the coffee process, adding ethanol increases solubility up to fivefold, enhancing the tincture's strength. Once the extraction is complete, the solution contains natural developer agents: mainly gallic acid in the case of walnuts, and caffeic and quinic acids in the case of coffee grounds.

A basic developer requires three components: water, a developing agent, and an activator. Sodium carbonate is a common activator in ascorbic acid–based developers, as it is easy to source, safe, and affordable. A sustainable alternative is wood ash, which contains about 40 percent potassium and sodium carbonates. It can be used directly or after refining into potash. To make potash, mix wood ash with water and let it stand for a day or two. Then filter the liquid and either evaporate it completely to obtain a fine, grayish powder (which is the potash), or use the filtered liquid to dilute the tincture, obtaining an active infusion. An active infusion of black walnut or spent coffee grounds is a fully functional developer, ready for photographic use.

Elemental Instructions

Black Walnut Tincture Developer	Black Walnut Infusion Developer	Coffee Grounds Extraction Developer
300 ml of tincture	100 ml of tincture	300 ml coffee grounds extraction
20 g sodium carbonate or 50g of potash	200 ml of water	20 g sodium carbonate or 50g of potash
	20 g sodium carbonate or 50g of potash	

Control image, same on both sides. 15 sec exposure Simple developer 1 min Fixer 7 min	
Left: black walnut developer Right: Spent coffee grounds developer	
15 sec exposure Black walnut tincture developer 6 min Fixer 7 min	15 sec exposure Spent coffee grounds developer 20 min Fixer 7 min
60 sec exposure Black walnut infusion developer 15 min Fixer 7 min	270 sec exposure Spent coffee grounds developer 20 min Fixer 7 min

All tests were performed using photo paper and an orphaned glass negative from the beginning of the 1900s, part of the Uruguayan Photo Club collection. Exposed by contact using a Durst enlarger with the lens aperture open wide and the head 18 cm over the table plane. Simple developer is a working solution obtained by mixing 300 ml of water, 20 grams of sodium carbonate and 5 grams of ascorbic acid.

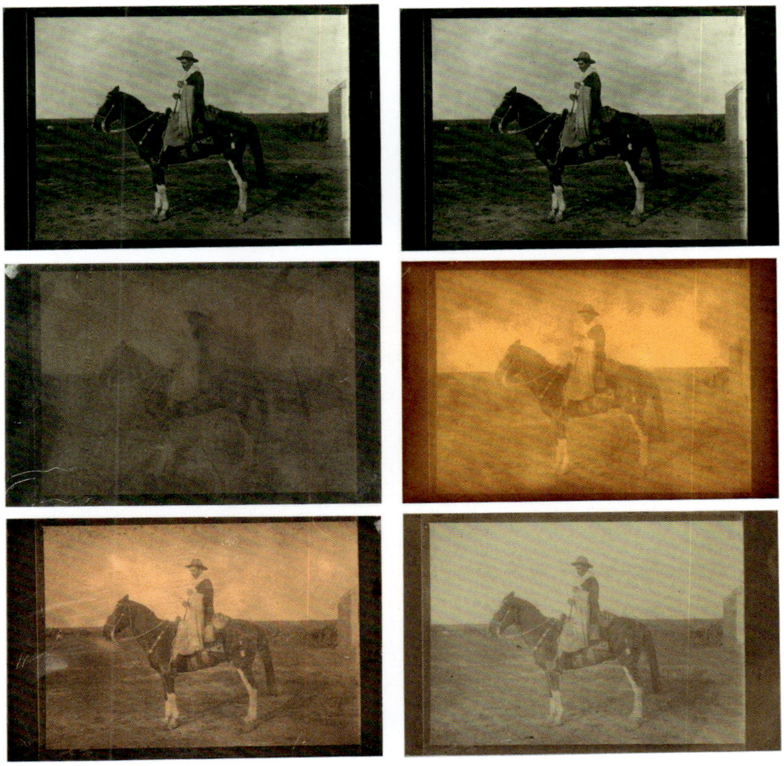

Conclusion

Reconnecting photography to the local, to its time, and to the materials we have around us is not just about chemistry—it is about freedom. When we step away from industrial formulas and start testing what is close by, what has been discarded, what grows near us, we begin to build a practice that is not only more sustainable, but also more meaningful. Slowing down, walking, experimenting—these are acts of resistance. The results are far from a standard black-and-white print—especially in the tincture, where black walnut adds a deep, dark cast over the image. These initial tests show it works. With further experimentation, the character of this developer can only evolve. This is not just about replacing one ingredient with another. It is about breaking the program, challenging the systems that shaped photography into a predictable product. Becoming a sustainable photographer today means becoming a kind of nomad alchemist, someone who listens to the seasons, works with what is at hand, and finds beauty in the imperfect and the unstable. It is not always easy, but it is possible. And, more than that, it is necessary.

Archival Collections

Robert Hunt to William Henry Fox Talbot, August 12, 1842. *Talbot Correspondence Project.* https://foxtalbot.dmu.ac.uk/letters/transcriptFree-text.php?keystring=Infusion+&keystring2=&key-string3=&year1=1807&year2=1877&pageNum-ber=1&pageTotal=6&referringPage=0

Bibliography

Andres R. Edwards, *The Sustainability Revolution: Portrait of a Paradigm Shift* (New Society Publishers, 2005).

Kelly Egan, "Handmade Films and Artist-Run Labs: The Chemical Sites of Film's Counterculture," NECSUS – *European Journal of Media Studies* 7, no. 2 (Autumn/Winter 2018): 173–190.

Vilém Flusser, *Towards a Philosophy of Photography*, trans. Anthony Mathews (London: Reaktion Books, 2000).

Macarena Gómez-Barris, *The Extractive Zone: Social Ecologies and Decolonial Perspectives* (Durham and London: Duke University Press, 2017).

William McDonough and Michael Braungart, *The Hannover Principles: Design for Sustainability* (Charlottesville: Expo 2000, 1992).

Timothy Morton, *All Art is Ecological* (Harmondsworth, UK: Penguin Classics, 2021).

John Towler, *The Silver Sunbeam* (New York: Morgan and Morgan, 1969).

Wikipedia contributors, "You Press the Button, We Do the Rest," Wikipedia. https://en.wikipedia.org/wiki/You_Press_the_Button,_We_Do_the_Rest#:~:text=%22You%20Press%20the%20Button%2C%20We,desire%20to%20take%20great%20pictures. last modified 2025 (accessed April 22, 2025).

Scott Williams, "Caffenol: An Alternative Film Developer," Class article, Rochester Institute of Technology, last modified 1995, https://repository.rit.edu/article/1124/ (accessed August 09, 2025).

Dom Winton, "Smuggling & Tariffs: The Complicated World of Coffee Imports to Origin Countries," *Coffee Intelligence*, last modified August 17, 2022, https://intelligence.coffee/2022/08/world-of-coffee-imports-origin-countries/ (accessed April 28, 2026)

Genevieve Yue, "Kitchen Sink Cinema: Artist-Run Film Laboratories," *Film Comment* (March 30, 2015).

PART 3
Native Sovereignty

CHAPTER 8
Corn Pollen Futures: Photography, Environmental Memory, and Indigenous Sovereignty on the Navajo Nation

Will Wilson (Diné)

> They were told to choose between two yellow powders. One was yellow dust from the rocks, and the other was corn pollen. The [Diné] chose corn pollen, and the gods nodded in assent. They also issued a warning. Having chosen the corn pollen, the Navajos were to leave the yellow dust in the ground. If it was ever removed, it would bring evil. (Eichstaedt, 1994: 47)

This origin story remains central to my practice as a photographer, storyteller, and citizen of the Navajo Nation. Across my career, I have returned again and again to this metaphoric crossroads—a choice between life and death, between relationality and extraction. Whether I have always been aware of it or not, this work has been my life's calling. From my early childhood on the reservation and my experiences at the Tuba City Bureau of Indian Affairs boarding school, to my middle school forensics speech on nuclear non-proliferation, to wandering the ruins of the Rare Metals uranium mill just outside of town, I have long been circling the environmental wounds in our landscape. My photographs seek to illuminate, intervene in, and transform these wounds into acts of witness, memory, and possibility.

In 2004, I began Auto Immune Response (AIR) (figs 1–3), a photographic and performative project in which I portray a Diné protagonist wandering a post-apocalyptic landscape. Drawing from Diné creation stories, AIR considers survival in the wake of environmental collapse and explores what Indigenous remediation might look like. The central figure of AIR is a recurring character, "the protagonist," who appears in large-format panoramic landscapes wearing protective prostheses performing ritual acts, including environmental monitoring and corn pollen offerings. The series culminates in the creation of the *Auto Immune*

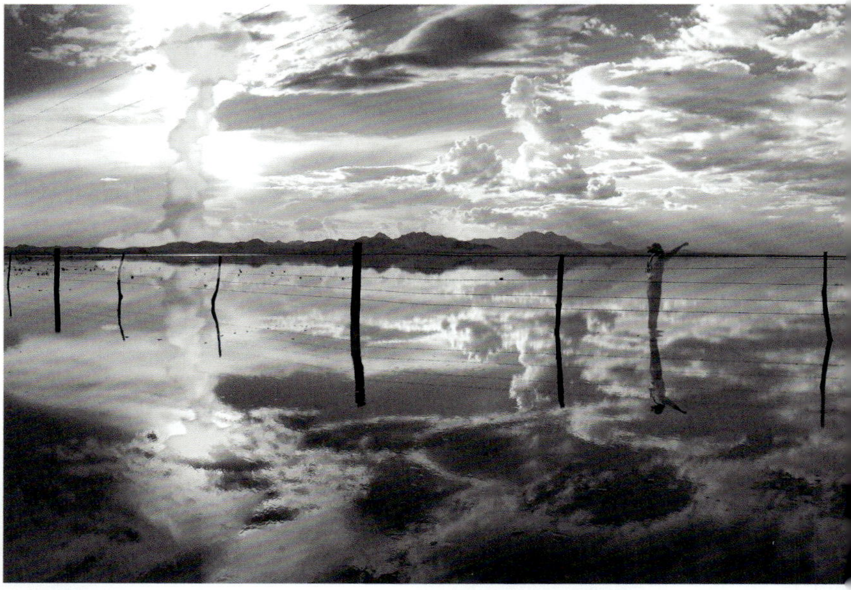

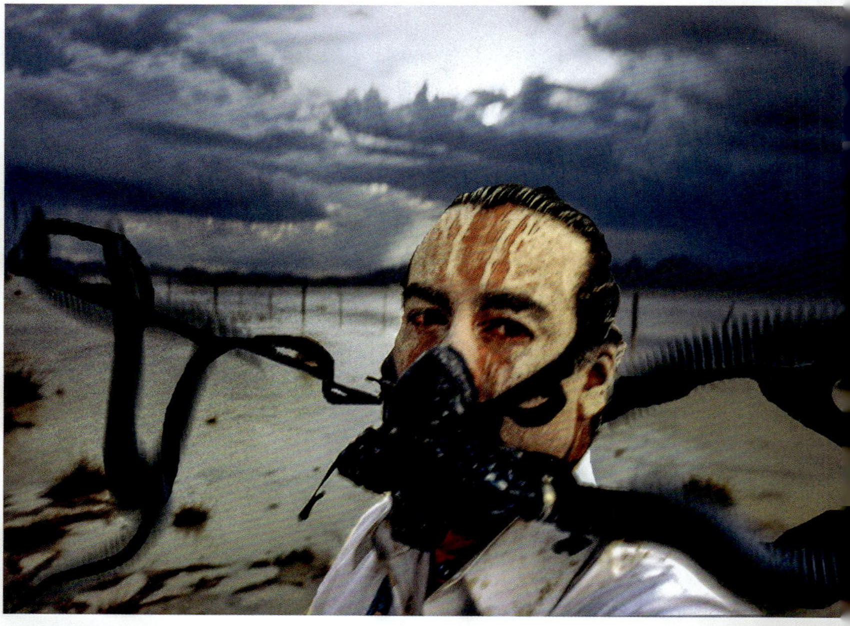

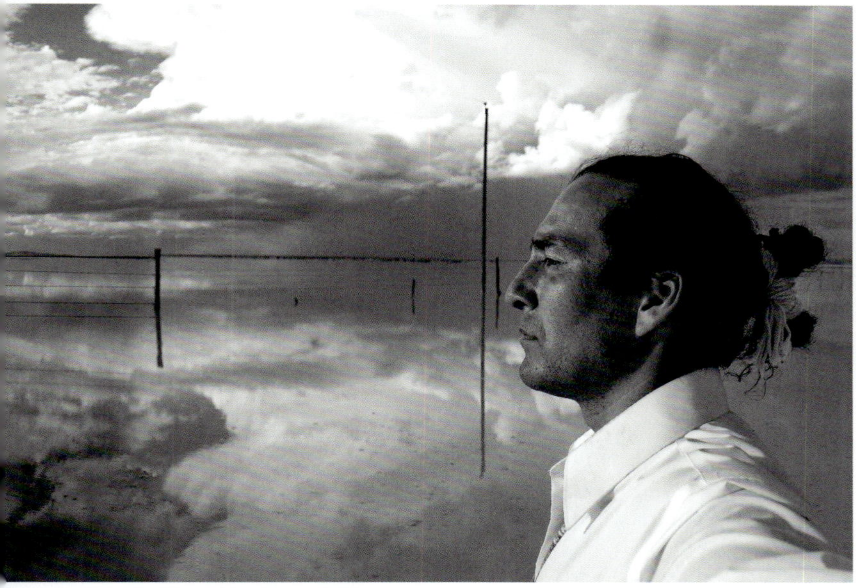

Fig. 1.

Will Wilson, *Auto Immune Response 1,* 2004, archival pigment print, variable dimensions. © Will Wilson.

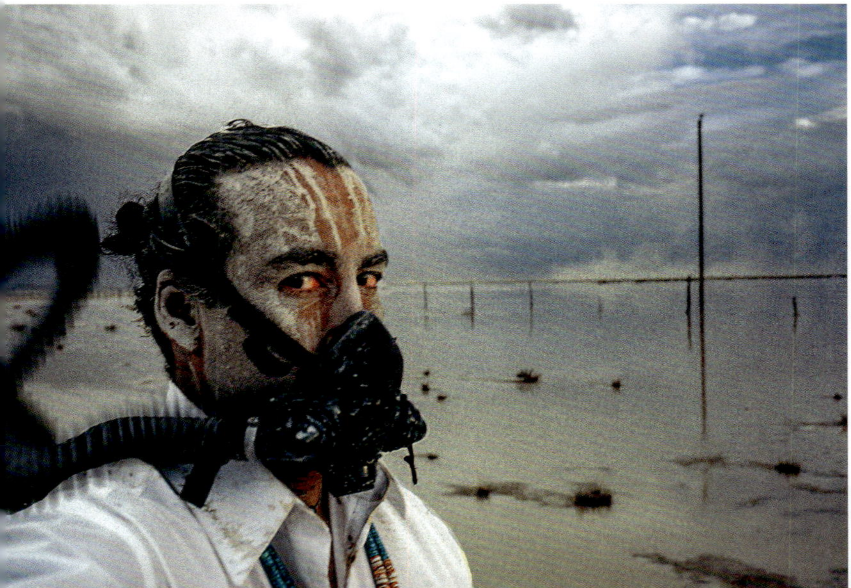

Fig. 2.

Will Wilson, *Auto Immune Response 5,* 2004, archival pigment print, variable dimensions. © Will Wilson.

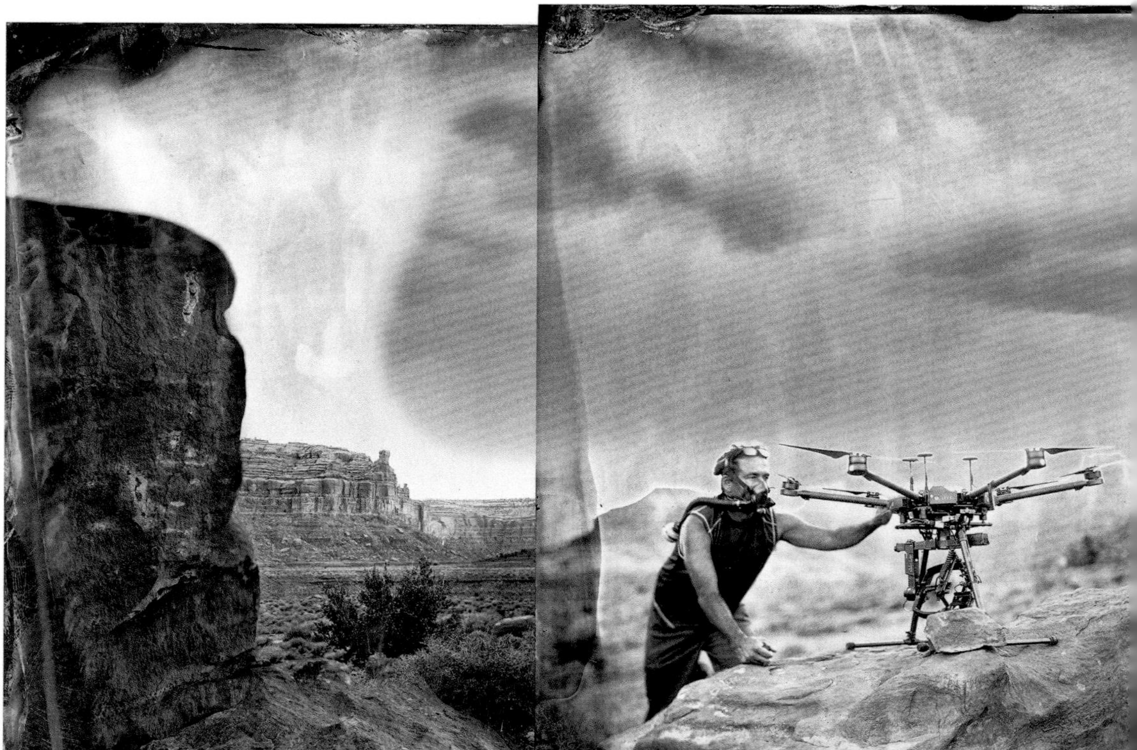

Fig. 3.

Will Wilson, *AIR/SURVEY,* 2020, archival pigment print, variable dimensions. © Will Wilson.

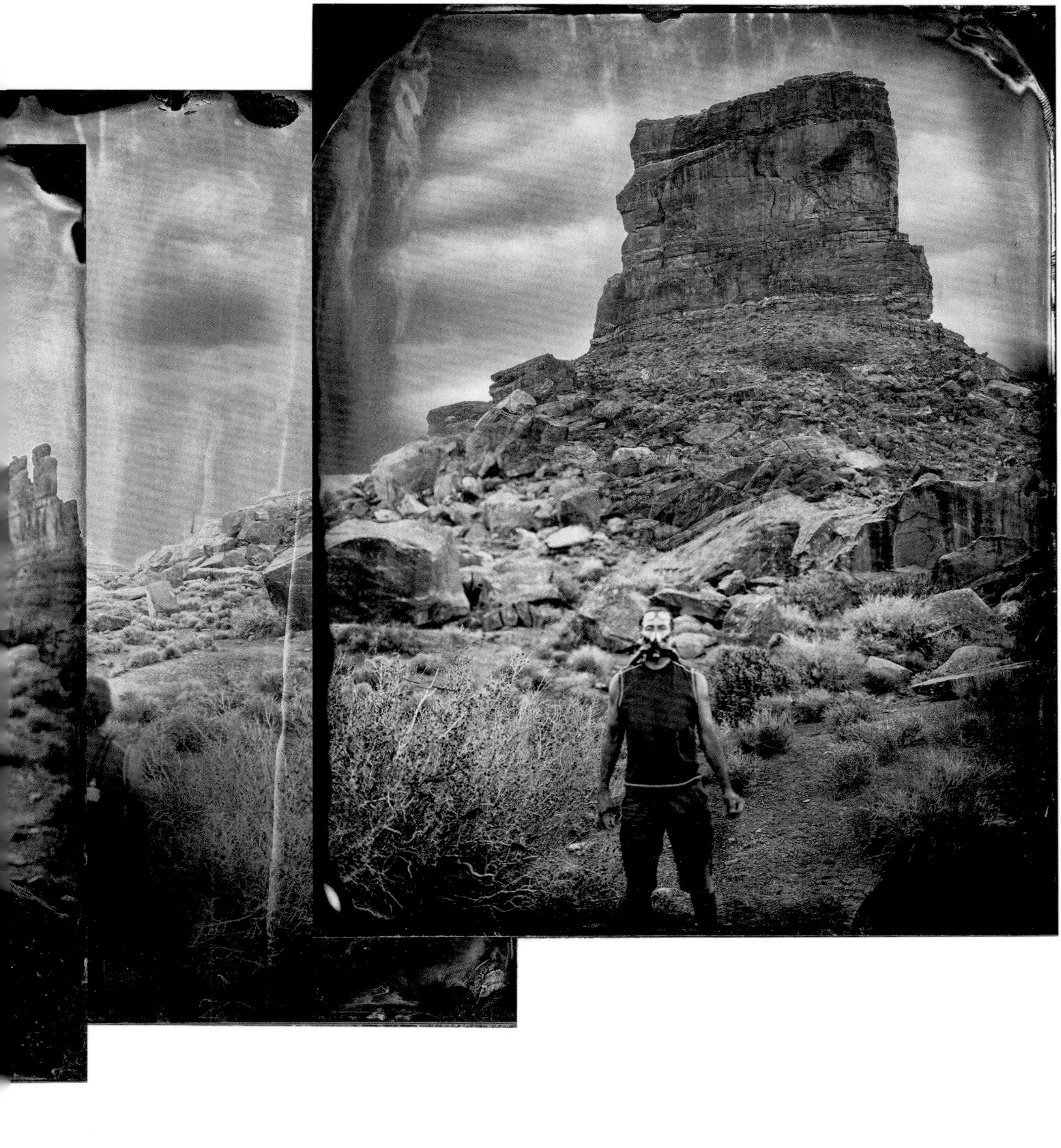

Response Laboratory (AIR LAB) (fig. 4), a site-specific hogan or greenhouse installation that cultivates Indigenous food and phytoremediation species. These plants, which can draw heavy metals out of contaminated soil, are both practical and symbolic: They embody Diné futurity and resilience, and the possibility of healing.

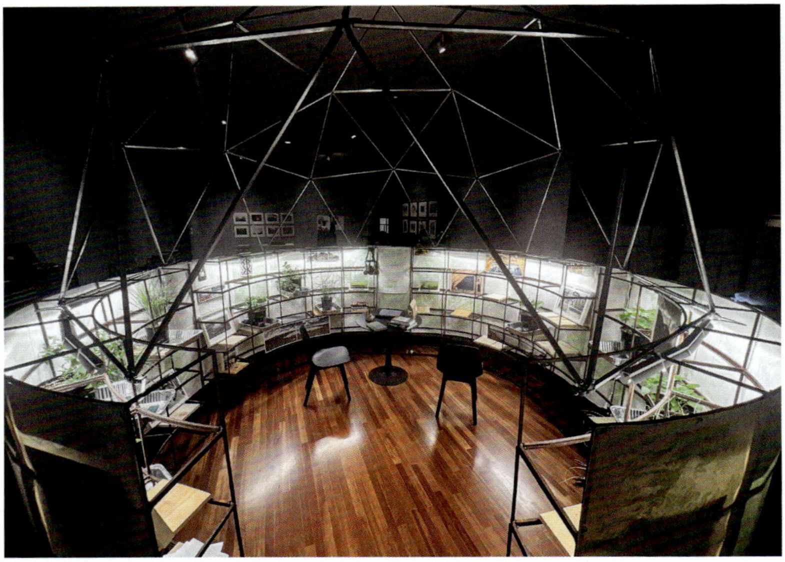

Fig. 4.
Will Wilson, *AIR LAB, Phytoremediation,* 2021, archival pigment print, variable dimensions.
© Will Wilson.

AIR laid the conceptual groundwork for my current project, *Connecting the Dots for a Just Transition* (CTDs) (figs 5–9), a multi-modal, community-based investigation into the history and ongoing legacy of uranium extraction on Diné lands. Where AIR imagined the future through speculative allegory, CTDs grounds its intervention in the present, mapping more than 500 abandoned uranium mines (AUMs) across the Navajo Nation. Launched in 2019, CTDs combines drone-based aerial photography, archival research, storytelling, and app development to create a critical Indigenous cartography—a new way of seeing and knowing place from within. Drone photography, historically tied to surveillance and militarization, is reimagined here as an instrument of counter-survey. My images hover above radioactive waste sites, Uranium Mill Tailings Radiation Control Act (UMTRCA) disposal cells, and contaminated rivers. Places such as Rare Metals (fig. 6), Mexican Hat (fig. 7), Shiprock (fig. 8), and Church Rock (fig. 9) become visible as wounds in the landscape—scars that connect Cold War history, extractive capitalism, and the lived experiences of Diné communities. The drone allows us to see what is otherwise hidden or buried, to visualize environmental violence from an Indigenous perspective.

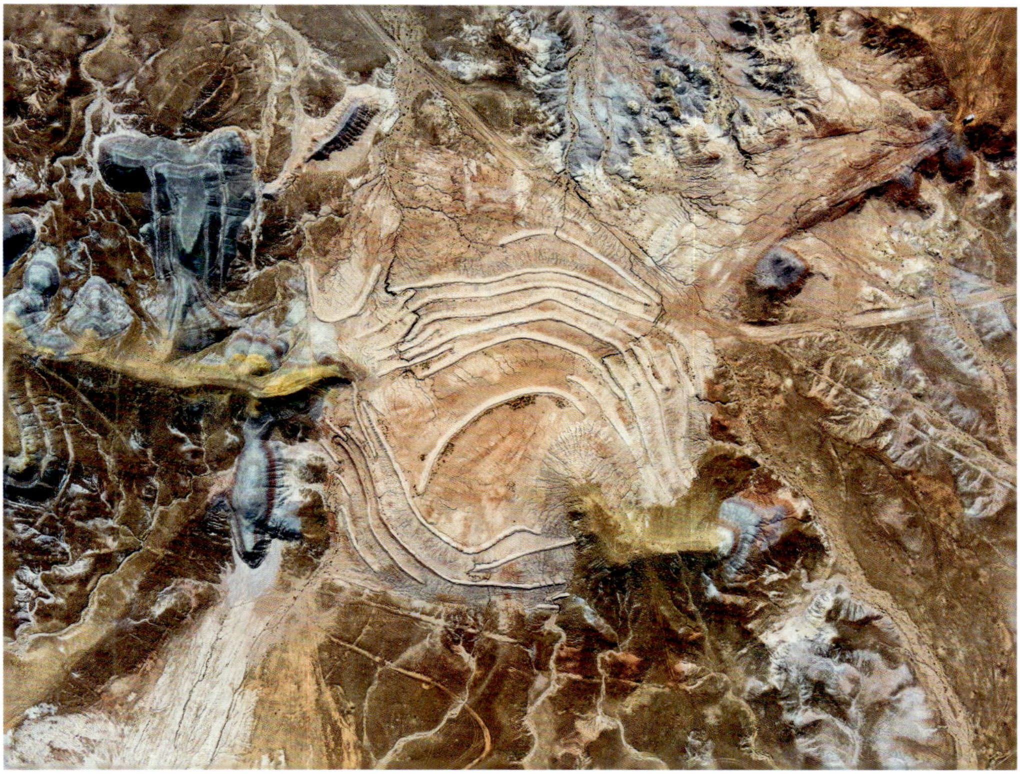

Fig. 5.

Will Wilson, *Cameron Complex (121,125,127) (Bird's-Eye View) AUM, Cameron & Coalmine Mesa, AZ, Navajo Nation, 35.877690 N / −111.392520 W,* 2020, archival pigment print, variable dimensions. © Will Wilson.

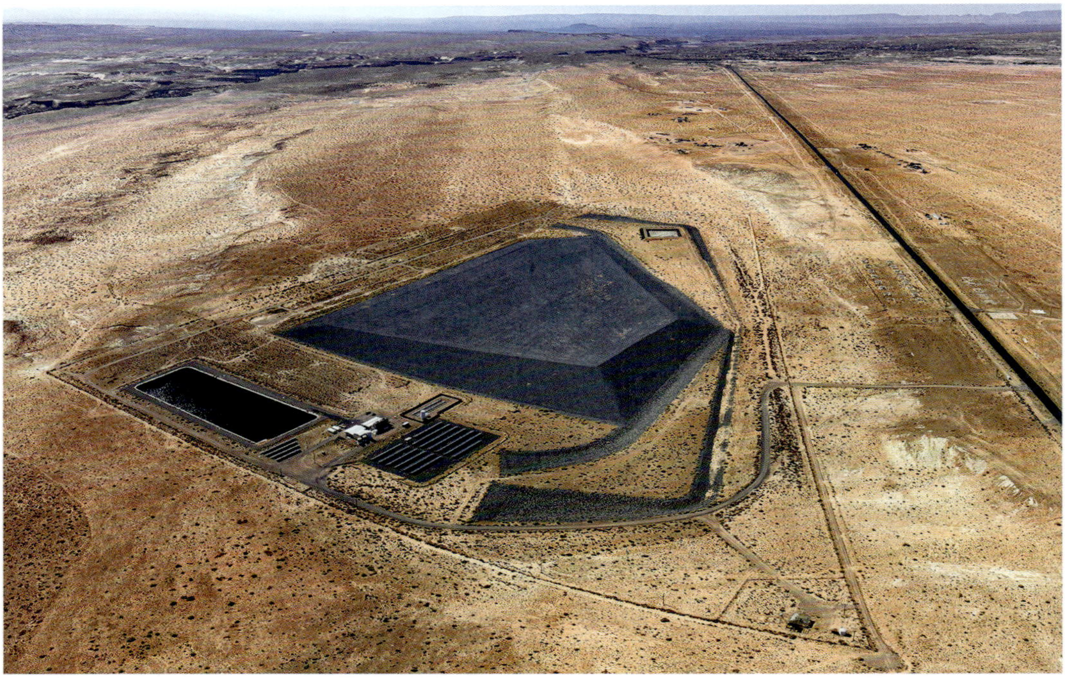

Fig. 6

Will Wilson, *Rare Metals Disposal Cell, Tuba City, AZ, Navajo Nation, 36.14527890580825 N / −111.13477888114164 W,* 2020, archival pigment print, variable dimensions. © Will Wilson.

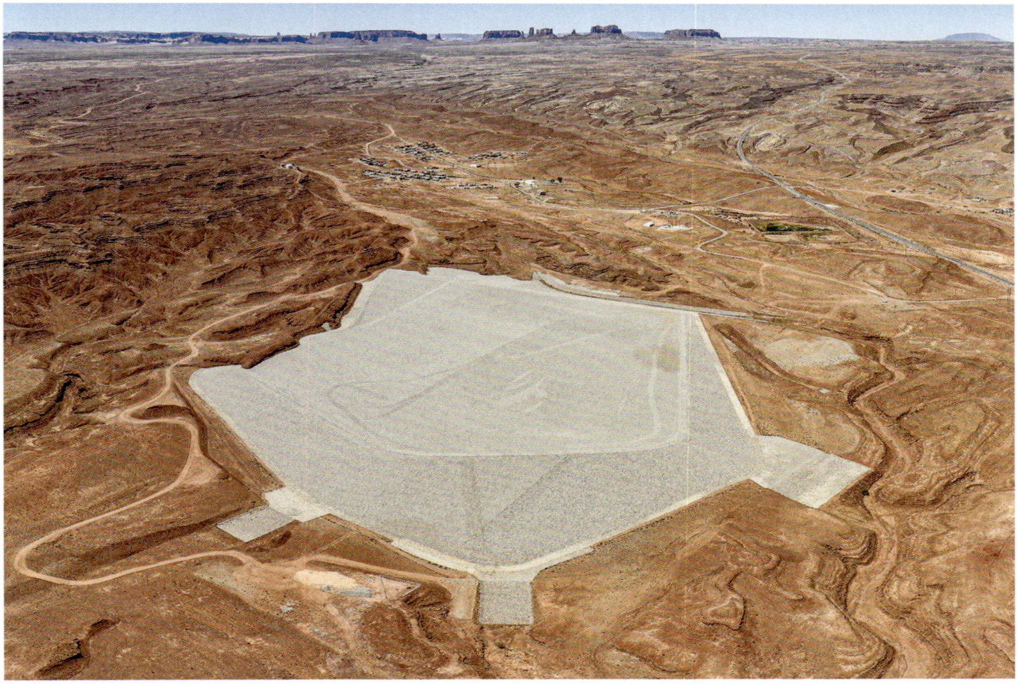

Fig. 7.

Will Wilson, *Mexican Hat Disposal Cell, Halchita, UT, Navajo Nation, 2019, 37.133608802895914 N / -109.87459125505639 W,* 2019, archival pigment print, variable dimensions. © Will Wilson.

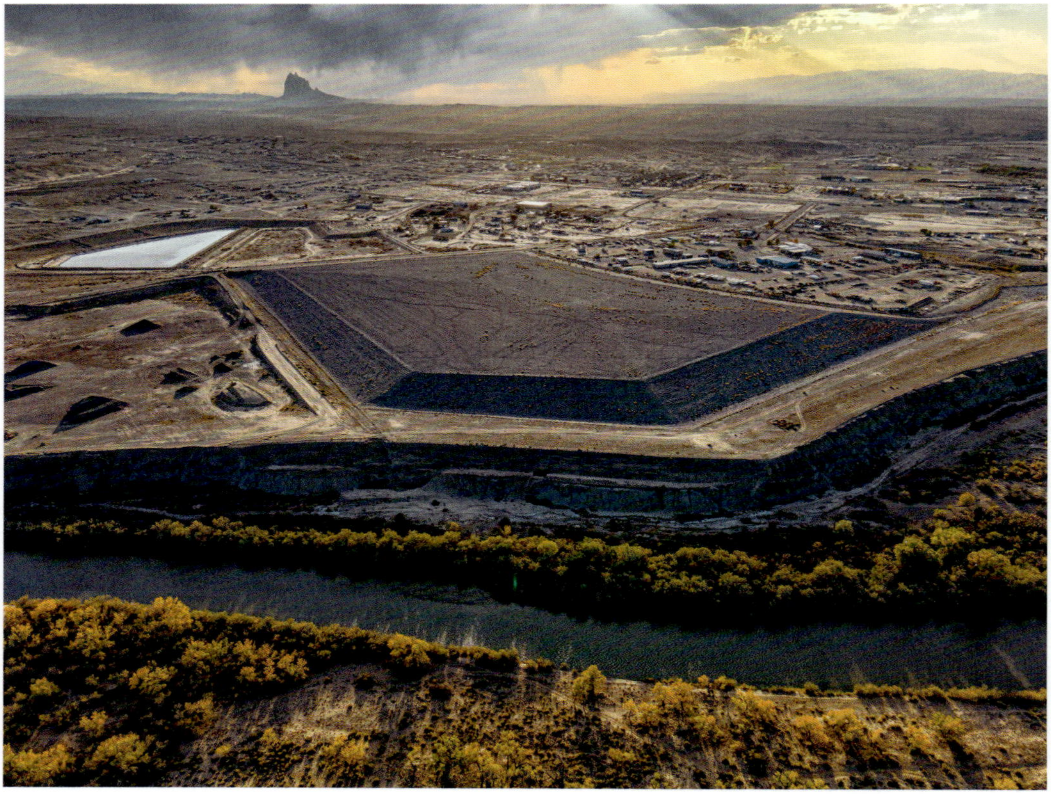

Fig. 8.

Will Wilson, *Shiprock Disposal Cell, Shiprock, NM, Navajo Nation, 36.769690596277414 N / –108.68547605691704 W,* 2023, archival pigment print, variable dimensions.

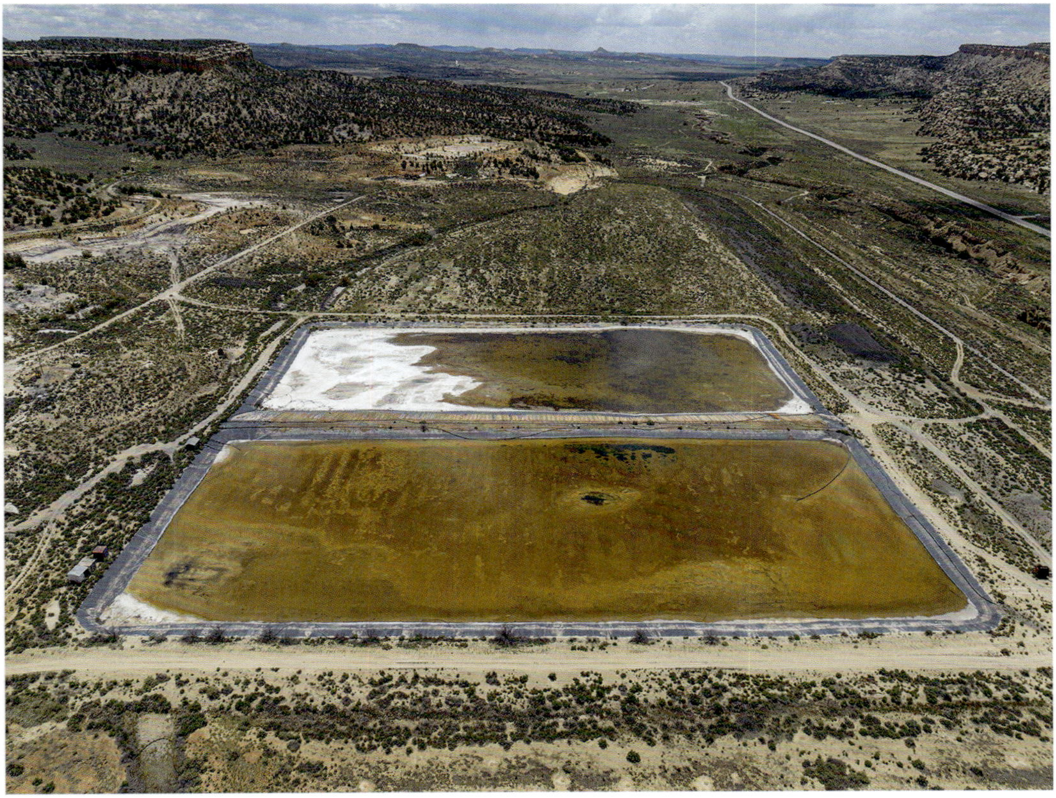

Fig. 9.

Will Wilson, *Church Rock Mill Spill Remediation Ponds AUM, Coyote Canyon, Navajo Nation, 35.64810921016795 N / −108.50126295589726 W,* 2022, archival pigment print, variable dimensions. © Will Wilson.

Yet these are not photographs of wastelands. CTDs refuses the dominant visual regime that renders Indigenous territories as empty, ruined, or disposable. These are images of home. Diné Bikeyah is not a passive backdrop to history, but an active, relational being. Our land remembers where our umbilical cords are buried, where our relations have lived and died. These landscapes hold memory, and they demand accountability.

CTDs expands beyond visual documentation. It includes a location-aware mobile app that provides users with data about AUMs, integrating Navajo Nation/USEPA site screening reports with on-the-ground photography. This tool, designed in collaboration with community stakeholders, enables users to situate themselves within a network of extraction, making visible what was once obscured. It is a means of empowerment, education, and advocacy. The mobile app will pinpoint the user's location and illustrate their position in relation to the more than 500 identified and assessed AUMs on Diné Bikeyah. A database linked to these mapped positions will enable the delivery of important data associated with the AUMs (for more information, please visit www.dinéconnect.org). The project also engages directly with community-based learning and organizing. In partnership with Diné College, CTDs has launched the Reframing Indigenous Remediation speaker series; held photography workshops; and provided resources to train emerging Diné documentarians, artists, and scientists. These efforts point toward a Just Transition—a vision of economic, cultural, and environmental restructuring rooted in Indigenous knowledge systems and sovereignty. Photography, in this context, becomes a generative practice. It is not just about representing the world, but about reshaping it. In CTDs, the camera becomes a tool for environmental justice, cultural resurgence, and intergenerational healing. The images connect the past to the present and the present to the future, asserting Diné presence and possibility across time.

My more recent photographic series, *Hubris on the Land* (2025) (fig. 10), extends these questions by juxtaposing aerial images of UMTRCA disposal cells with iconic works of land art in the American West by the artists Robert Smithson, Nancy Holt, Michael Heizer, and James Turrell. Many of these monumental interventions were created with little to no acknowledgment of Indigenous presence or sovereignty, reflecting a settler-colonial gaze that framed the land as empty and available for artistic occupation. By placing these artworks in visual dialogue with sites of extraction and contamination, *Hubris on the Land* (fig. 10) challenges the aestheticization of the American West and calls attention to the ecological consequences of both artistic and industrial interventions.

Together, these bodies of work compose an ecological visual history of Diné resistance, vision, and survivance. They function as counter-archives, unsettling the settler gaze and remapping the visual

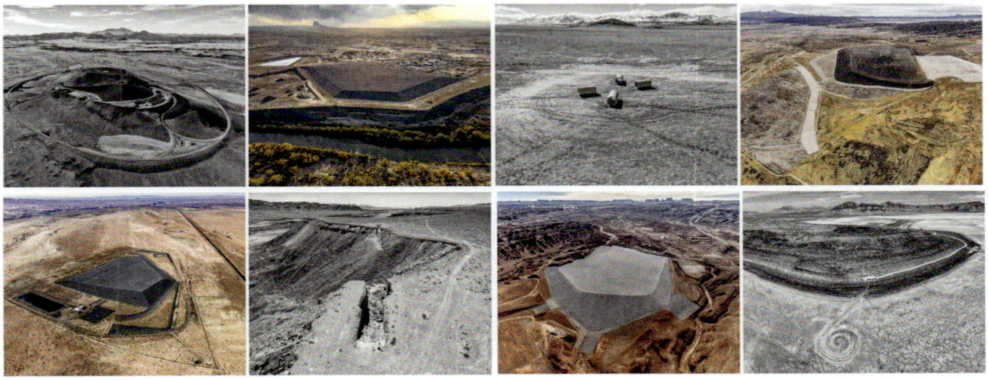

and material terrain of the Southwest. My aim is to tell a different story—one rooted in reciprocity, relationality, and remembrance. As scholars Janet C. Berlo and Jessica L. Horton have noted, recent Western turns toward "new materialisms" often echo Indigenous epistemologies that have long recognized the agency and intelligence of land, water, and non-human life. (Horton and Berlo, 2013: 17–18) For Diné people, these are not theoretical constructs but everyday truths. Our survival has always depended on maintaining respectful relationships with the beings that sustain us. AIR, AIR LAB, and CTDs honor this worldview by documenting harm, yes, but also by envisioning repair.

Fig. 10.
Will Wilson, *Hubris on the Land,* 2025, archival pigment print, variable dimensions. © Will Wilson.

In the end, the work circles back to that foundational choice between the yellow dust and the corn pollen. It is not merely a story from the past; it is a question we must answer again and again. Through image, action, and community, I choose corn pollen. I choose beauty, balance, and the enduring possibility of Indigenous futurity.

Bibliography

Janet Catherine Berlo and Jessica L. Horton, "Beyond the Mirror: Indigenous Ecologies and 'New Materialisms' in Contemporary Art," *Third Text* 27, no. 1 (2013): 17–28.

Peter H. Eichstaedt, *If You Poison Us: Uranium and Native Americans* (Santa Fe: Red Crane Books, 1994).

CHAPTER 9
Āina and Accession Photographs of Queen Kapi'olani and Queen Lili'uokalani

Emily Cornish

When photography arrived in the Kingdom of Hawai'i in the 1840s, it was subjected to Hawai'i's ecology. The earliest extant photograph of Kīlauea volcano, taken by Hugo Stangenwald in 1855 (fig. 1), provides evidence of the environment's impact on this medium. In this photograph, steam rises from vents along Halema'uma'u crater, evoking a ghostly and atmospheric impression of Pele's famous home instead of a sharp and well-defined rendering of the caldera. Stangenwald's photograph highlights how high humidity, fluctuations in sunlight, and dust dancing on tradewinds posed unique technical challenges for early photographers wishing to capture Hawai'i's storied landscapes. Given photography's connection to colonial exploration, surveillance, resource mapping, and extraction, Stangenwald's daguerreotype raises interesting questions about how Hawaiian *'aīna* (land) might have been resistant to the urge to visually possess, catalog, and market different environments.

Later in the century, Queen Kapi'olani and Queen Lili'uokalani created their own depictions of Hawaiian land within photography. Unlike Stangenwald's, their approach was not representational, and it drew upon portrait photography to visualize their complex connections to Hawaiian *'āina*. Kapi'olani and Lili'uokalani were frequent patrons of photography studios in Hawai'i, with the earliest portraits of both women dating to the 1850s. They also frequented studios in the United States and England during their travels abroad. It is not a stretch to say that both queens were familiar with the conventions of the Victorian portrait studio, opening up space for them to leverage the homologues and the formulaic aspects of this kind of photography in order to articulate a sovereign relationship to land, founded on relational and reciprocal interactions between humans and the non-human world rather than transactional ones. The depictions of this relationship within their portrait photographs relied on *kaona*, a Hawaiian aesthetic device,[1] to layer different visual and material symbols, poetry, metaphor, and genealogy to portray their relationships and responsibilities to Hawai'i's environments. This essay analyzes how Queen Kapi'olani and Queen Lili'uokalani used compositional choices alongside foreign materials in their accession photographs (figs 2 and 3) to create powerful images that conveyed their relationships to *'āina* as potent expressions of their political authority and legitimacy during the 1880s and 1890s.

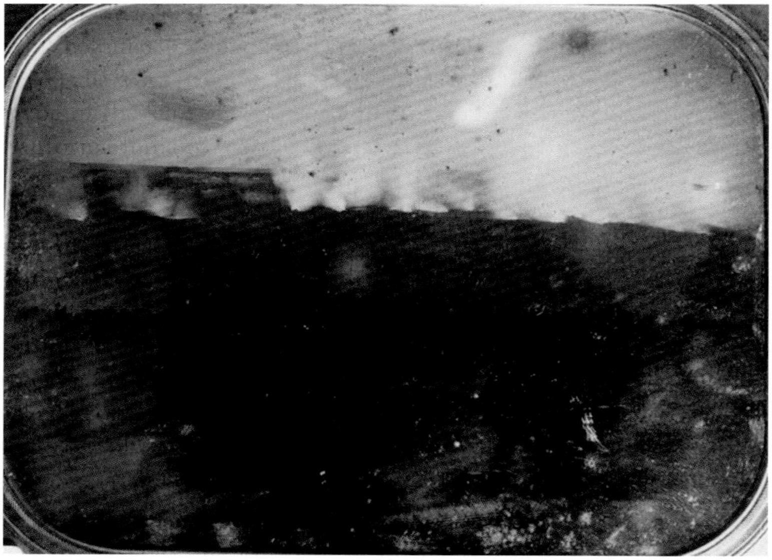

Fig. 1.
Hugo Stangenwald, *Kīlauea volcano*, 1855. Daguerreotype. Courtesy of the Hawaiian Mission Houses Historic Site and Archive.

Histories of photography in Hawai'i have tended toward narratives of White settler triumph in the island kingdom and the ocular possession of Indigenous land and people through ethnographic and touristic imagery. Many scholars, including Jane Desmond, Patricia Johnston, and Joyce Hammond, have produced studies detailing how depictions of Hawaiian women in photographic media generally enabled the visual colonization of Hawai'i by portraying said women as attractive, exotic, non-threatening, and inviting stand-ins for the land itself. (Desmond, 1999: 459–501; Hammond, 2001: 1–32; Johnston, 2002: 188–225) Concurrent with such projects, Native Hawaiians actively embraced photography and embedded it within their culture and traditions.

One goal of this essay is to shake up the lens and expand the interpretative possibilities surrounding photography and ecology during this period. Another goal is to add greater nuance to narratives of Indigenous encounters with nineteenth-century photography by analyzing the accession photographs of two Hawaiian queens. These portraits are neither exoticizing nor ethnographic in their form and content. They are indicative of the aristocratic air that permeated photographs of the Kalākaua era, evincing a high level of self-control and an awareness that self-representation and bodily displays within photography convey powerful messages about individuals and societies. (Maxwell, 1999: 198)

The first section of this essay analyzes the coronation photographs of Julia Kapi'olani, who became *Mo'i Wahine* (queen) of Hawai'i when her husband, David Kalākaua, won the 1874 election for the throne. Queen Kapi'olani's coronation portraits combined foreign materials with Indigenous motifs and concepts to legitimize her and her husband's rule by visualizing her genealogical connection to the *'āina* through the incorporation of kukui nut motifs in her crown and gold-embroidered palapalai ferns in her coronation dress. Queen Kapi'olani's coronation photograph, along with King Kalākaua's, was intended to project their authority and royal prerogatives to a global audience.

The next section analyses how Lili'uokalani used her accession portrait, taken in 1891, to foreground the important relationship between herself and Hāloa ke kalo (the taro plant, one of the children of Wākea), the elder sibling of humanity. The photograph visualizes Lili'uokalani's political authority derived from this relationship, which was rooted in Hawaiian cosmology. Additionally, this connection positions Lili'uokalani within her accession portrait as a key source of protection and nourishment for her land and people. This chapter concludes by examining the various lives of this photograph, first as a gift from Queen Lili'uokalani to the US Naval Secretary in 1916, and, finally, as a painted mural in the Kalihi-Palama neighborhood on O'ahu.

Queen Kapi'olani's Coronation Portraits, 1883

King David Kalākaua's ascension to the throne, on February 12, 1874, marked a new era in the Hawaiian Kingdom, with the shift in power from the Kamehameha dynasty to the Kalākaua dynasty. One year prior, Major General Scholfield and Brigadier General B. S. Alexander were instructed by the United States Navy to scope out Hawai'i's ports for their defensive capacities and the islands' commercial potential. This marked the beginning of the US government's interest in extraterritorial expansion in Hawai'i as a form of official policy and included a renewed interest in Pu'uloa, commonly known as Pearl Harbor. Hawai'i was facing, in Kalākaua's own words, "a peculiar situation" involving a majority, though decreasing population of *Kānaka Maoli* (Native Hawaiians) and an increasing settler populace who were interested in controlling the nation's politics and economic resources. (quoted in Kamehiro,

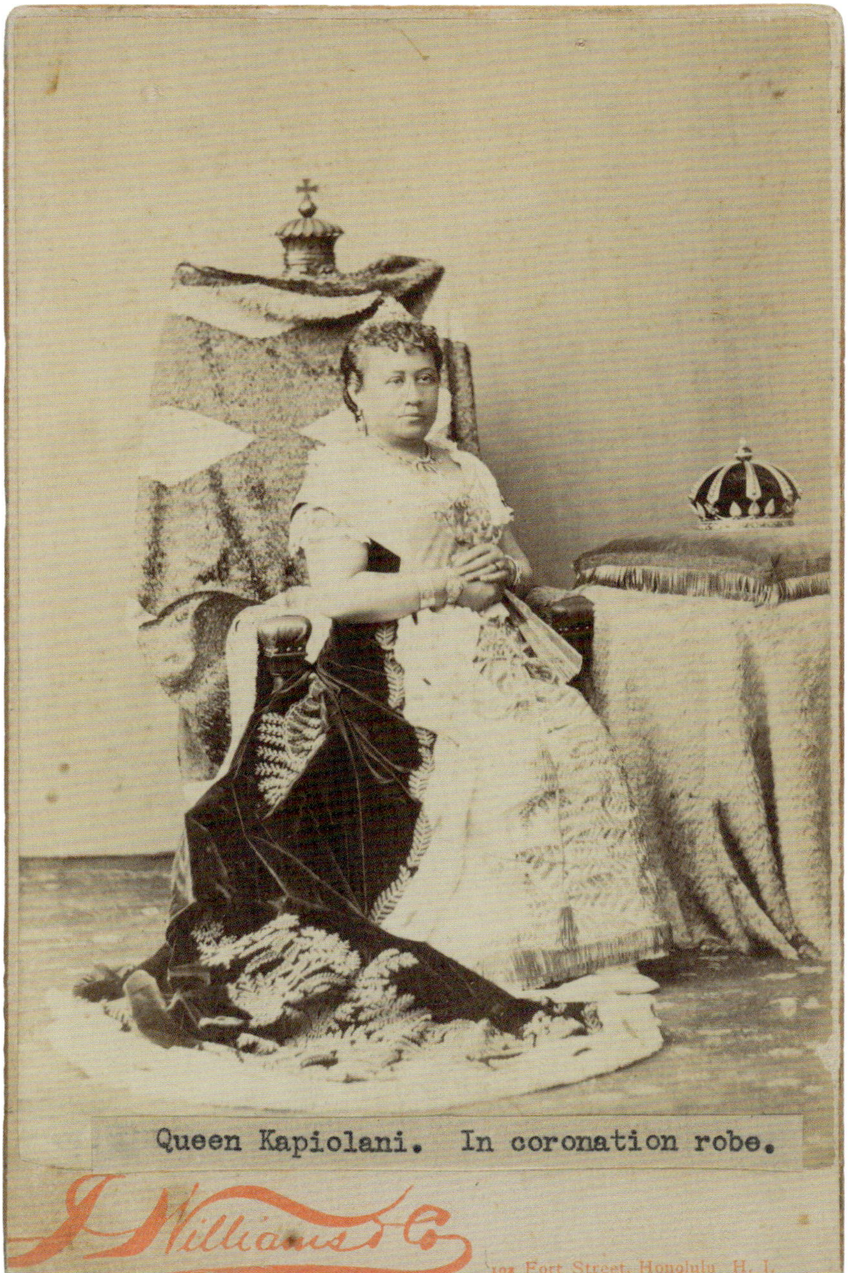

Fig. 2.

J. J. Williams, *Queen Kapi'olani in Coronation Robe*, circa 1883, Cabinet Card. Courtesy of the National Portrait Gallery, Smithsonian Institution. Gift of the Bernice Pauahi Bishop Museum.

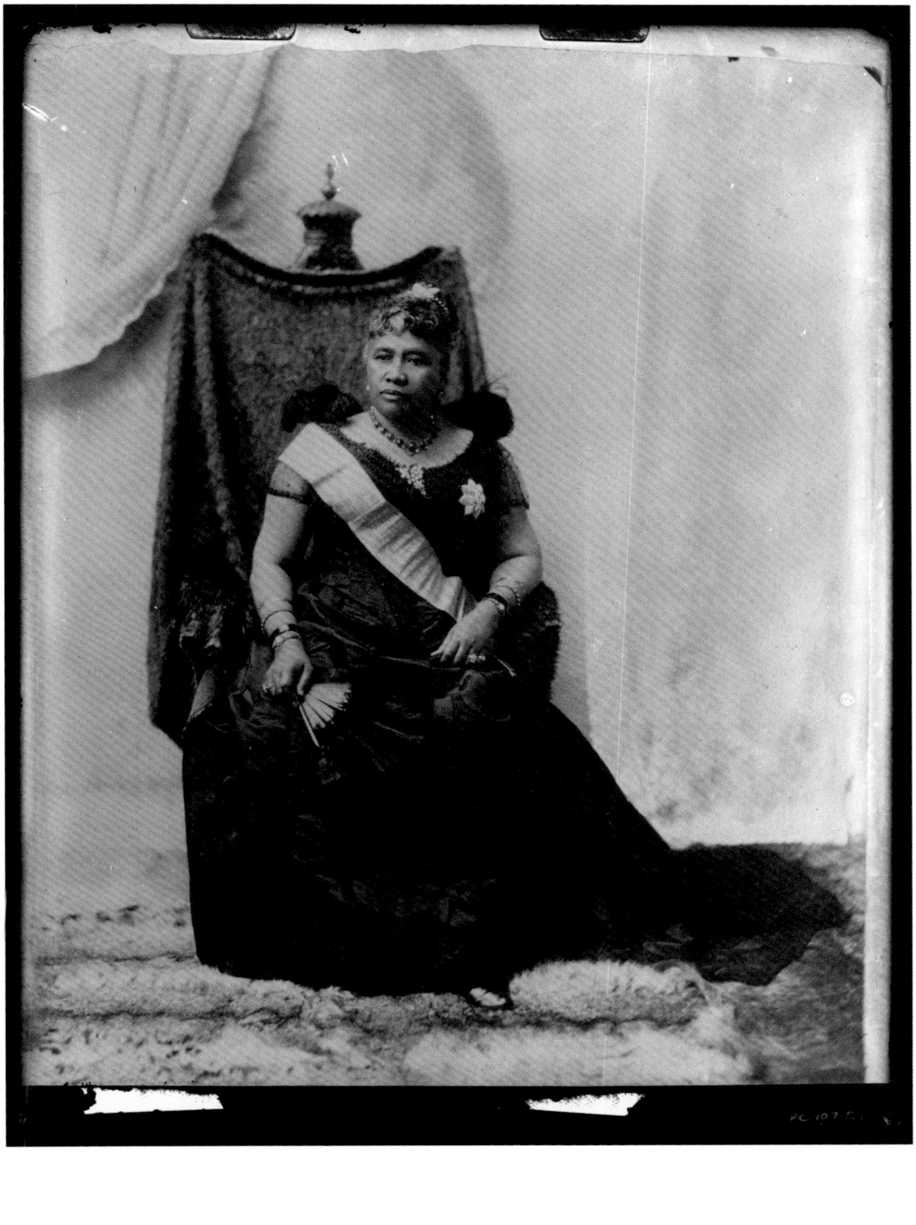

Fig. 3.

J. J. Williams, *Queen Lili'uokalani enthroned*, c. 1891. Glass plate negative of the original. Courtesy of the Hawai'i State Archives.

2018: 86) Tied to this peculiar situation were questions of how best to steward Hawai'i's environment and resources for the wellbeing of this multi-cultural and growing nation. Following the election, Kalākaua and Kapi'olani needed to secure and authenticate their position as the ruling family of Hawai'i in the face of ongoing discontent from Emmaites and increased imperial interest in the island kingdom by the United States and pro-American settlers.[2] With these issues in mind, Kalākaua adopted the motto *ho'oulu lāhui*, often translated as "to increase the nation." More than a motto, it was used by King Kalākaua and Queen Kapi'olani to transform the role of Hawaiian elites in caring for their land and people into a "distinctly modern national" enterprise. (Caldeira, 2018: 13)

Looking at the nuances contained in definitions of *ho'oulu* and *lāhui* helps us understand how Kapi'olani and Kalākaua approached this monumental undertaking and the interrelationship between human beings and *'āina* implicit in this task. The 1865 Hawaiian dictionary provides several definitions for *ho'oulu,* including "to grow as a plant," "to increase in any way," "to become stronger," "to grow or increase as good or evil in a community," "to grow up," "to grow in size or strength," and "to cause to spring up, as seeds." (Andrews, 1865: 16)[3] Modern definitions of *ho'oulu* include "to grow," "to sprout," "to propagate," and "to protect." (Combined Hawaiian Dictionary, 2020) The 1865 dictionary defined *lāhui* as "to prohibit," "to forbid," "to lay a *kapu,*" "to proclaim a law or ordinance," and "as a time of coming together; hence, an assemblage, a company; a union of many." (Andrews, 1865: 321, 328.)[4] Over time, these definitions have coalesced into the following modern interpretations: "nation," "race," "tribe," "species," and "to assemble or gather." *Ho'oulu lāhui* connects Kalākaua and Kapi'olani's political, social, and economic concerns and policies to long-standing human relationships with one's land and environment.

Ho'oulu lāhui became expressed in the visual culture of this dynasty, which produced a robust array of images and objects that spoke to Hawai'i's wellbeing and future as something to be carefully cultivated, incorporating Indigenous revitalizations and prudent uses of foreign ideas, objects, and technology. A vast array of visual culture, including photography and performances of state, was appropriated in service of this agenda. Finally, *ho'oulu lāhui* and its expression in Hawaiian visual culture of the period stands in stark contrast to settler notions and representations of nationalistic increase. For instance, Manifest Destiny, the ideological stance that US territorial expansion and imperialist extraction were both inevitable and divinely mandated, is represented in photography through such things as sublime landscapes, mines, and railroads. Typically, within such photographs, Indigenous connections to and presence on the land are minimized and Indigenous people are posed to conform to settler narratives or omitted altogether.

On February 12, 1883, nine years to the day after King Kalākaua and Queen Kapiʻolani ascended the throne, they held an official coronation. In addition to this ceremony, they sat for a series of portrait photographs to commemorate the occasion. The portraits were reprinted and distributed in various formats, including as cabinet cards (fig. 2), indicating these images' wider circulation. Life-size versions were even sent by King Kalākaua to the Women's Industries Centenary Fair in Sydney in 1888. Extant versions of these photographs show Kapiʻolani standing or seated near her throne, with her coronation regalia, crown, and *ʻahuʻula* (feather cloaks) prominently displayed. Queen Kapiʻolani was photographed in a variety of poses, with both full-length and cropped versions of the images being printed. Her crown, coronation dress, and two *ʻahuʻula* featured in these portraits reflect the deep connections between the reigning *Moʻi Wahine* and the Hawaiian *ʻāina.*

In Figure 4, Queen Kapiʻolani stands parallel to her crown. The fan she holds creates a line that leads the viewer's eye to the cushion upon which the crown sits. According to the February 17, 1883, issue of the *Pacific Commercial Advertiser*, Queen Kapiʻolani's crown was "precisely similar [to the king's], except that it is a trifle smaller. Each crown contains 521 diamonds, 54 pearls, 20 opals, 8 emeralds, 1 large carbuncle, and 6 kukui jewels." The photograph's composition successfully indicates the relationship between queen and crown while skirting any issues of decorum related to Kapiʻolani's status as an *aliʻi* (chiefly or royal) woman. It would have been inappropriate for the photographer or any attendants of lesser rank to pass their hands over Kapiʻolani's head, the most sacred part of her body, in order to place and adjust the crown during the photoshoot.[5] Instead, Kapiʻolani appears in all versions of her coronation photographs wearing a tiara, which forms a tasteful triangular point atop her head and potentially serves to cover and protect the most sacred part of the queen's head. (Kaeppler, 1982: 93)

The kukui nut jewels on Kapiʻolani's crown were "a deep reddish black, highly polished," and highly significant both within the crown and, by extension, within the coronation photographs. (*Pacific Commercial Advertiser*, February 17, 1883)[6] Precious gems, such as diamonds, were fairly commonplace symbols of European royalty or even empire. Kukui, or candlenut, constituted a uniquely Hawaiian material used to denote Kapiʻolani's royal status. Kukui has a range of uses and significance, including as medicine and food. As Hawaiian historian David Malo recounts in his *moʻolelo* (history) of Hawaiʻi, the oil derived from the nut of this tree was used in torches as an important source of light. (Malo, 2020: 93) The act of burning such torches by day was a marker of elevated status reserved for members of the Kalākaua family. (Kamehiro, 2009: 51) The light emitted from these torches could be associated with the god Lono, whose domains included the clouds, the

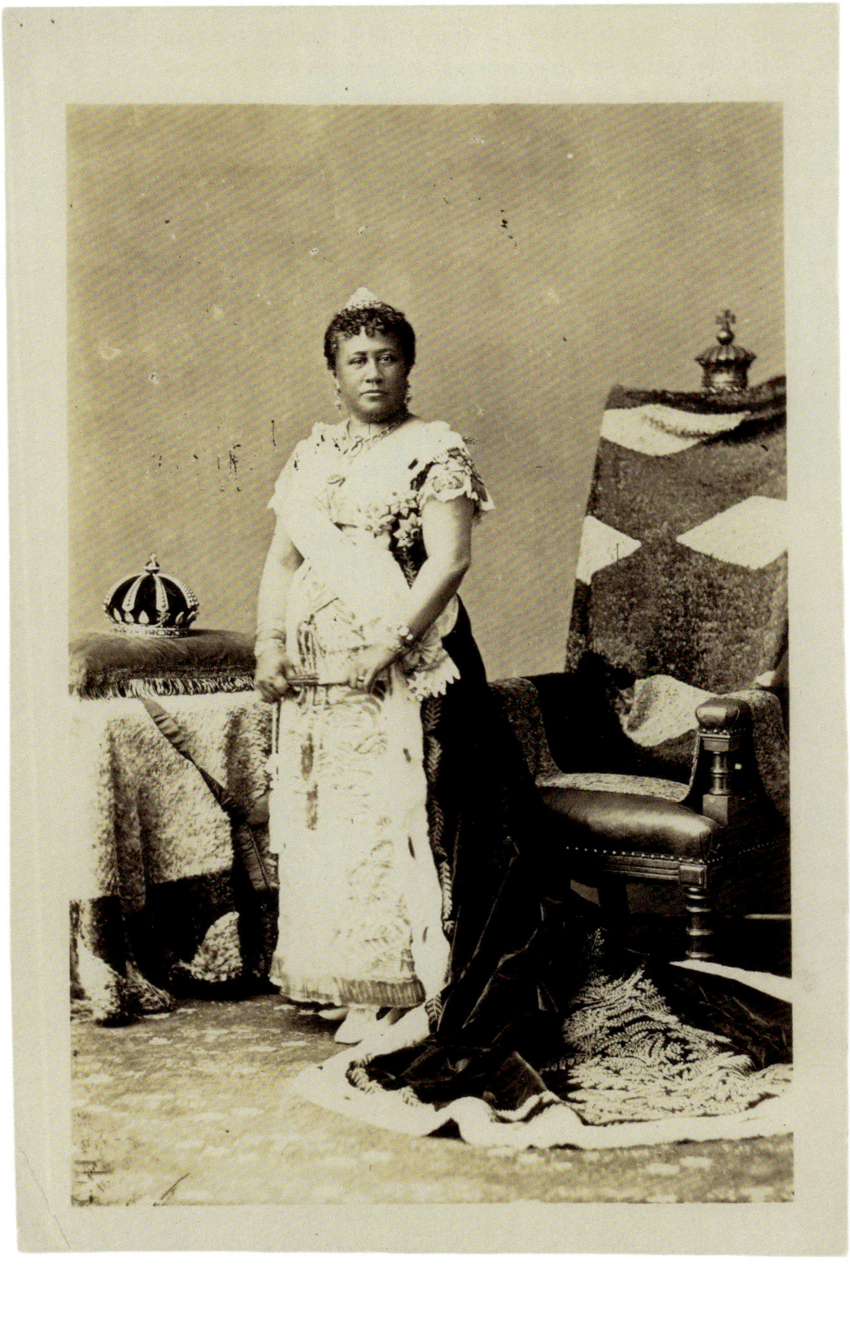

Fig. 4.

J. J. Williams, *Kapiʻolani coronation portrait (standing)*, 1883. Courtesy of the Hawaiʻi State Archives.

winds, the sea, and agriculture. He is also the patron god of the Makahiki season, which corresponded to the rise of the Pleiades and marked the change in weather in Hawai'i. Kapi'olani was a descendant of this deified ancestor through her mother's side. As such, the representation of kukui in Kapi'olani's crown, and its connection to light, coalesces into a visually complex reference to the bright and shining nature of Kapi'olani as the current queen within this photograph.

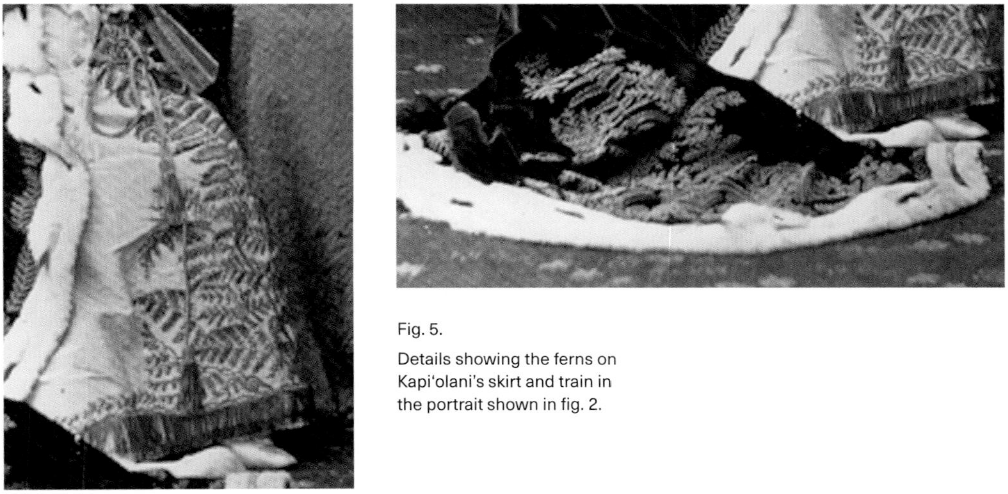

Fig. 5.
Details showing the ferns on Kapi'olani's skirt and train in the portrait shown in fig. 2.

The appearance of the embroidered palapalai ferns (fig. 5) running the length of Kapi'olani's skirt and train stands out within her coronation photographs and presents another significant connection between Queen Kapi'olani and the god Lono in this photograph, which further informs how this image materializes Kapi'olani's political authority and legitimacy. Queen Kapi'olani chose to dress in modern Victorian fashion for her coronation, designed by Charles Furneaux and made in London from white satin and gold-embroidered palapalai ferns. This garment combined local custom and motifs—to indicate Kapi'olani's rank and sacred status—with global fashion trends. In Figure 2, Queen Kapi'olani's train is arranged in such a way as to show off the fern embroidery. Additionally, Kapi'olani angles her body to display the embroidery on the front of her gown, giving viewers a sense of the ferns growing from the floor, up and over her body, visually evoking the entwined nature of human and plant life. Malo recounts that palapalai was used as a food source, especially during times of famine. (Malo, 2020: 94, 116) It was also a material component found in older forms of *ki'i* (various forms of images including photographs and older statues), especially ones created during Makahiki. (Malo, 2020: 224–237) Kapi'olani's display of this fern within her coronation portrait captures the multifaceted messages and ideas about her and her family's political authority and the interrelationship between that authority and Hawai'i's ecology.

Photography would have aided in projecting these ideas beyond the immediacy of the coronation ceremony and the queen's local actions on behalf of her people. Kapi'olani was heavily involved in distributing food, paying for funerals, and providing other forms of support to her people. (Caldeira, 2018: 13) In 1890, she founded the Kapi'olani Maternity Home to ensure the care of Hawaiian mothers and babies.[7] Kapi'olani's actions form a crucial component in legitimizing her and her husband's rule. Beyond being a recognizable symbol of the monarchy, Kapi'olani's crown and the palapalai embroidery on her coronation dress imbue this photograph with associations to complex connections between *'āina* and plants that provide light and nourishment, support for their kingdom. The inclusion of these objects within this image can be interpreted as part of the complex social, political, and cultural formation that was *ho'oulu lāhui* and ultimately works to make a pointed statement about Kapi'olani as a queen who upholds her responsibilities to her people.

Eight years later, Lili'uokalani also selectively mixed Hawaiian and European symbols of royal power within her accession photograph, suggesting that she drew on Queen Kapi'olani's coronation photographs as models for her own portrait. The following section will analyze how Lili'uokalani deployed her self-fashioning within her accession photograph to materialize her relationship to *Hāloa ke kalo* as foundational to her sovereign authority.

Lili'uokalani Enthroned, 1891

In November of 1890, Lili'uokalani was appointed regent so that her brother, King Kalākaua, could travel to San Francisco for his declining health. Roughly two months later, on January 29, 1891, the USS *Charleston* was sighted near Diamond Head with its flags at half-mast. Unbeknownst to Queen Kapi'olani and Lili'uokalani, who was acting as regent, the king had died while abroad, in San Francisco, on January 20, 1891, and Lili'uokalani was now queen. The USS *Charleston* was returning the remains of her brother David Kalākaua to the kingdom. No coronation ceremony was held, as the court and the country launched into funeral rites for their late king. Queen Lili'uokalani's reign was a turbulent one, plagued by internal and external threats to her political authority. These threats included fallout from the McKinley Tarriff Act of 1890, which effectively crashed the economy of the Hawaiian Kingdom by imposing tariffs on Hawaiian sugar. John L. Stevens, who served as the US Minister to Hawai'i during her reign, was eager for the US imperial expansion to extend to the Pacific Islands, even waxing poetic about Hawaiian land and resources in the US press.[8] Queen Lili'uokalani experienced ongoing ripple effects from the 1887 "Bayonet Constitution" which facilitated the repeated dismissal of her cabinets by the legislature, especially toward the end of 1892, leading to the destabilization of her government. At one point, her reign was threatened by the National Reform Party, which lobbied for an end to the monarchy and for Hawai'i to become a republic of its own. Finally, a secret annexation club comprised of pro-American businessmen that was created during her reign engaged in treasonous activities that culminated in the coup on January 17, 1893.[9]

Queen Lili'uokalani's accession photograph was produced shortly after she ascended the throne in 1891. The glass plate negative of this portrait (fig. 3) from the Hawai'i State Archives is indicative of the portrait's reproduction and wider dissemination. The photograph visually suggests she understood how the challenges facing Hawai'i interconnected with land and resources. In his account of her accession, Ralph Kuykendall made a similar observation about Lili'uokalani's conception of political authority and action, stating she was: "proud of her noble lineage and conscious of the authority, the power that had been exercised by high chiefs in earlier times, and she was determined to maintain the traditional prerogatives of the ruling chief." (Kuykendall, 1967: 380)

Her bodily display of *kalo* (taro) on the dress she wore for the portrait indicates that she saw her genealogical connection to *kalo* as a crucial sanction of her authority and actions as sovereign. In this photograph, Queen Lili'uokalani is seated on an *'ahu'ula*-covered throne and appears bedecked in jewels and royal orders, including, notably, the Royal Order of Kalākaua, which was created to commemorate her brother's accession to the throne. The tonal qualities of the photograph highlight the different textures of the objects included, such as the individual feathers of the cloak at her back and the rippling material of her dress. Close scrutiny of her gown in this portrait reveals delicate swirling volutes that form a *kalo* leaf pattern. *Kalo* is an important plant in Hawai'i, with a variety of practical and ceremonial uses, from food and medicine to its prior use in the funerary rites of ruling chiefs. (Malo, 2020:114, 185) In her seminal work on Kalākaua-era visual culture, art historian Stacy Kamehiro refers to *kalo* as the "staff of life." (2009: 51) Before Hawai'i's ecosystems were heavily disrupted by settler-colonial agricultural production and extraction, *kalo* was a cornerstone of the Hawaiian diet. These changes began with such things as the harvesting of sandalwood and later expanded to crops, such as sugarcane, pineapple, and rice, and continue in the present with luxury developments and resorts. A *New York Times* article from 2019 that discussed the revival of *kalo* in Hawai'i aptly noted, "history calls this progress, the subduing of nature to industry and the needs of swelling population—and islands are often the first to pay the price for it." (Mishan 2019)[10] As such, the appearance of *kalo* within Queen Lili'uokalani's accession photograph draws a significant connection between political authority and those who nourish the land and its people.

In the cosmogenic chant known as the *Kumulipo, kalo* is the elder sibling of all *Kānaka Maoli.* Hāloa ke kalo was the first child of Wākea (the sky father, one of the progenitors of humanity) and his daughter Ho'ohōkūkalani. However, this child was stillborn, and from the spot where Hāloa ke kalo was buried sprouted the first taro plant. Their second child, Hāloa ke Kanaka, was charged with tending his elder sibling, i.e., the *kalo* plant. (Lili'uokalani, 1897; Malo, 2020: 324)[11] In addition to recounting the creation of the world, Pualani Kanaka'ole Kanahele notes that the *Kumulipo* provides a foundation for present action. (1997) This was certainly true for Lili'uokalani and Kalākaua during the nineteenth century. As part of a political and cultural strategy to establish his family's genealogical right to rule, Kalākaua frequently alluded to his lofty genealogy, which could be traced to deified ancestors in the *Kumulipo* in a variety of artforms, including his publication of a Hawaiian translation of the chant in 1889.[12] Notably, Lili'uokalani published the first English translation in 1897. This edition of the text was published in Boston by the Lee and Shepherd publishing house. The title page notes that the queen began her translation "during her imprisonment at 'Iolani Palace and afterward at Washington Place Honolulu was completed at Washington DC May 20, 1897." During this time Lili'uokalani was traveling around the United States presenting her and the Hawaiian people's case against the annexation of Hawai'i to the United States, all of which suggests that the text of her translation was meant to function as a critique of US imperial aggression toward Hawai'i and a proclamation of her sovereignty rooted in her genealogy.[13]

The *Kumulipo* includes themes of change, churning, and the preservation of life, both human and non-, reemphasizing that the connection between people and *'āina* is not one purely based on extraction and profit but is instead familial and sacred. (Kanaka'ole Kanahele, 1997) The text's elaboration of the responsibilities and interactions between people and their environment, combined with its connection to Lili'uokalani's lofty genealogy, would have made the *kalo* motif a powerful point of reference in her official accession portrait. Within this photograph, Lili'uokalani's dress metaphorically wraps her body in *kalo* leaves, potentially indicating her *kapu* status. Similarly to Kapi'olani's coronation photographs, this creates a visual correlation between Lili'uokalani's body, *kalo*, and the right or authority to rule in Hawai'i. This point takes on greater significance given that on January 17, 1893, Queen Lili'uokalani's rule was cut short by a coup spearheaded by a group of settler oligarchs, including Sanford B. Dole. Dole, who was the cousin of James Dole of the Dole Pineapple Company, was named president of the provisional government that was put in place because of the coup and was then named governor of the territory post-annexation.

Far from being an ending, the overthrow of Queen Liliʻuokalani marked a new phase in her fight for Hawaiian sovereignty. This photograph has an enduring connection to that struggle and continues to be reproduced in different media, including paintings, protest fliers, T-shirts, posters, and pareo (a kind of sarong or wrap-around skirt popular in Hawaiʻi and other areas of the Pacific). I therefore conclude by analyzing two examples that demonstrate different ways Queen Liliʻuokalani's accession photograph has been "re-codified" and aligned with various forms of protest and struggles for sovereignty in Hawaiʻi during the twentieth and twenty-first centuries.

The first example is a gelatin silver print (fig. 6) from the collections at the Library of Congress. This particular version contains an inscription that reads "To the Hon. Josephus Daniels, Secretary of the Navy, from Liliuokalani, Hawaii, October 1916." This inscription (fig. 7) indicates the photograph was as a gift to the man who oversaw the US Navy, which claimed territorial bases in places such as Hawaiʻi. In 1919, Daniels would travel to Hawaiʻi, where his wife participated in the ceremonial opening of the Wai-Momi (Pearl Harbor) Dry Dock, a place deeply entangled with the history of this Hawaiian dynasty and US settler colonialism in the Pacific. Queen Liliʻuokalani's decision to give a print of her accession photograph—an image that announces her sovereign right to rule over Hawaiʻi in ways that were legible in both Hawaiian and Euro-American visual contexts—to Daniels, a man who was a functionary of the settler government that first helped to usurp her throne and then annex her country, *reveals the political complexity of her position.* While we cannot be certain of her exact intentions behind this gift, the image itself suggests certain analytical possibilities that align with her actions as queen before and after the illegal overthrow, as a significant expression of her sovereign authority and through its depiction of the *kalo* leaf, which heralds her genealogical connection and responsibilities to Hawaiʻi.

Using this photograph in a political gift exchange of this kind further reflects how Liliʻuokalani employed adornment and images to protest and critique her overthrow (1891) and the annexation of Hawaiʻi to the United States (1893).[14] For example, both during and following her reign, Liliʻuokalani often dressed in black to symbolize she was in mourning. (Terao-Guila, 2021: 154; Liliʻuokalani, 1897: 280) Queen Liliʻuokalani ended her reign in a state of uncertainty that called for mourning the tangible loss of sovereignty that she and Native Hawaiians experienced

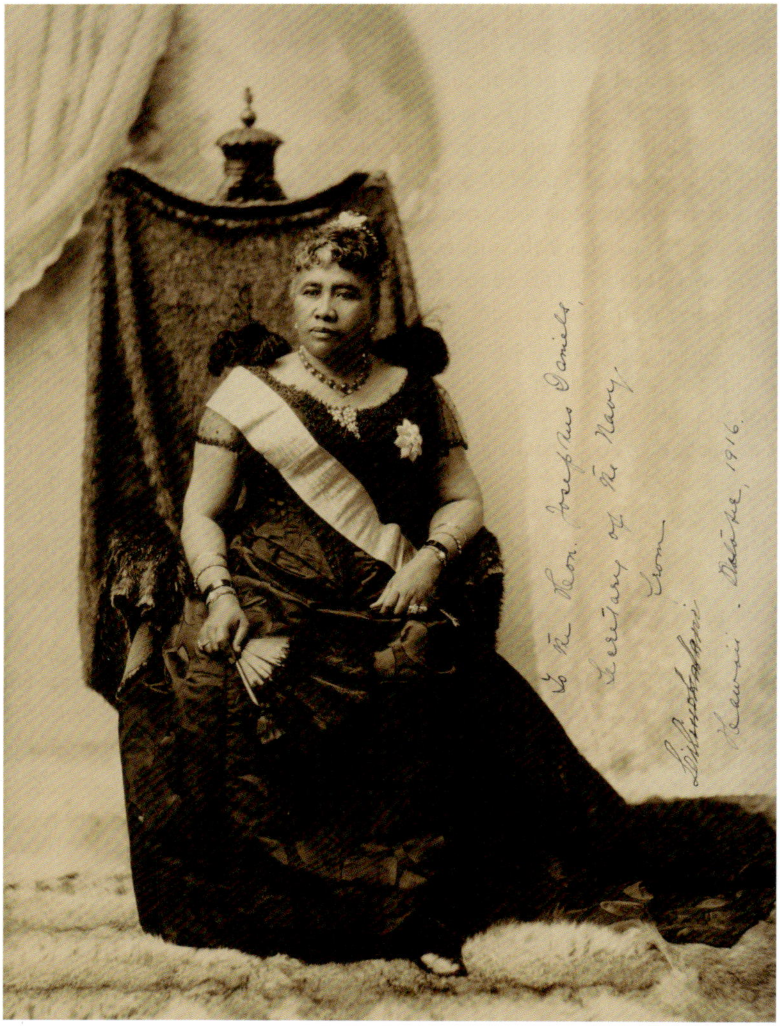

Fig. 6.

J. J. Williams, *Queen Lili'uokalani enthroned, inscription: "To the Hon. Josephus Daniels, Secretary of the Navy, from Lili'uokalani, Hawaii, October 1916,"* c. 1891. Gelatin silver print. Courtesy of the Library of Congress.

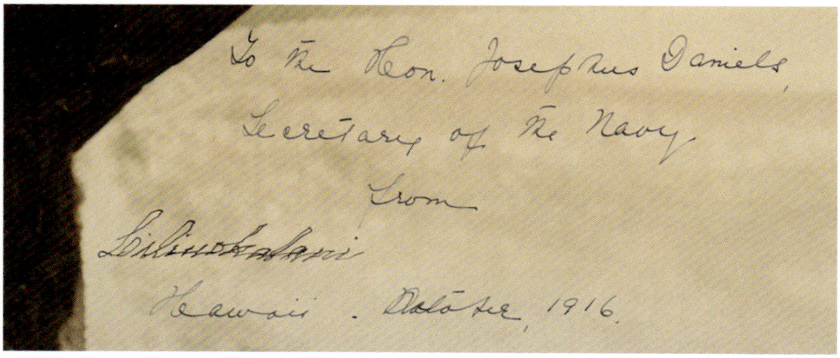

Fig. 7.

Detail of text from Queen Lili'uokalani's accession portrait shown in fig. 6.

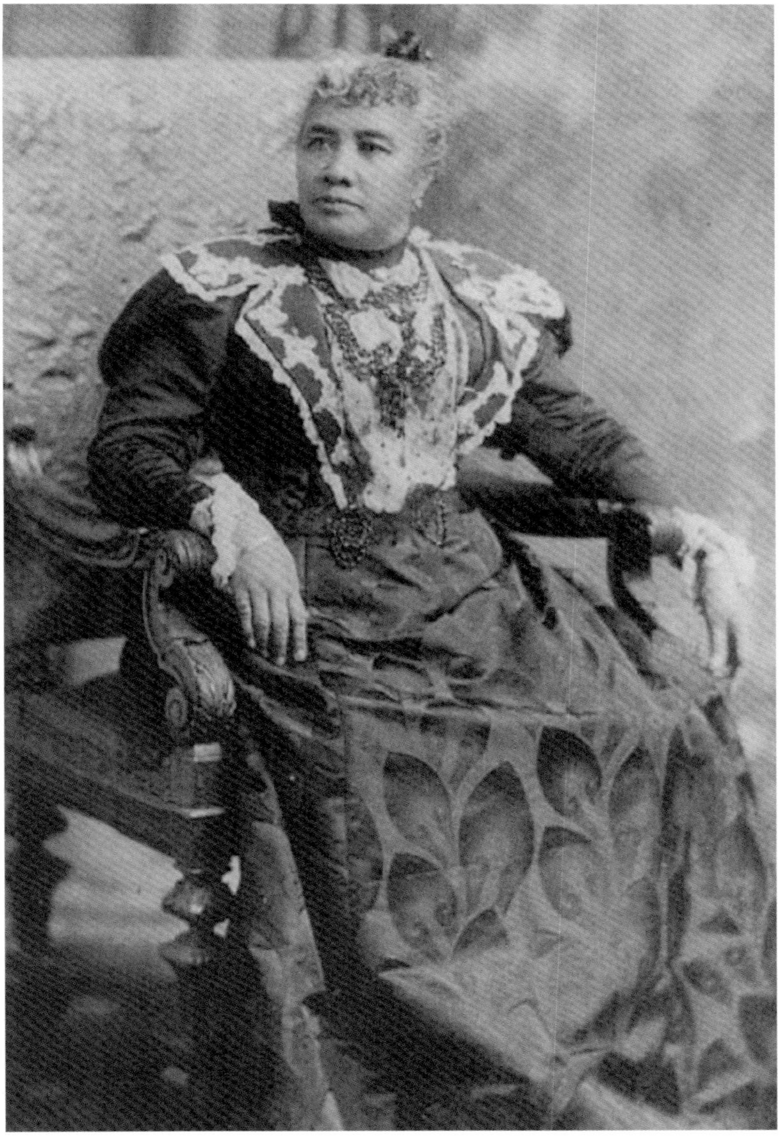

Fig. 8.
Photographer unrecorded, *Queen Lili'uokalani*, c. 1897. Courtesy of the Hawai'i State Archives.

as a result of the illegal overthrow of the Hawaiian monarchy. After 1893, she appeared in multiple photographs depicting her enthroned, in poses eerily reminiscent of her accession portrait, including another photograph with the *kalo* motif visible on her dress (fig. 8) This suggests that after 1893, Liliʻuokalani deliberately cultivated portrait photographs that depicted her still-enthroned yet in mourning, indicative of a state of unrelinquished sovereignty that she was invested in keeping alive and visible for Native Hawaiians and the American government.

The mural *Ola Ka Wai, Ola Ka Honua* (As the Water Lives, the Earth Thrives) (fig. 9), offers a re-interpretation of the accession portrait, extending this analysis into the present. The mural makes the connection between Liliʻuokalani and different facets of *ʻāina*—mountains, water, sea life, and plants—explicit. In the mural, painted in 2011 by a cohort of twenty-five graffiti artists in the Kalihi-Palama neighborhood on Oʻahu, the artists recreated Liliʻuokalani's original accession portrait. However, instead of placing Liliʻuokalani on an ʻahuʻula-draped throne, they depicted her on a seat that visibly transforms into a lush green landscape. Her blue cordon becomes a stream flowing down and across her body. Estria Miysharo, one of the contributing artists, interprets this part of the mural as Liliʻuokalani continuing "to pour her heart and help" *Kānaka Maoli* in the present. (Tamaira, 2017: 10) Her famed butterfly brooch, which is barely perceivable in the original photograph, appears animated in the mural, poised on her head as if ready to take flight.

In her article "Walls of Empowerment," Mārata Tamaira argues that the creation and placement of *Ola Ka Wai, Ola Ka Honua* on Koaka Street in the Kalihi-Palama neighborhood was an act that "overcame the ideological 'wall' of US settler colonialism by transforming the physical walls of the built environment into sites of indigenous empowerment." (Tamaira, 2017: 2) She further notes that walls are enmeshed with the current pursuit of development in Hawaiʻi, which cannot be disentangled from contemporary forms of mili-tourism, which continue to impact Hawaiʻi's ecology. (Tamaira, 2017: 5–6) We can read boundaries and delineations in both historic and contemporary photographs of such things as sugarcane and pineapple plantations and resorts. We might also ask: What do such walls keep out? What is physically and conceptually beyond their frames?

While not the central focus of *Ola Ka Wai, Ola Ka Honua,* taro are also present in the mural. Taro was once a prominent feature of the landscape in this area of Kalihi and was cultivated through a system of irrigated terraces known as *loʻi kalo,* which creates a jarring juxtaposition between the industrialized urban sprawl of today and the verdant scene in the mural. (Tamaira, 2017: 8) Unlike in the photograph, in the mural these plants are no longer contained within the boundaries of the queen's dress. The taro plants are visible to the right through irrigated

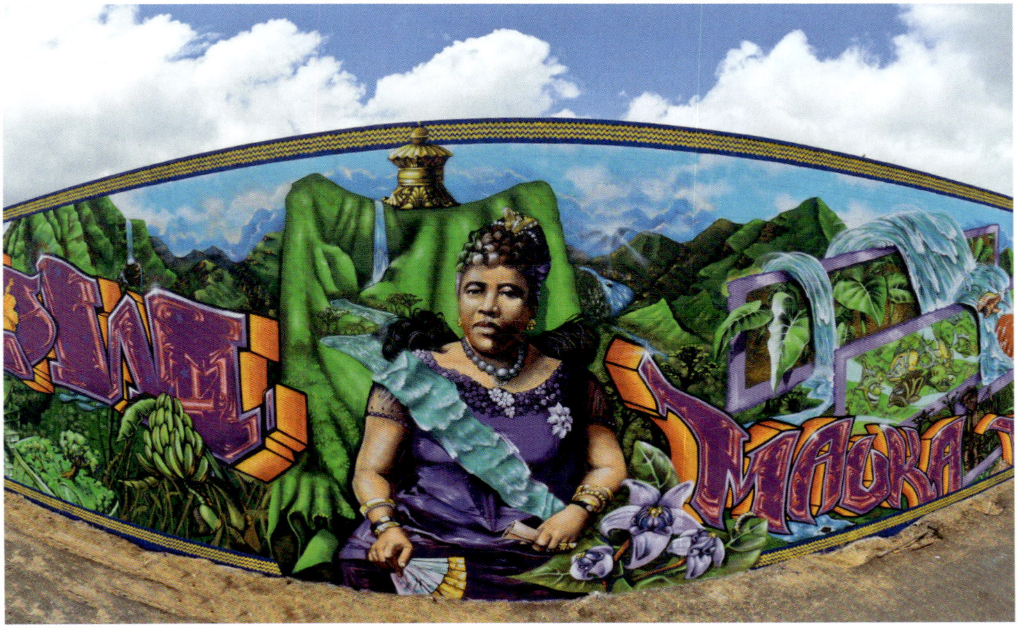

Fig. 9.
Estria Miyashiro, John "Prime" Hina, and others, *Ola Ka Wai, Ola Ka Honua (As the Water Lives, the Earth Thrives)*, 2011. Mural. Estria Foundation, Kalihi, O'ahu. Photograph courtesy of Smarthistory.

terraces. The right-hand side of this image juxtaposes past methods for land stewardship with the futuristic vision on the left. On the left side of the mural, plant life melds with the urban sprawl of Honolulu, in much the same way as Queen Liliʻuokalani melded important concepts about the relationship between Hawaiʻi's leaders and *ʻāina* in her accession photograph. The mural not only reflects the diversity of organisms that comprise Hawaiian *ʻāina* as inseparable and intersecting aspects with the image of Queen Liliʻuokalani, but figures these as critical to sovereign futures in Hawaiʻi.

Queen Liliʻuokalani and Queen Kapiʻolani used a conventional form of photography to play upon different visual, cultural, and political registers for power. Their accession portraits became a way for them to visually affirm that sovereignty in Hawaiʻi arises from deep ancestral connections to *ʻāina* and that, further, it is maintained by reciprocal interaction with the natural world. Photography has frequently participated in settler-colonial projects that framed land as empty, available, or extractable, minimizing Indigenous ecological knowledge and relations to place. Certainly plenty has been said about how the role it has played in creating the "American West," even allowing that mythic space to materialize at great distances from the continental United States. Placing the portraits of Queen Kapiʻolani and Queen Liliʻuokalani in dialogue with these histories invites us to consider how photographic depictions of our environments are neither passive nor merely part of the backdrop but foreground land as lived, governed, and ethically bound.

Notes

1. See Nicole Furtado and Stacy Kamehiro's essay *Kapulani Landgraf, ʻauʻa* for a thorough definition of *kaona* and its function in Hawaiian visual culture.
2. The term Emmaites refers to members of the Queen Emma political party, which started in 1874 and disolved in 1882. The party was vehemently anti-American and critical of the Reciporicty Treaty Kalākaua brokered with the United States in 1875.
3. *Hoʻo* is a causative prefix added to verbs. As such, most definitions for the term hoʻoulu derive from the verb ulu, with the exception of "to cause to spring up," which stems from the word hoo.
4. The meaning of *lāhui* derives from several words in the Andrews dictionary, including *laa, lahu,* and *lahui.*
5. See Kamehiro, 2009: 51–53, for a discussion of how this issue was skirted during the coronation ceremony which reinforces my interpretation of the placement of Kapiʻolani's crown in this photograph.
6. Research into whether they choose to use real kukui nut or gems that mimicked this form are ongoing. Whatever the case, Stacy Kamehiro points out that the incorporation of kukui in the royal crowns drew a visual and material equivalence between kukui and precious gems, such as diamonds (Kamehiro, 2009: 51).
7. This organization is still in operation today, as the Kapʻiolani Medical Center for Women and Children.
8. See, for example, Stevens, 1893.
9. While John L. Stevens was not an official member of this organization, he was certainly aligned with it and supported its efforts.
10. *Kalo* is still grown and consumed in Hawaiʻi. At one point, more than 35,000 acres of taro were cultivated in the island archipelago. Today it is estimated that the number is around 310 acres. *Kalo* farmers today are also subject to the fickle practices of industrialized agriculture and the commodification of such products as *poʻi* (a dish made with taro). Individuals and organizations, such as the non-profit Kakoʻo ʻŌiwi, are currently working to restore wetland ecosystems in the area of Kaneohe on Oʻahu through the cultivation of taro grown in irrigated terraces (*loʻi kalo*). *Loʻi kalo* help restore nuturients to the soil and mitigate the effects of land-based run-off on esturaries and inland coral reefs.
11. *Hāloa* translates to long breath. It denotes both children of Wākea and Hoʻohōkūkalani and is sometimes used as a poetic epithet for *lauloa taro*, one of the many native varieties of taro, named for its long leaves.
12. For more on Kalākaua and the genealogical trial he underwent when he became king, see Kamehiro, 2009.
13. Robin L. Bott makes a similar case. In Bott, 1997: 140–156.
14. In 1893, the US celebrated the hundred-year anniversary of the arrival of Columbus in the Americas with the Colombian Exposition, also known as the Chicago World's Fair. In 1898, the year the US annexed Hawaiʻi as a territory, Puerto Rico, Guam, and the Philipines were ceded to the United States as part of the Treaty of Paris.

Bibliography

Lorrin Andrews, *A Dictionary of the Hawaiian Language to Which is Appended an English-Hawaiian Vocabulary and a Chronological Table of Remarkable Events* (Honolulu: 1865).

Robin L. Bott, "Hawaiʻi's Story by Hawaiʻi's Queen." *I Know What is Due to Me: Self-fashioning and Legitimization in Queen Liliuokalani's Hawaii's Story by Hawaii's Queen, Remaking Queen Victoria* (Cambridge: Cambridge University Press, 1997).

Leah Caldeira, "Visualizing Hoʻoulu Lāhui," in *Hoʻoulu Hawaiʻi: The King Kalākaua Era*, ed. Healoha Johnston (Honolulu: Honolulu Museum of Art, 2018), 11–35.

Combined Hawaiian Dictionary (University of Hawaiʻi at Hilo, Day, 2020), *Wehewehe.org*, https://wehewehe.org/ (accessed April 28, 2026).

Jane Desmond, "Picturing Hawaiʻi: The 'Ideal' Native and the Origins of Tourism, 1880–1915," *Positions: East Asia Cultures Critique* 7, no. 2 (1999): 459–501.

Joyce D. Hammond, "Photography, Tourism and the Kodak Hula Show," *Visual Anthropology* 14, no. 1 (2001): 1–32.

Nicole K. Furtado and Stacy L. Kamehiro, "Kapulani Landgraf, 'Au'a," in *Smarthistory*, March 26, 2025, https://smarthistory.org/kapulani-landgraf-au-a/ (accessed April 28, 2025).

Patricia Johnston, "Advertising Paradise: Hawai'i in Art, Anthropology, and Commercial Photography," in *Colonialist Photography: Imag(in)ing Race and Place*, eds. Eleanor M. Hight and Gary D. Sampson (London, New York: Routledge, 2002), 188–225.

Adrienne Kaeppler, "Genealogy and Disrespect: A Study of Symbolism in Hawaiian Images," *RES Anthropology and Aesthetics* 3 (1982): 82–107.

Stacy Kamehiro, *The Arts of Kingship: Hawaiian Art and National Culture of the Kalākaua Era* (Honolulu: University of Hawai'i Press, 2009).

Stacy Kamehiro, "Worlding the Kingdom of Hawai'i: The Art of International Relations," in *Ho'oulu Hawai'i: The King Kalākaua Era*, ed. Healoha Johnston (Honolulu: Honolulu Museum of Art, 2018), 79–112.

Pualani Kanaka'ole Kanahele, "Forward," to *The Kumulipo: An Hawaiian Creation Myth* (Honolulu: Pueo Press: 1978 / 1997), n.p.

Ralph Kuykendall, *The Hawaiian Kingdom-Volume 3: The Kalālaua Dynasty, 1874–1893*, (Honolulu: University of Hawai'i Press, 1979).

Lili'uokalani, trans., (1897), *The Kumulipo: An Hawaiian Creation Myth* (Honolulu: Pueo Press: 1978 / 1997).

David Malo, *The Mo'olelo of Hawai'i of Davida Malo*, Vol. 2, Hawaiian Text and Translation, ed. Charles Langlas, Jeffrey Lyon (Honolulu: University of Hawai'i Press 2020)

Anne Maxwell, "Colonial Photography and Indigenous Resistance in Hawai'i: the Case of the Last Royal Family," in *Colonial Photography and Exhibitions: Representations of the 'Native' and the Making of European Identities* (London: Leicester University Press, 1999), 192–223.

Ligaya Mishan, "On Hawaii, The Fight For Taro's Revival," *New York Times T Magazine*, November 17, 2019, 122.

John L. Stevens, "A Plea for Annexation," *The North American Review*, 157(445) 1893.

"The Coronation: Processions, Salutes, Decorations etc," *The Pacific Commercial Advertiser*, (February 17, 1883), https://chroniclingamerica.loc.gov/lccn/sn82015418/1883-02-17/ed-1/seq-2/ (accessed 25 April, 2025).

Mārata Tamaira, "Walls of empowerment: reading public murals in a kanaka maoli context," *The Contemporary Pacific* (2017): 1–36.

Agnes Terao-Guila, *Hawaiian Women's Fashion: Kapa, Cotton, and Silk* (2019; Printed in the Republic of Korea, 2021).

Chapter 10
Bedrock Mortar vs. Photograph: Nevada City Rancheria Nisenan Cosmovision and the Malakoff Hydraulic Mine

Jordan Reznick

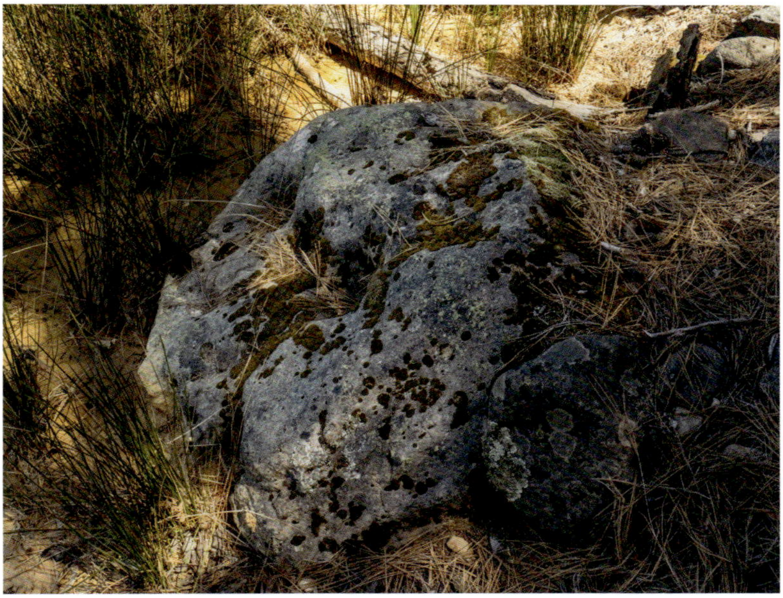

Fig. 1.
Jordan Reznick, *Bedrock mortar at Malakoff Diggins State Historic Park*, 2024. Digital image. Author's collection.

During the summers of 2024 and 2025, I visited a Nisenan bedrock mortar (fig. 1) at Malakoff Diggins State Historic Park in the Sierra Foothills north of Nevada City, California—all within the homelands of the Nevada City Rancheria Nisenan Tribe. The rediscovery of the mortar in 2022 had halted an environmental remediation project intended to resolve the historic mine's continued leeching of metals into the Yuba River below. The grinding stone's presence indicated a Nisenan relationship to the mined land that required their consultation before the project could continue. I sometimes visited the mortar with Scott Green, Tribal Liaison for California State Parks, but mostly I visited it alone, alternately sitting with the stone and traversing surrounding trails. I hoped to understand its relationship to historic photographs of the North Bloomfield Gravel Mining Company's Malakoff Diggins hydraulic mine made during the mine's height of operation between 1871 and 1883.

While the State Park's project was on pause, the Tribe and its many allies, including Green, archaeologist Mark Selverston, archaeologist Chris Ward, and I, worked to support the Tribe's wishes that the mortar be empowered to tell its story to non-Native audiences. The Tribe's vision for the stone charted a more complete path to repair than the state's initial plan for environmental remediation. Since 2023, I have also spent countless hours in conversation and community with Spokesperson for the Tribe Shelly Covert; other tribal members; and Covert's non-Nisenan executive assistant, Ember Amador. I presented the research below for feedback to members of the Tribal Council and, at the Tribe's request, to non-Native local audiences in Nevada City and Grass Valley. It represents one component of Nisenan-allied support for the Tribe's efforts in relation to the stone. Selverston and Ward's archaeological research on the stone's history and attributes endorsed the Tribe's request to have the stone moved to the rim of the mining pit. Green facilitated the stone's relocation to a place in the park easily accessible to Tribal Elders and visitors. In tandem with these efforts, the photo-historical contribution I present here aims to situate the stone's story within an Indigenous-centered framework of justice.

This chapter pursues the ways photography, environmental law, and Indigenous Knowledge intersect in the visual record of hydraulic gold mining, an extractive process that used high-pressure water to dislodge auriferous gravel from mountainsides, mutilating many California lands. I examine two series of photographs of Malakoff Mine—the most infamous of such mines in California. The first series, by prominent California landscape photographer Carleton E. Watkins, primarily marketed the mines to overseas investors. In contrast, the second series, by John A. Todd, served as legal evidence in the historic *Woodruff v. North Bloomfield Mining Company* case that ended hydraulic mining.[1] Against the visual languages that alternately celebrated and condemned Malakoff, I foreground the Nisenan's recognition of the land as kin, storyteller, witness, and wellspring of legal Protocols. The Nisenan's cosmovision (their spiritual and physical worldview), rooted in ancestral Tribal Knowledge and history, offers an interpretive framework for understanding these images as evidence of the connected terrors of genocide and ecocide. I present a Nevada City

Nisenan-centered interpretation of Watkins's and Todd's images that broadens the cultural-historical framework through which landscape photographs might be interpreted and connects them to Native-allied frameworks of justice.

Malakoff's Legal and Photographic Histories

Historian Powell Greenland dubbed Malakoff Diggins the "quintessential California hydraulic mine" because it so exemplified hydraulic mining's titanic development and undoing. Its environmental impact was correspondingly colossal. In 1866, the North Bloomfield Gravel Mining Company—a corporation formed by San Francisco financial elites—purchased extensive disused claims in and around North Bloomfield and began consolidating water rights and constructing a drainage tunnel to enable large-scale hydraulic mining. By the close of 1874, it had completed a 7,800 foot-long tunnel through the bedrock leading from Malakoff Hill to the South Fork Yuba River. The company obtained a lion's share of the water rights for the San Juan Ridge area above the mines and purchased the massive Rudyard Dam. To deliver water to the mines, North Bloomfield constructed 157 miles of ditches, dammed the water from both the North and South Yuba Rivers, and built at least 5 miles of flumes—projects that used approximately 2 million feet of lumber. Monitors fitted with newly invented high-powered nozzles aimed jets of water at the Malakoff hillside to dislodge the geological strata extending from the topsoil surface to the bedrock below. Monitors then guided the loosed rock into sluices, where gold was separated from gravel with the application of highly toxic mercury. Depleted gravel was dumped into a tunnel extending 8,000 feet from the mining pit to the Yuba River below. In 1874, North Bloomfield adopted hydro-powered electric lights to enable mining twenty-four hours a day. In this manner, a mountain ridge was transformed into a canyon 7,000 feet long, 3,000 feet wide, and 600 feet deep. (Greenland, 2001; 69, 109, 150, 198–226; quote at 198; Hult-Lewis, 2011: 158–162; Selverston et al., 2023: 25)

The hydraulic mining industry was embroiled in constant litigation. North Bloomfield, along with approximately thirty other major hydraulic mining operations in the region, cast an estimated 1.6 billion cubic yards of debris (roughly a third of it from Malakoff alone) into the 150 mile stretch of tributaries connecting to the San Francisco Bay, causing regular catastrophic floods to plague adjacent farmlands and towns. A resulting feud simmered between the mining community centered around Nevada City and the downstream farming community centered around Marysville. Numerous lawsuits failed to curb hydraulic mining

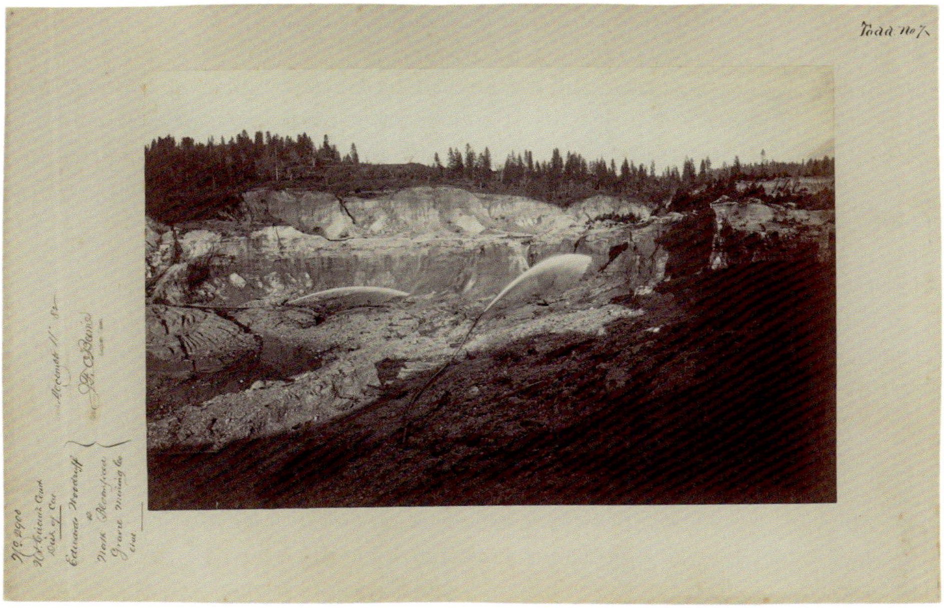

Fig. 2.

John A. Todd, *Hydraulic Mining, Exhibit 7*, 1882–1883. Albumen print, mount 14 × 22 in./36 × 56 cm. San Francisco, California Historical Society, PC-RM-Todd.

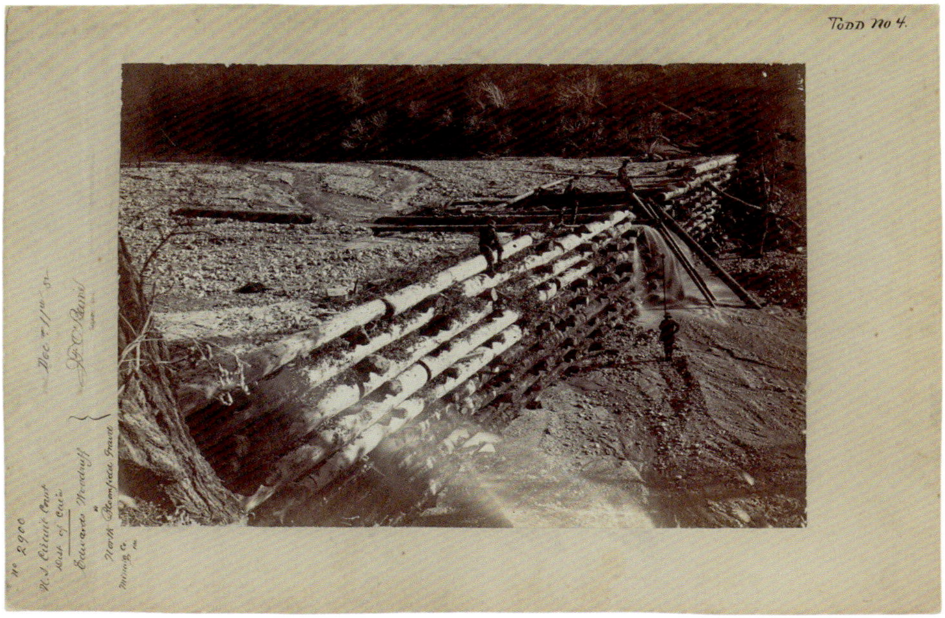

Fig. 3.

John A. Todd, *Men Posed on Log Dam in River Bed*, 1882–1883. Albumen print, mount 14 × 22 in./36 × 56 cm. San Francisco, California Historical Society, PC-RM-Todd.

until farming interests consolidated their complaints into a single suit against North Bloomfield, the largest and most notorious of the mining companies, as defendant. *Woodruff v. North Bloomfield Mining Company* was filed at the Federal Circuit Court for the District of California in 1882. Justice Lorenzo Sawyer issued an 1884 injunction against North Bloomfield that effectively ended hydraulic mining in California—and thus halted most California gold mining operations. (Woodruff, 1884) The case's success in curbing ecological impact that extended beyond a landowner's terrain helped chart the beginnings of what would become environmental law. (Ziebarth, 1984: 274–279; Greenland, 2001: 56, 227–268, 274–275; Vigars, 2016: 235–249)

Todd made seventeen photographs of the mine and the destruction it caused downstream watersheds, in 1882–1883 (figs 2 and 3), commissioned by the prosecution in the *Woodruff* case. (Hult-Lewis, 2011: 153) His images represent an early case in which landscape photographs appeared in a US courtroom. (see Mnookin, 2015) In contrast, Watkins made seventy mammoth photographs of Malakoff and its related resources in 1871 and circa 1878 under commission by North Bloomfield, who planned to woo overseas investors. (figs 4 and 5) Watkins's images, though not apparently produced as evidence in the case, provide a worthy contrast to Todd's images because of their divergent ends: Watkins's photographs emphasize the mine's positive attributes and long-term economic viability.[2] Both photographers were tasked with conveying the colossal scale of the mining operation and accordingly focused primarily on the northwest area of the diggings, near where the height of the manmade cliff is most severe. However, most telling of the photographers' contrary aims are the pictures that supplement those of the mines. Impressive dams and flumes upstream frame Watkins's diggings, while devastated farms, dams, and rivers frame Todd's. One set of images describes the pinnacle achievement of industrial mining. The other describes its ruinous consequences.

As photo historian Christine Hult-Lewis chronicles, although Watkins did not make images for the case, the nascent visual language of law fundamentally shaped his work. By the time he photographed Malakoff in 1871, he had cut his teeth as a photographer making several of the first landscape photographs ever presented in US courts. The requirement that he testify under oath to the objectivity of his views provided him with an education in the cultural strictures of visual fact. Hult-Lewis demonstrates that Watkins translated these courtroom lessons to mining commissions, combining his mastery at constructing a sense of factual "evidence" with his power to create stunning compositions. Hult-Lewis aptly describes Watkins's images taken at Malakoff as early instances of the "technological sublime," when, in the face of industrial destruction of nature, California's pastoral beauty yielded

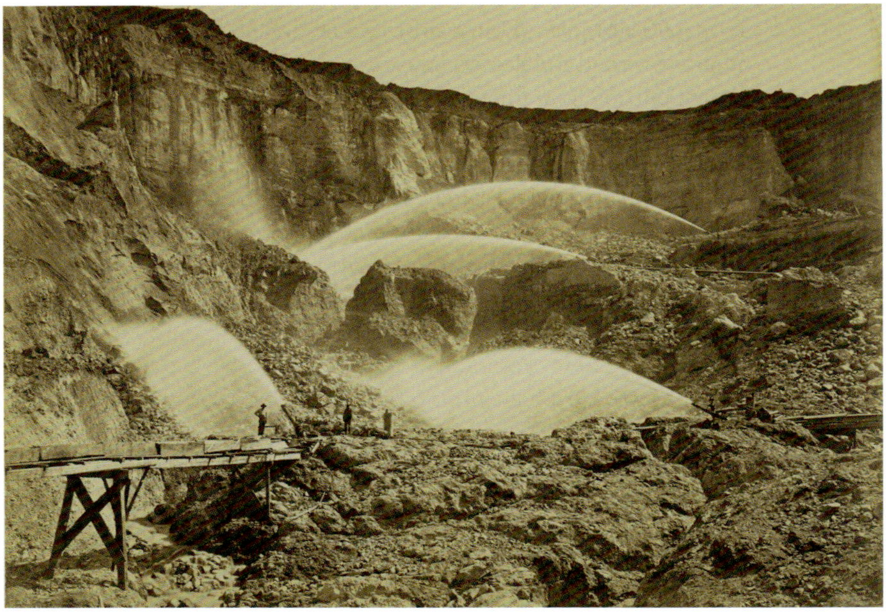

Fig. 4.

Carleton Watkins, *Malakoff Diggings, Nevada County, California,* ca. 1875. Albumen print, 15 3/16 × 21 7/16 in./38.6 × 54.5 cm. J. Paul Getty Museum. Courtesy of Getty's Open Content Program.

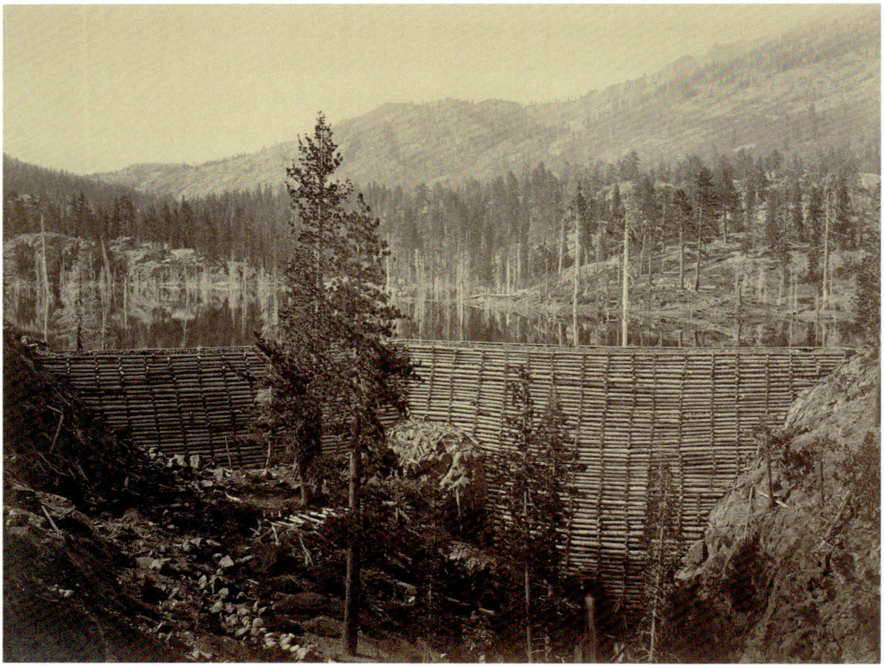

Fig. 5.

Carleton Watkins, *Rudyard (English) Reservoir, Central Dam, Nevada County*, ca. 1878. Albumen print, 15 3/16 × 21 7/16 in. Los Angeles, J. Paul Getty Museum. Courtesy of Getty's Open Content Program.

to an aesthetics of fright mixed with pleasure. (Hult-Lewis, 2011: 23, 39–151, 165–170, 218; see also Scott, 2017: 205–207)

Todd's images do not achieve the level of Watkins's in their technical mastery or aesthetic affect. They are smaller in size and partially obscured by shadow, each providing a frank, distant view of the northwestern area of the mine, with little visual reference to provide a sense of scale. Judge Sawyer's ruling does not indicate that photographs proved valuable in substantiating the prosecution's case. He concluded that "the amount of debris discharged into the rivers by these operations can only be duly appreciated by actual observation," describing the scene as "the highest degree weird and startling." (Woodruff, 1884: 4) While Watkins's images of the mining pit better approximate such a sense of shock, neither photographer portrayed what firsthand accounts describe. One reporter wrote of water that was "beaten" into a "cloud of red foam" as the monitors sprayed it with the force of "a wicked, vicious, unutterable indignation." The "incessant roar" was broken only by "rattling and slipping noises" of boulders being wrested from the land. (*Nevada Daily Transcript*, 1879) No sense of such terrific force appears in either photographer's images of the gentle arcs of water that stretch across the canyon. While each image portrays the extraordinary scale of the mines, the images translate only a fraction of the reality witnesses described. These visual records are fundamentally inept at helping us imagine what occurred at Malakoff Mine. The legal scholar Jennifer L. Mnookin refers to such images as semi-legible visual evidence, which requires expert testimony to decipher facts that a lay viewer cannot. Such images cannot "speak for themselves" but must be "made to speak" through expertise. (Mnookin, 2014: 65)

A Nisenan Perspective on Malakoff

The Nevada City Rancheria Nisenan Tribal account of Malakoff's history speak to histories not yet considered in interpreting Todd's and Watkins's images. The Nisenan, on whose homelands most of the Central Mines were located, were among the hardest hit by genocidal gold rush violence. (Madley, 2016: 69, 85–99, 179–185, 195, 220; Johnson, 2018: 133–187) I first learned of the mine's impact on the Tribe from Spokesperson Covert, who told me the story over takeout on a June evening in the backrooms of 'Uba-Seo, a non-profit gallery the Tribe runs in Nevada City. She described a family camping trip at the diggings in 1975, when she was nine years old. Covert wandered away from her aunts and cousins to discover her grandfather Dutch Rose alone, contemplating the enormous void of the defunct mine:

> I went over, and I was like, "Hi, grandpa," you know, like, "What are you doing?" And he had his hanky on his eyes. And he said, "Oh, you know, just, just remembering," and I said, "Oh, what are you remembering?" And he said, "Well"—and this is the thing that keeps landing on me is like—"Well, this is one of places" or "this is our place." And he meant literally, like where we were standing was a site. If he would have used archaeological terms or something he would have been like, "This is a center of, you know, Nisenan living in blah-blah-blah family or whatever, you know, for eons or whatever." But he was standing there, and he said, "But when the gold rush came, they used big hoses and they pointed them up at the mountain, washed the mountain down to get to the gold." He said, "But they turned the—they turned—but they put the hoses on the people too." And he was crying. (Shelly Covert, in discussion with the author, June 2023)

Covert and Rose's accounts situate the images within the Tribe's historical survival of genocide and ongoing grief.

Selverston identified the Nisenan bedrock mortar in a ca. 1878 Watkins photograph (fig. 6) and calculated the boulder's original home. He estimates the stone was dislodged between 1874 and 1876, sliding down the hill into the diggings, confirming what Rose told his granddaughter: "this is our place." California State Park archives reveal that the parks service had learned of this stone and several other village locations distributed throughout much of the area that is now Malakoff Diggins State Historic Park in archaeological reports commissioned since 1949. (State of California, 1949; Selverston et al., 2023: 44–51) It remains unknown how many villages resided within the area of the mining pit. The grinding stone is one of the few survivors who can speak to those villages' history.

Remnants of the earliest mining settlement established at Malakoff stand near the site from which the mortar was displaced, suggesting the Campoodie could have been displaced as early as 1854. (Mining Claim Location Record for Malakoff Hill, 1855; Selverston et al., 2023: 20–22) While the history of this particular Nisenan town's encounter with miners remains unknown, genocidal violence across the Central Mines area was profound. (Madley, 2016) When miners found gold near a Nisenan Campoodie, sometimes they forced the entire village to relocate. Sometimes they simply burned it down. (Johnson, 2018: 159, 170; Covert in discussion with the author, June 2024) Miners also murdered entire villages as retribution for rumored crimes or competition with whites at the mines. (Madley, 2016: 96–99; Johnson, 2018; 159–166) The high elevation of Malakoff suggests it may have been a seasonal home deserted when mining and widespread violence made annual migration untenable. However, by the 1880s, at least twenty Nisenan resided close

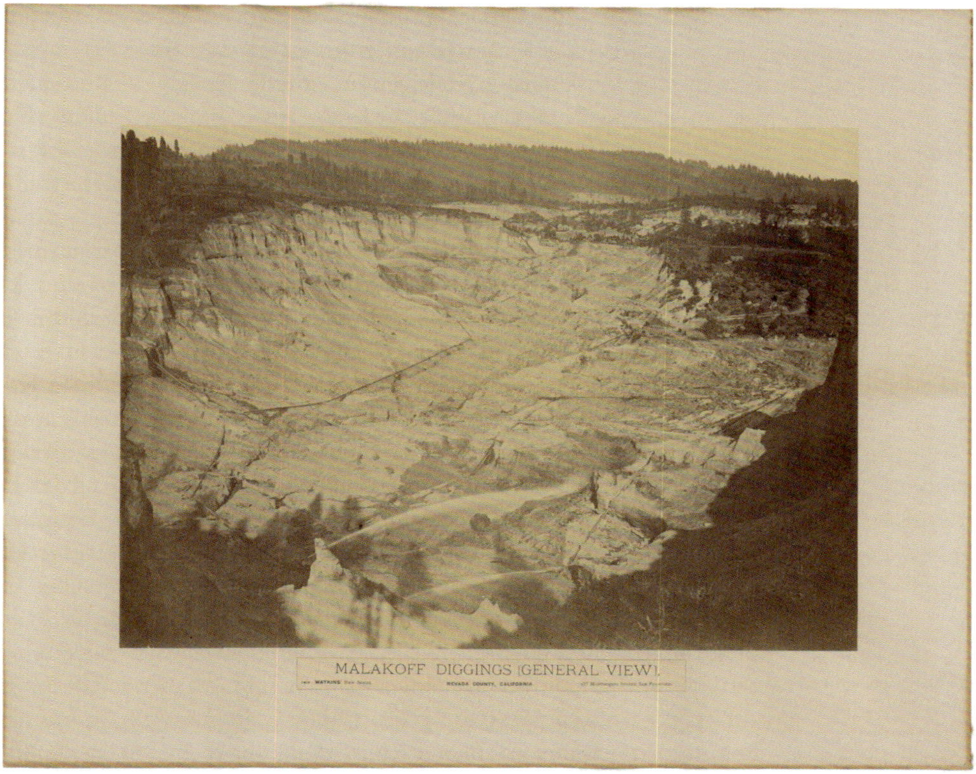

to the miners in a North Bloomfield neighborhood next to the slaughterhouse dubbed the North Bloomfield Campoodie. (Shelly Covert, in discussion with the author, June 2023; Selverston et al., 2023: 23)

The mortar's material testimony to history, however fragmented, frames Todd's and Watkins's photographs as historical pictures of a Nisenan place. Nevada City Rancheria Nisenan Tribal Council Elder Lorena Davis explains her experience of their homeland mortars: "I just feel the presence. I can almost see the women working and I can feel their presence." (Lorena Davis, in discussion with the author, June 2024) A grinding stone was the hub of daily life of a Campoodie. Women prepared meals in and around the mortar's holes. They told stories. They relayed gossip. Children played around their mothers, grandmothers, and aunties. Women imparted tribal epistemologies to younger generations, sitting on the lichen-covered stone's rough flesh. It was one of the most familiar sights, sounds, smells, and touches of daily life. (Simpson, 1977: 60, 65)

Archaeologists believe mortars were used by the Nisenan as early as 1,300 years ago. (Selverston et al., 2023: 62) Chair of the Nevada City Rancheria Nisenan Tribal Council Richard B. Johnson dates the Tribe's

Fig. 6.

Carleton Watkins, *Malakoff Diggings (General View)*, ca. 1878. Albumen print, 15 × 21 in. Bancroft Library, University of California.

use of mortars to at least 5,000 years ago. (Richard B. Johnson, in discussion with the author, June 2024) The Nisenan told ethnographer C. Hart Merriam in 1907 that mortar holes "have always been here," in line with California Indian oral histories that describe the co-emergence of California land and Native Californians during time immemorial, when World Maker made the land and the People in unison for one another. (Merriam, 1966: 320; Simpson, 1977: 35, 60; Akins and Bauer, 2021: 13–15)

How should we interpret images of Malakoff considering the grinding stone's witness testimony to histories that cannot be fleshed out in the scope of historical detail typical of photo-historical discourse? Native scholar and artist Hulleah Tsinhnahjinnie (Seminole/Muscogee/Diné) argues for counter-reading colonial photographs with Native ways of knowing that magnify their subjects' place in Native histories and cosmologies. (Tsinhnahjinnie, 1998) Below, I consider how discourse on landscape photography might, likewise, center the legitimacy of Native epistemologies in considering the historical and cultural contexts through which we make sense of photographs' histories.

Photography and Western Law in California

Since the post-structural turn in the history of photography, the photograph's earliest appearances at trials have provided photography studies with insights into how the visual conventions associated with objective fact constructed an aesthetic companion to state power. (see, for example, Sekula, 1986: 3; Burgin, 1987; Tagg, 1988) In North America, that discourse has so far overlooked the fact of purloined Indigenous land at the center of settler-colonial legal logics. As American studies scholar Monica Huerta explains, it was, crucially, within the context of counting Black bodies and Indigenous land as the property of White Americans that photography's conventions and Western property law formed interdependently around new notions of legal "objectivity." Colonial theft was itself regarded as objective fact—a neutral basis for the consideration of legal evidence. (Huerta, 2023: 1–10) It was immaterial to a US court of law whether the land that appeared in photographs was the unceded homeland of a Native tribe or nation. No courtroom photographic conventions developed for documenting or identifying land theft because no court would docket such a case.

This chapter takes up this absence as a site rich for inquiry. If the disregard for the sovereignty and lifeways of Native peoples was foundational to the legal procedures that developed around photography, Indigenous Knowledge supplies a necessary corrective as a mode of expert interpretation. The Tuscarora art historian Jolene Rickard argues that landscape photographs in particular demand Native interpretation because they so profoundly obscure the epistemological connections between landscapes and their ancestral caretakers. (Rickard, 1998) By situating photographs within epistemologies of justice native to the landscapes of their making, we might come to see the Malakoff images as witnesses to violence against the Nisenan and their homelands.

Consider for example two similar photographs of boulders: Watkins's 1861 *Rancho San Antonio: Near View of Rock Outcropping with Taller Formation at Left* (fig. 7) is an early example of legal landscape photography, presented as Exhibit 5 in the case *U.S. v. D.&V. Peralta*. Dugan Aguilar's (Mountain Maidu/Pit River/Walker River Paiute descended) 1998 *Aalam* (fig. 8) is a portrait of a living Nisenan storytelling stone. Watkins also made two additional corresponding photographs of rocks for the Peralta case. Each aimed to determine which boulder matched testimony describing the property boundary at the heart of a dispute between competing land claims in present-day Alameda County made under the California Land Act of 1851 (during which landowners had two years to prove that their land tenure preceded the US occupation

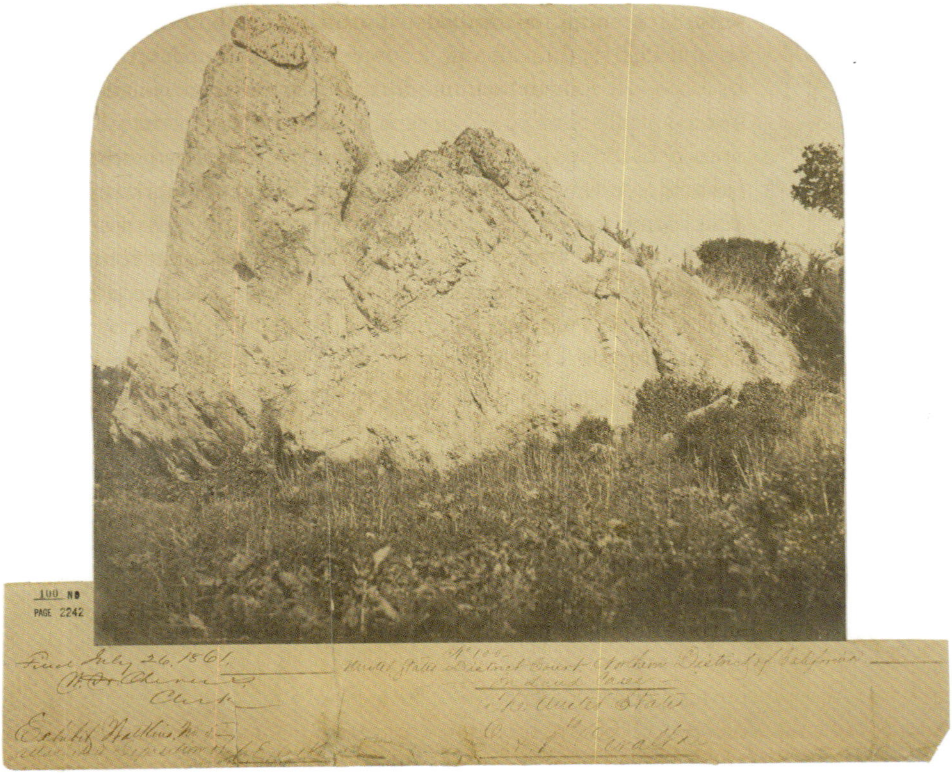

Fig. 7.

Carleton Watkins, *Exhibit 5. Rancho San Antonio: Near View of Rock Outcropping with Taller Formation at Left*, 1861. Salted paper print, approximately 34×42 cm. Bancroft Library, University of California Berkeley.

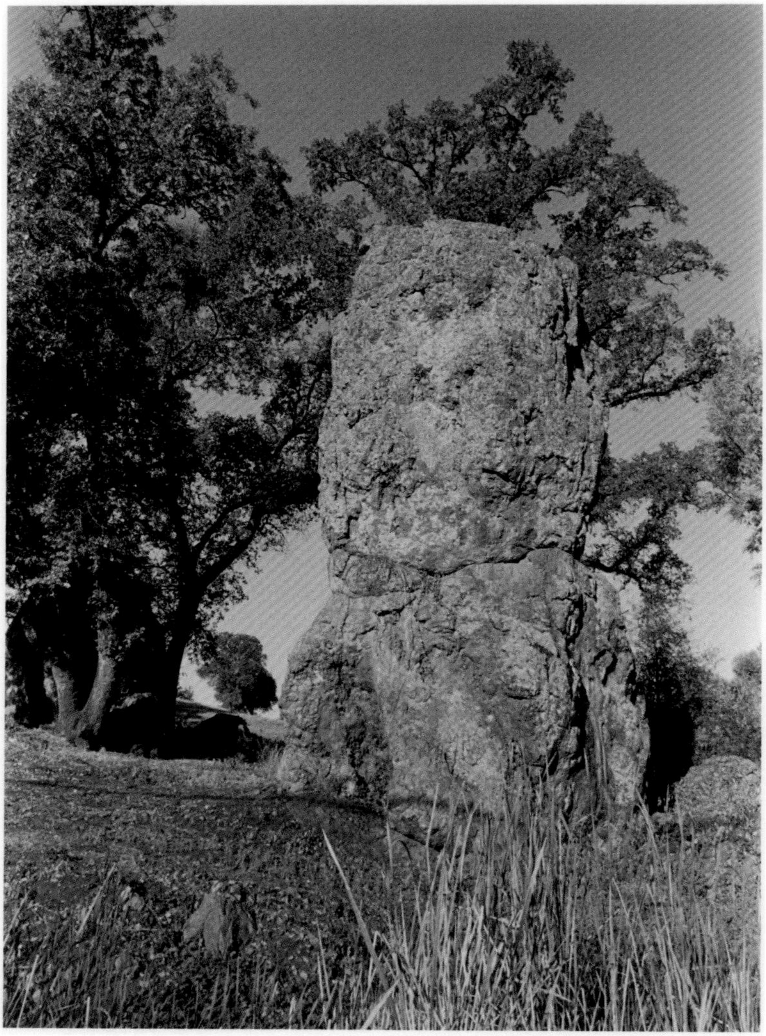

Fig. 8.

Dugan Aguilar, *Aalam,* 1998. Silver gelatin print, 14×11 in. Collection of the Oakland Museum of California, gift of the family of Dugan Aguilar. © Family of Dugan Aguilar.

of California) and the congressional act of 1860 (which permitted competing parties to contest claims made under the 1851 act). (Hult-Lewis, 2011: 72–74) Because Native land ownership was discounted by the United States, California Indians lost nearly all their homelands under these laws. (Heizer, 1989; Johnson, 2018: 146–147; Akins and Bauer 2021: 115–116) In the Peralta case, the court determined that it was indeed the stone pictured in Exhibit 5 that marked the property boundary in question, though, as Hult-Lewis details, the judge found the photograph's evidence insufficient and had to visit the site in person to determine the boundary. (Hult-Lewis, 2011: 76–86) Even so, few twenty-first-century viewers need to stretch their imaginations to regard the boulder in Watkins's picture as a property marker. Yet property and its boundaries are invisible abstractions. They belong less to the visual surface of reality that a camera records than to the tacit frameworks of Western law that viewers employ to understand an image's significance.

In contrast, Aguilar's portrait of Aalam serves as evidence within the context of Nisenan Knowledge. It is at Aalam where the story of Bear Woman and Deer Woman ends in a dramatic scene, when Deer Woman's two children leap atop the stone to escape the revenge of Bear Woman. Just when Bear Woman is about to claw them to death Aalam rises up, lifting the fawns to the Upper Meadow in Sky World. Bear Woman's claw marks remain visible on the stone's surface today. In some versions of the story, before the fawns' final ascent, they toss down pine nuts, manzanita berries, and two varieties of acorn—foods that would sustain the Nisenan and forever remind them of the story's moral messages. The story recalls the epoch between the creation of Turtle Island and the creation of the Nisenan, when animals peopled California and established its ethical and ecological protocols. (Powers, 1976: 342; Uldall, 1992: 24–30; Bibby and Aguilar, 2005: 62; Shelly Covert, in discussion with the author, February 2024) Told since time immemorial, it verifies Nisenan land tenure long before the Land Act of 1851. The oral history legitimating that fact is no less visible than that of the property boundary in Watkins's Exhibit 5.

The last time I visited Aalam, the stone remained in its ancestral place, serving as a regal marker of the entrance to an estate for sale for 4.7 million dollars. Aguilar's photograph evidences Indigenous Title as objectively as any photograph presented in a US court. Yet, imagine the Nisenan presenting *Aalam* in a US court to challenge the legitimacy of the sellers' deed. Imagine them presenting it to establish their land tenure in their current petition to reinstate their federal recognition.

This should encourage you and me to reconsider, not merely the construction of photographic objectivity, but how the embrace of alternative interpretations of photographic facts brings decolonized epistemologies into view. Why not prioritize interpretive epistemologies Indigenous to the land on which landscape photographs are made? Indigenous thought contains legal Protocols, developed over the course of thousands of years of collaboration with ecosystems, that outclass Western advancements in environmental law in many respects. (Temper, 2019) These philosophies—narrated in stories, songs, and objects that do not easily translate into the flat lines of Western logic—require developing modes of photographic interpretation that match their manner of truth telling. According to Jo-ann Archibald and colleagues, valuing Indigenous stories "traverses new theoretical, methodological, and pedagogical realms where Indigenous stories, experiences, and understandings are the core of the meaning-making process." (Archibald et al., 2022: 13) How might the intergenerational storied Nisenan Knowledge which derives from the land itself supply additional context for interpreting the Todd images presented in Judge Lorenzo Sawyer's courtroom?

The Testimony of Feathers, Furs, and Beads

The destruction pictured in Todd's images becomes legible through the metonymic logics of Indigenous oral and material cultures—modes of testimony through which Native North American peoples have long articulated their sovereign relationships to land in negotiations with occupying colonial governments.[3] For the Nisenan, this practice is visible in the regalia worn by the Native signatories in the sole surviving daguerreotype of the signing of one of eighteen treaties negotiated between California Indians and the United States. (fig. 9) The Native subjects were retrospectively identified as Maidu headmen, but they are more likely Nisenan, who are frequently misidentified as Maidu because of anthropologists' historical errors in categorization. (Johnson, 2018: 1–3) Anthropologist Thomas C. Blackburn more recently identified the image as likely one of the lost Robert H. Vance gold rush daguerreotypes, but mistakenly placed it at Bidwell's Ranch. However, photo historian Peter Palmquist notes that the northernmost points of Vance's gold rush daguerreotyping trip were Yuba City and the Feather River—fifty miles south of Bidwell's Ranch. It is much more probable that Vance made the daguerreotype on July 18, 1851, at Camp Union, near the Yuba River, at the treaty signing by ten Foothill Nisenan headmen. (Palmquist, 1978: 166, 173; Blackburn, 2016: 317; Johnson, 2018: 145–146)

Their regalia adds considerable meaning-making dimension to the image. It not only indicates which Nisenan district each headman represents, but also references the Protocols governing Nisenan Country. The materials that make up the regalia are some of the most rare and valued articles of the Nisenan world: wildcat furs, trade beads, and powerful feathers. They stand apart from other types of Nisenan regalia in their capacity to indicate rank and status. The regalia materials are not symbolic of their eminence, but, instead, material evidence of it—rooted in the land. Each specific configuration of materials is the pride of the headman's own Campoodie. Each man is the wealthiest of his town, but his wealth is a measure of his service to his community. Headmanship is first and foremost a position of responsibility for the wellbeing of the people. A Campoodie's pride in its headman's regalia stems from the fact that people provide him wealth to the extent that he provides them the conditions to thrive. In this chain of reciprocity, the responsibility of the headman for his people was, in turn, reciprocal between the people and the land. People who hold steadfast to Protocols and at peace with

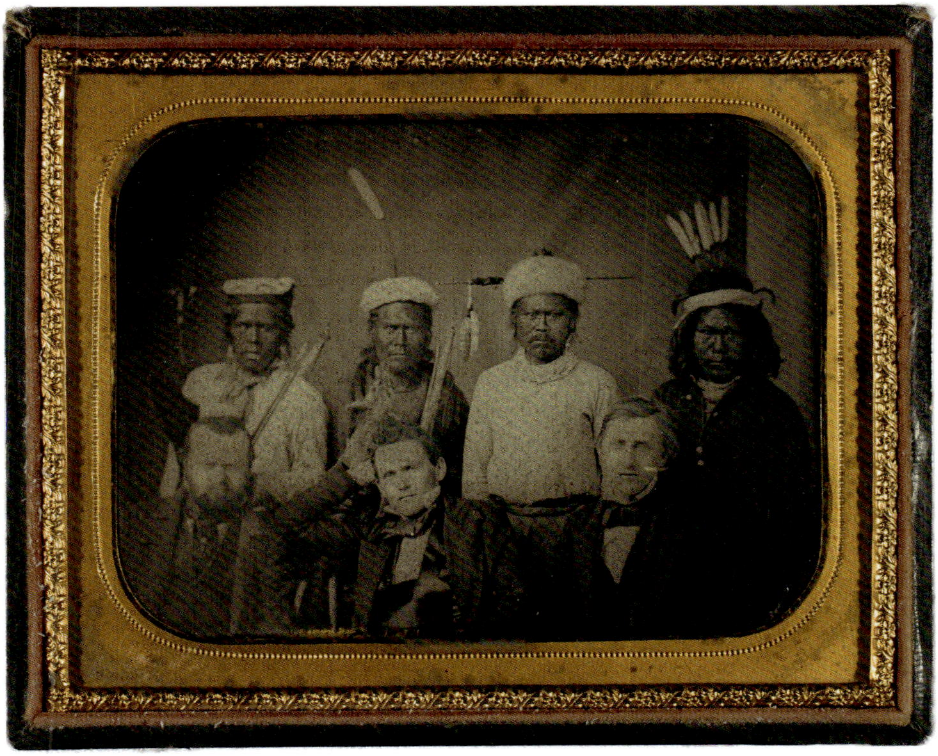

one another can caretake the land to make it richly biodiverse. They have good luck in the hunt. Their peace and bountiful resources also result in fruitful trading relationships with other tribes. The lush products of the land and trade that each headman wears testify to the peace and wellbeing of the people and land in his care. (Shelly Covert, in discussion with the author, June 2024; Richard B. Johnson, in discussion with the author, June 2024)

Fig. 9.
Maker unknown (probably Robert Vance), *Four Maidu (probably Nisenan) Headmen and Treaty Commissioners*, ca. August 1851. Daguerreotype, half plate, 3 3/8 × 4 5/8 in. Courtesy of the International Museum of Photography at George Eastman House, Rochester.

Thus each headman presents himself as a sovereign diplomat to a government with no such grounded legitimacy. Prior to traveling to Camp Union, the headmen painstakingly selected twelve square miles of land between the Bear and Yuba rivers named in the treaty, on which they could suitably sustain their lifeways as Nisenan Peoples (in the treaty's words) "forever." (*Treaty with the Das-pia, Ya-ma-do, etc.*, 1851; Kappler 1927) They stand before the camera weighted with life-and-death responsibility to their Campoodies. To a nineteenth-century Nisenan person, the feathers, furs, and beads they wear spoke all of this. If the regalia were permitted in court, each object could be deciphered for the specific philosophies and histories that affirm the Nisenan's sovereignty and Indigenous Title. (on regalia in the western courtroom, see Temper, 2019)

The photograph provides a guide to how a Nisenan cosmovisual reading of photographs might function: Visual evidence is not symbolic but grounded in the specific ethical ecological practices that result in the material presence of what lies within the camera's frame. In the same spirit, the Protocols that inform the interpretive framework for visual and material evidence is not abstract or arbitrary, even if the storytelling that describes those Protocols seems fantastical to Western ears. The Protocols come from the land. They are learned from the land. They are quite simply the law of the land.

The profound relatedness to land with which the headmen assert their sovereign authority makes it all the more poignant that, in 1852, the California Senate declared that the treaties "committed an error in assigning large portions of the richest mineral and agricultural lands to the Indians, who did not appreciate the land's value" and persuaded the US Senate to reject all eighteen treaties, place them under an "injunction of secrecy," and instead support the state in removing the Native population and sanctioning genocidal violence. (Heizer, 1989; Madley, 2016: 167–171; Johnson, 2018: 148–150) The daguerreotype stands today as what the anthropologist Audra Simpson (Kahnawà:ke Mohawk) describes as "a knotty reminder" of "other orders, other authorities, and an earlier time that has not fully passed." (Simpson, 2017: 22) As such, the daguerreotype reveals photography's early entanglement with a legal order capable of documenting Indigenous sovereignty while simultaneously mobilizing the law to extinguish it.

The Testimony of Stone, Acorn, and Seeds

Where regalia signal rank and status in a portrait, a landscape testifies to the Tribe's adherence to the Protocols that govern everyday life. In Nisenan cosmology, Protocols center on the acorn as the primary wellspring of all life. The Nisenan Creation Story begins with a handful of soil placed on Turtle's back. Earth was, on its first day of existence, dry and barren. On the second day, World Maker brought the first acorn down from the Upper Meadow and nestled it into the ground. This act prompted the first rains, which formed the rivers and brought earth to life, causing the first oak tree to grow, and with it the promise of everlasting sustenance for the Nisenan on the condition that they follow World Maker's instructions about how to tend to acorn.[4] This doctrine of Original Instructions guides the ecological stewardship of Nisenan homelands. Indigenous scholars describe Original Instructions as pedagogical and intellectual traditions for collaboratively developing ecological knowledge based in long-term observation and research. (Kimmerer, 2015: 180–181) They outline practical ethical directives for hunting, harvesting, human relationships, and ceremonies that foreground the interdependence of humans, plants, and animals, with the goal of intergenerational cultivation of the land's fertility. They warn of the catastrophic results of indulging human impulses to consume and stress the centrality of self-restraint as a condition of citizenship in the interspecies community of Earth. (Kimmerer, 2015; Nelson, 2017: 254)

In Nisenan epistemology, land is alive. Stones are storytellers. Memories are emplaced. Knowledge is relational. Law is learned from the land. Nisenan Original Instructions include the annual tending of oak trees and other favored culinary crops through burning, sowing, harvesting, and food preparation. (Simpson, 1977: 35; Anderson, 2005: 262, 269–270) Every animal in the Nisenan world has a song to safeguard it. (Richard B. Johnson, in discussion with the author, June 2024) The acorn has dances and songs to celebrate each of its stages of life, from the very first acorn flower that appears in the spring to the Big Time celebration surrounding the acorn harvest in the fall. (Simpson, 1977: 46, 70, 88) These instructions have been passed down from the very first Nisenan Peoples to every generation since time immemorial. The preparation of acorn and other foods connects women to the Original Instructions, ancestors, and descendants, providing a sense of one's place in a community of humans and more-than-humans. (Simpson, 1977: 35, 60; Garroutte, 2003: 114–117), or, as Covert puts it: "To really be Nisenan is between the Nisenan and the land." (Shelly Covert, in discussion with the author, June 2024)

The presence of the bedrock mortar at Malakoff supplies material testimony to the fact of the Original Protocols of the Nisenan that governed daily Campoodie life at the places disappeared into the mine's pit. The women who prepared acorn and other cuisine at the Malakoff Campoodie could have read the Creation Story from its six cups. (Simpson, 1977: 10, 35) Grinding was one of many daily acts that constantly maintained the Original Instructions. It is difficult to convey to twenty-first-century readers the degree to which such stories and Protocols animated the meaning of even the most mundane daily tasks. As Covert describes, every aspect of life was permeated with matrixes of Traditional Knowledge. Were a Nisenan person to feel insincerely connected to that fabric, going through the motions of life would feel strangely hollow, like reading aloud the words of another language without knowing their meaning. (Shelly Covert, in discussion with the author, June 2024) Tribal Elders today recount that, even after genocidal trauma inhibited storytelling, their parents' knowledge of the land was so great that they survived poverty under which their White neighbors perished. (Lorena Davis, in discussion with the author, June 2024) Every rotation of the pestle that carved the cups of the Malakoff mortar is an immutable record of that knowledge. As Green puts it, "This rock is a storyteller." (Green, 2024) Even if the mortar was not traditionally a storytelling stone, today it is. If presented in court, it could, in fact, be used to read the Laws of Nature governing Campoodie life.[5]

The mortar's presence provides a framework with which to assess the extent of the loss described in photographs of a barren canyon of gravel where Nisenan life thrived for millennia, until only recently. Regardless of the reign of US law over California, Nisenan laws based in long-term environmental research provide a perceptual framework that contrasts to Western law's view of land as private property. The Nisenan knew in 1884—and long before—what European Americans were still ignorant of. In the words of the Ecuadorian constitution, which enshrines the Rights of Nature based upon Indigenous cosmology: "Nature [...] has the right to integral respect for its existence and for the maintenance and regeneration of its life cycles, structure, functions and evolutionary processes." (Constitution of the Republic of Ecuador, article 71) The mountain is connected to the river, which is connected to the valley below. When one entity is violated, the entire ecosystem breaks down. Within this framework Todd's photographs are a vision of the Creation Story's warnings: If nature is not respected, the world will turn topsy turvy.

Justice for the End of the World

The Creation Story cautions that when people neglect the Original Instructions and evil takes over, life will cease and the land will become as bare as the first day of creation. From a Nisenan perspective, Malakoff photographs corroborate the fact that the caretaking of nature's life cycles has all but ceased. Malakoff appears as a hub of cataclysm, ending Campoodie worlds together with many others: the worlds of trees, meadows, rivers, fish, and countless more-than-humans. The parched earth mirrors the dryness of Earth's first day. Time appears to reverse—the land dries, rivers evaporate, valleys desiccate. The Creation Story says Turtle must wait for World Maker to return and make Earth anew. (Simpson, 1977: 23) When the grinding stone tumbled from its original village home, some Nisenan may have felt that time had come. As Nevada City Rancheria Nisenan Tribal Chair Richard B. Johnson describes of his grandparents, who survived the era of hydraulic mining, witnessed the murder of nearly everybody they knew, and lost every home they made: "They had nothing. They would never have nothing. There was no future." (Richard B. Johnson, in discussion with the author, 2024) Though both Nisenan culture and land would survive the connected devastations of gold mining and genocide, the immensity of their sense of loss, like the tears in Grandpa Rose's eyes, are important contexts for assessing Todd's images. To grasp the scale of ruin from a

Nisenan perspective provides a means to appreciate the monumental strides the Tribe has made in cultural renewal, as well as the long road to justice that remains ahead.

The Campoodie at Malakoff was only one of many Nisenan towns built on hills near Nevada City that were rapidly reduced to a few remaining piles of red clay. Johnson estimates that by 1864, the roughly 7,000 Nisenan who had lived in hundreds of villages at the outset of the gold rush had been reduced to just 500 people. With the passage of the Dawes Act in 1887, the Nisenan entered a four-decade struggle to retain seventy-five acres of their homeland, later designated the Nevada City Rancheria. In 1959, federal agents sold the land against the Tribe's will, and five years later, they terminated its federal recognition—ostensibly to force assimilation into mainstream American society. (Johnson, 2018: 6, 188–216)

After decades of work building relationships with local residents and government officials, the Tribe is today a prominent and respected voice in Nevada City civic life. In 2024, it repurchased and began restoring one of its former villages near Nevada City, Yuliça, with substantial support from the local community. As of this writing, the Tribe also continues its sixty-year legal struggle to have its federal recognition restored.

In September 2025, tribal members directed the return of their grinding stone to the land and a California State Parks crew relocated it to the rim of the diggings, where elders may visit it and interpretive texts will soon frame Nisenan history for the public. The mortar supplies necessary context for Watkins's photographs, which appear throughout the site to help visitors imagine Malakoff's past. It tells the story these images both document and obscure: As a Nisenan place, Malakoff is a testament to the violence of colonization on Native lands and peoples. The Nisenan persist robustly. The work of repairing the harms of the gold rush, however, remains unfinished.

Notes

1. A third series of eleven photographs by W. A. Clinch for the subsequent 1899 case *US vs. North Bloomfield Mining Company* will not be discussed here. (See Scott, 2017: 201–207)

2. Watkins's 1871 series was made to show North Bloomfield's potential to British investors, but it is not known why the ca. 1878 images were commissioned. (Hult-Lewis, 2011: 172–174, 210–212)

3. For example, the Two-Row Wampum belt, a treaty between the Hodinöhsö:ni' and the Dutch, believed to be the first treaty between Europeans and Native Americans, represents Indigenous and European nations rowing canoes on their own paths. It outlines a set of legal, social, and ecological protocols by which Native peoples and Europeans could share overlapping territories in everlasting peace. (Hill, 2019: 118)

4. Note that there are many versions of the Creation Story. (Simpson, 1977: 22, 35; Covert, in discussion with the author, February 2024). On acorns in California Indian cosmovision, Original Instructions, and ecological practices more broadly, see also Akins and Bauer, 2021: 17–24.

5. Selverston and colleagues did not find residual physical evidence of acorn in the cups of the rediscovered mortar; instead, they found pine nuts (first provided by Deer Woman's fawn not far away, at Aalam) and six other grasses, seeds, roots, and nuts. It remains possible that acorns were prepared here or at other mortars nearby. Johnson explains that mortar cups were washed with hot water between each use, which would have washed acorn away. Covert explains that because of the bitterness of acorns, women may have prepared them in separate cups. Eyewitness accounts state that boulders weighing tons were blasted into fragments that could be washed through the tunnel, suggesting that what remains may be part of a larger bedrock mortar. (*The Nevada Daily Transcript*, 1879: 2; Selverston et al., 2023: 51–65; Shelly Covert, in discussion with the author, June 2024; Richard B. Johnson, in discussion with the author, June 2024)

Archival Collections

Record for Malakoff Hill, Nevada County, CA. Mining Claims Collection, County Clerk Recorder's Office, 1855.

State of California Department of Parks and Recreation, Archeological Site Survey Record, Malakoff State Historic Park Collection, California State Parks Archives, 1949.

Bibliography

Damon B. Akins and William J. Bauer, *We Are the Land: A History of Native California* (Berkeley: University of California Press, 2021).

M. Kat Anderson, *Tending the Wild: Native American Knowledge and the Management of California's Natural Resources* (Berkeley: University of California Press, 2005).

Jo-ann Archibald, Jenny Bol Jun Lee-Morgan, and Jason De Santolo (eds), *Decolonizing Research: Indigenous Storywork as Methodology* (London: Bloomsbury Academic, 2022).

Brian Bibby and Dugan Aguilar, *Deeper than Gold: Indian Life in the Sierra Foothills* (Berkeley: Heyday Books, 2005).

Thomas C. Blackburn, "A New Look at Some Old Data: The Nisenan Photographs of Alexander W. Chase," *Journal of California and Great Basin Anthropology* 36, no. 2 (2016): 311–325.

Victor Burgin (ed.) *Thinking Photography* (London: Macmillan Education, 1987).

Eva Marie Garroutte, *Real Indians: Identity and the Survival of Native America* (Berkeley: University of California Press, 2003).

Powell Greenland, *Hydraulic Mining in California: A Tarnished Legacy* (Spokane, WA: A. H. Clark, 2001).

Robert F. Heizer, *The Eighteen Unratified Treaties of 1851–1952 between the California Indians and the United*

States Government. (Berkeley: California Indian Library Collections Project distributor, 1989).

Susan M. Hill, *Clay We are Made of: Haudenosaunee Land Tenure on the Grand River* (Winnipeg, MB: University of Manitoba Press, 2019).

Monica Huerta, *The Unintended: Photography, Property, and the Aesthetics of Racial Capitalism* (New York: New York University Press, 2023).

Christine Hult-Lewis, *The Mining Photographs of Carleton Watkins, 1858–1891, and the Origins of Corporate Photography* (PhD diss., Boston University, 2011).

Richard B. Johnson, *History of Us: Nisenan Tribe of the Nevada City Rancheria* (Santa Rosa: Comstock Bonanza Press, 2018).

Charles Joseph Kappler, *Indian Affairs: Laws and Treaties*. Vol. 4. 2nd ed. (Washington, DC: United States Government Printing Office, 1927).

Robin Wall Kimmerer, *Braiding Sweetgrass: Indigenous Wisdom, Scientific Knowledge and the Teachings of Plants* (Minneapolis: Milkweed Editions, 2015).

Benjamin Madley, *An American Genocide: The United States and the California Indian Catastrophe, 1846–1873*. (New Haven, CT: Yale University Press, 2016).

C. Hart Merriam, "Ethnographic Notes on California Indian Tribes," in *Reports of the University of California Archaeological Survey: Vol. 3*, ed. Robert F. Heizer (Berkeley: University of California Archaeological Research Facility, 1966).

Jennifer L. Mnookin, "Semi-Legibility and Visual Evidence: An Initial Exploration," *Law, Culture and the Humanities* 10, no. 1 (2014): 43–65.

Jennifer L. Mnookin, "The Image of Truth: Photographic Evidence and the Power of Analogy," *Yale Journal of Law and the Humanities* 10 (2015): 1.

Melissa K. Nelson, "Getting Dirty: The Eco-Eroticism of Women in Indigenous Oral Literatures," in *Critically Sovereign: Indigenous Gender, Sexuality, and Feminist Studies*, ed. Joanne Barker (Durham: Duke University Press, 2017), 229–260.

The Nevada Daily Transcript, "A Great Gravel Mine," July 30, 1879.

Peter E. Palmquist, "Mirror of Our Conscience: Surviving Photographic Images of California Indians Produced Before 1860," *Journal of California Anthropology* 5, no. 2 (1978): 162–178.

Stephen Powers, *Tribes of California* (Berkeley: University of California Press, 1976).

Jolene Rickard, "The Occupation of Indigenous Space as 'Photograph'," in *Native Nations: Journeys in American Photography*, ed. Jane Alison (London: Barbican Art Gallery, 1998), 57–71.

Conohar Scott, "Photographing Mining Pollution in Gold Rush California," *Photographies* 10 no. 2 (2017): 189–209.

Allan Sekula, "The Body and the Archive," *October* 39 (1986): 3.

Mark D. Selverston, Bruce Owen, and Doshia Dodd, "Cultural Resources Evaluation of Context 143 of CA-NEV-551/H, A Bedrock Milling Feature in the Malakoff Diggins Hydraulic Mine Complex, Nevada County, California," Report on file at Sonoma State University (2023).

Audra Simpson, "The Ruse of Consent and the Anatomy of 'Refusal': Cases from Indigenous North America and Australia," *Postcolonial Studies* 20, no. 1 (2017): 18–33.

Richard Simpson, *Ooti: A Maidu Legacy* (Millbrae, CA: Celestial Arts, 1977).

John Tagg, *The Burden of Representation: Essays on Photographies and Histories* (Amherst: University of Massachusetts Press, 1988).

Leah Temper, "Blocking Pipelines, Unsettling Environmental Justice: From Rights of Nature to Responsibility to Territory," *Local Environment* 24, no. 2 (2019): 94–112.

Hulleah J. Tsinhnahjinnie, "When is a Photograph Worth a Thousand Words?," in *Native Nations: Journeys in American Photography*, ed. Jane Alison (London: Barbican Art Gallery, 1998), 41–55.

Hans Jørgen Uldall, *Nisenan Texts and Dictionary* (Berkeley: California Indian Library Collections distributor, 1992).

Kaitlin N. Vigars, "Buried Beneath the Legislation It Gave Rise to: The Significance of Woodruff v. North Bloomfield Gravel Mining Co.," *Boston College Environmental Affairs Law Review* 43, no. 1 (2016): 235–249.

Woodruff v. North Bloomfield Gravel Mining Co. et al, No. 2900 (9th Circuit January 7, District of California, 1884).

Marilyn Ziebarth, "California's First Environmental Battle," *California History* 63, no. 4 (1984): 274–279.

CHAPTER 11
Everywhere is West

Mercedes Dorame (Tongva), with an essay by Emilia Mickevicius

Across a diffuse ground of shallow water, a bed of seagrass gently sways, its individual threads swirling into common orientations as if dreaming in the same direction (*Seagrass Emerging—Shooxar*; fig. 1). The water magnifies the seagrass's vivid hues of lime and ocher. Some of the sinewy strands breach the surface, their exposed lengths appearing like a series of scrawled marks. Above the grass, the water reflects a burst of overcast daylight, collapsing sea and sky onto the same plane. It has always been like this, the photograph attests: light, water, and vegetation sustaining one another and giving each other form.

Elsewhere, a pink portal beckons (*Algae Portal—Shooxar Tukuupar*; fig. 2). An ovular pool teeming with kelp and algae opens onto an arrangement of clay-colored stones that sit atop a bed of crushed shells. The stillness of the scene is an illusion, temporary: The depth and shape of the pool are in perpetual flux, shifting with the tide. Nearby, we encounter another alchemical surface in the form of glistening asphaltum. Its bulbous, undulating forms, inky teal like the night, scatter the overhead sunlight. Microscopic and immense, it is a nebula that births a thousand stars.

The photographs in Mercedes Dorame's series *Everywhere is West* picture land, but the word landscape is inadequate to describe them. For they are not hierarchical views or prospects, classifications that connote power and dominion as well as stasis. Rather, they posit an alternative temporality bound to tides, seasons, geology, and Dorame's Gabrielino-Tongva ancestors. Refuting binaries of night and day, instantaneity and duration, and earthly and celestial realms, they function as thresholds: visual gateways onto worlds that hum according to their own eternal, cyclical rhythms. Witnessed and cared for by the Tongva and Chumash for generations in kincentric harmony, and in the present by Dorame, these meeting points of land, sea, and sky become invitations to regard the natural world with reverence for its inherent wisdom—indeed, to consider it as kin. (Martinez, 2018; ahtone, 2021)[1]

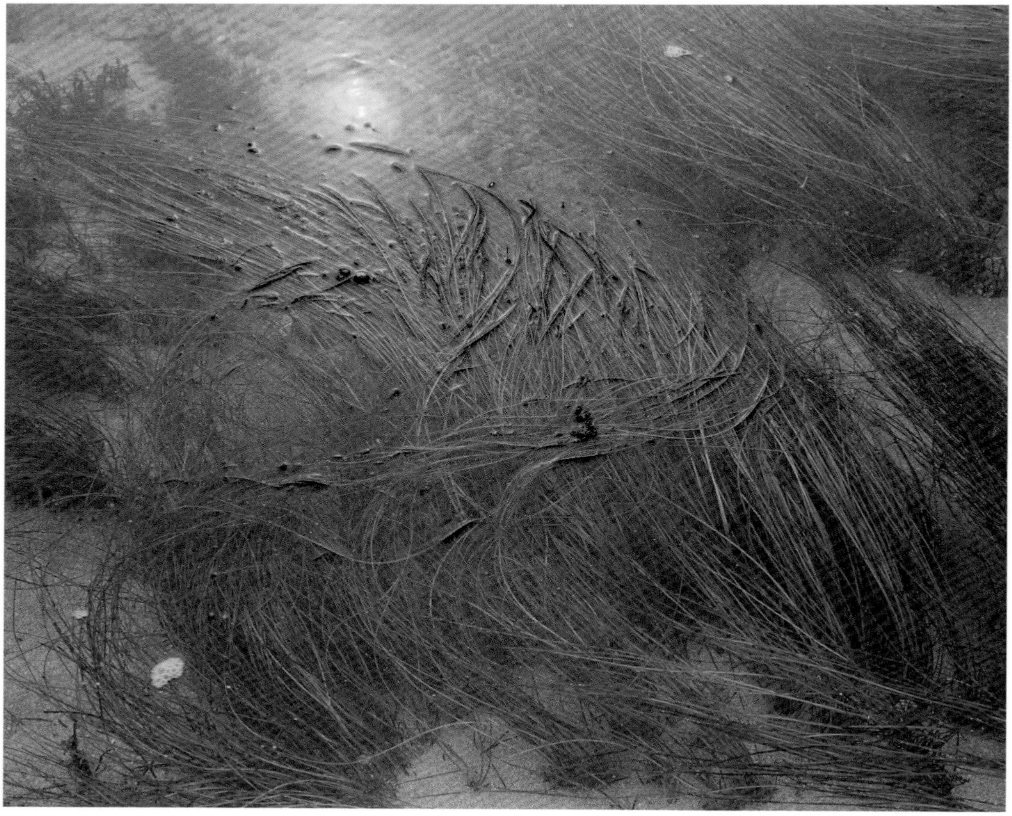

Fig. 1.

Mercedes Dorame, *Seagrass Emerging – Shooxar*, 2023. Inkjet print, dimensions variable.

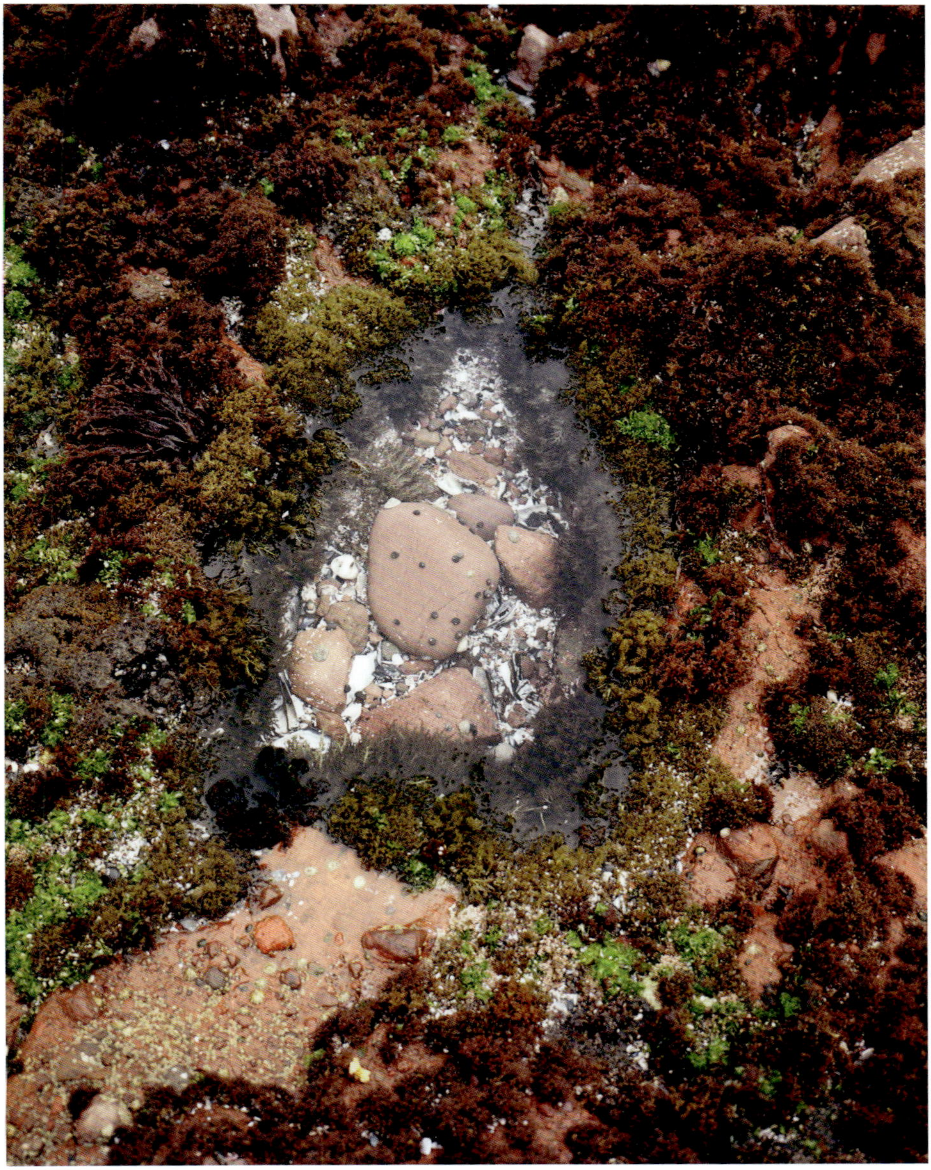

Fig. 2.

Mercedes Dorame, *Algae Portal – Shooxar Tukuupar,* 2022. Inkjet print, dimensions variable.

Dorame has long used photography to engage what she has called the "problematics of visibility" for First Peoples of coastal California, who have endured centuries of erasure by a settler-colonial culture that has subjected their ancestral homelands to concrete and commodification. Training her lens on places that hold cosmological importance for her Tongva community, and on the elements of the natural world that have sustained their lifeways for millennia, she forges a dialogue with her ancestors and asserts their continuous presence and stewardship of the land and water. This commitment to engaging unseen stories and working against the perception of land as absent of human presence propelled her contribution as a member of the inaugural cohort of Unseen California, which galvanized her to create this body of work. (*Language,* 2024)[2]

Growing up in Tovangaar (Los Angeles), Dorame often oriented herself by searching for the ocean. But surrounded by water on the Channel Islands of Pimugna (Catalina Island) and Limuw (Santa Cruz Island), she found that every direction read as West, and therefore as home. (*Language,* 2024)[3] Twenty-four miles off the city's coast, the artist encountered a landscape governed by the rhythms of the tide. The water's cyclical encroachment and retreat continuously shifted what was visible, obscuring or revealing tar, soapstone, and abalone. Tuning into the terrain and drawing upon research into her ancestors' journeys to the islands by *ti'at* (plank-built canoe; in Chumash, *tomol*), she created photographs that honor and animate this vital flux.

A defining way in which Dorame departs from representational convention and reorients the viewer to the land is by having us look down to look up. She summons water to appear alternately transparent and opaque as it reflects the sky; this characterizes not only *Seagrass Emerging* but also the viscous, moon-reflecting sheen of the water's surface in *To Give Moonlight—Pahaatkomok Mwaar* (fig. 3). Elsewhere, the ground itself takes on the appearance of sky, as in the case of the tar in *Asphaltum Seeps—Shaanat Chuuynok'e I* (fig. 4)—a substance, it bears noting, that Dorame's ancestors used to caulk their boats.

The artist's elimination of the horizon in the majority of her images, especially in those she presents in square format, also disperses our attention radially and directs it downward at the earth. Moving with Dorame across the terrain, our eyes trained on the ground, we imaginatively inhabit the searching gaze of her ancestors, who were propelled by their sophisticated knowledge of the islands' tidal rhythms and the myriad uses of its vegetation. In this way, Dorame's photographs elaborate a relationship to time that Chiricahua Apache scholar Nancy Marie Mithlo has described as "moving between registers" rather than adhering to strict, linear chronology. (Mithlo, 2020: 49)[4] The images are multigenerational: They layer deep geological time, the historical presence of

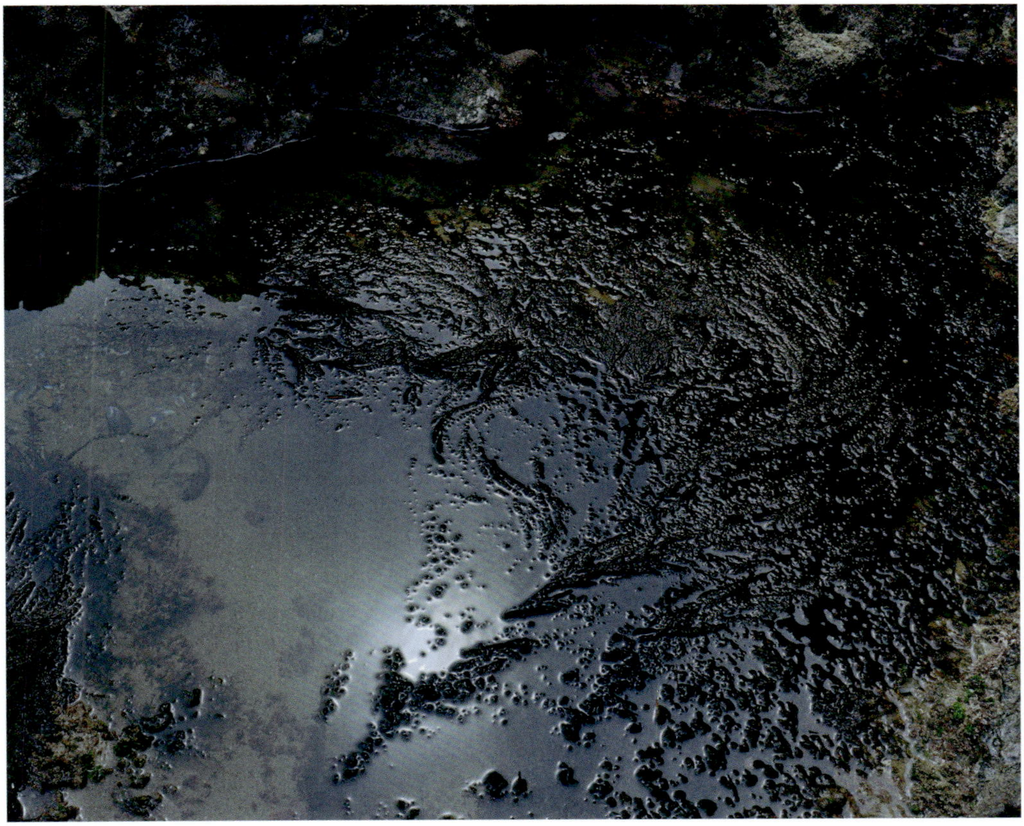

Fig. 3.

Mercedes Dorame, *To Give Moonlight – Pahaatkomok Mwaar,* 2023. Inkjet print, dimensions variable.

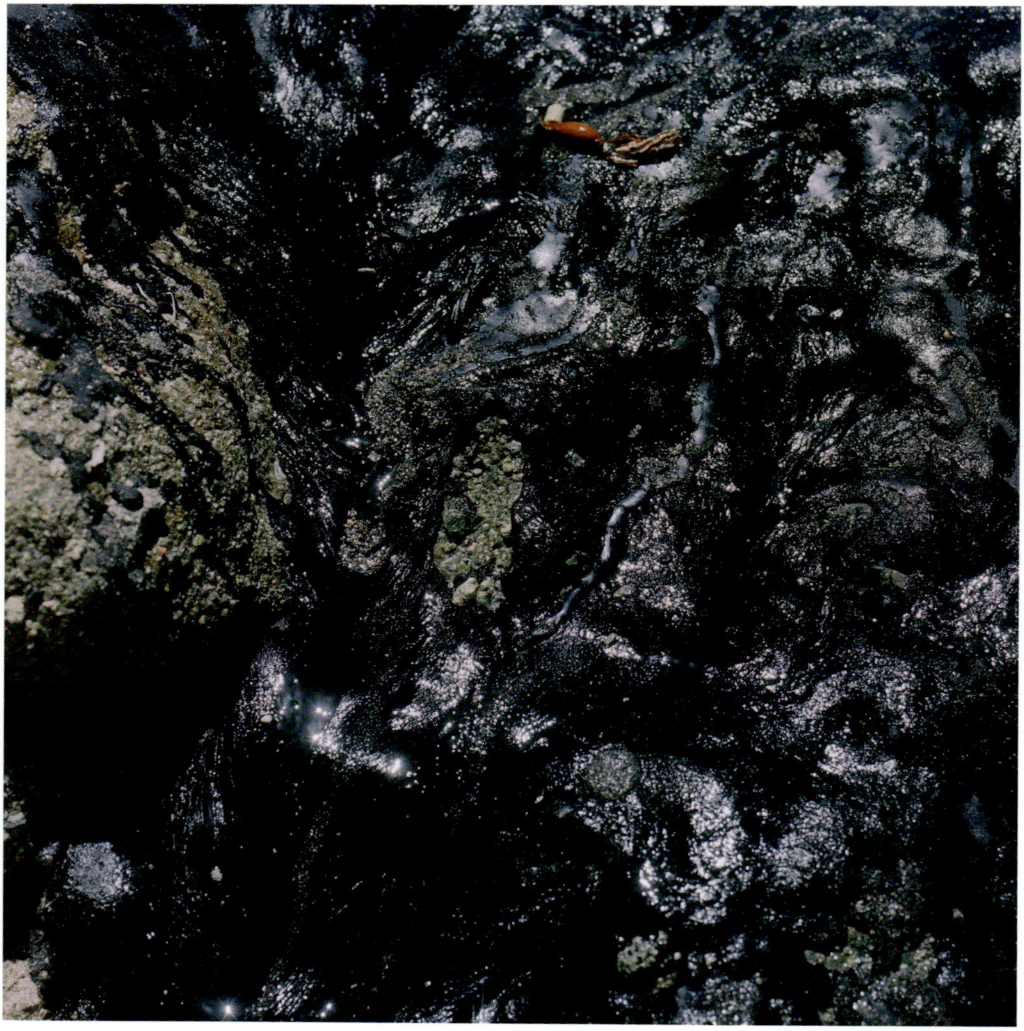

Fig. 4.

Mercedes Dorame, *Asphaltum Seeps – Shaanat Chuuynok'e I,* 2023. Inkjet print, dimensions variable.

ancestors, and the artist's own embodied gaze that perceives and records the dances of light reflected off the protean pools. They express history as an "experiential continuum," as scholar Daniel Wildcat (Yuchi member of the Muscogee Nation of Oklahoma) has described, with "time understood as space mediated." (Wildcat, 2006: 151)

Though *Everywhere is West* vibrates with Dorame and her ancestors' ecological knowledge, the photographs also uphold a certain ineffable quality of the islands' terrain that might be described as the impossibility of its complete knowability—in the sense not of romantic mystification, but rather of acknowledgment of its self-sustaining intelligence. Dorame grants the land and the water, and the life they sustain, a complexity of being and capacity to evolve that western culture generally ascribes only to humans. In *Dissolving Mountainside—Kotiit 'Akuupyavet* (fig. 5), for example, fog obscures a distant hill whose upward slope of shrubby chaparral takes on the ephemeral shape of a wave or a fin. Dorame presents the mountain and fog as mutually constituting one another visually as they nevertheless resist clarity. Like the tide pools—indeed, like people—they continuously change and shift into being, no single moment quite the same as the next.

Particularly as a member of a tribe that remains federally unrecognized and without sovereign land, Dorame ultimately harnesses photography as "reclamation," Mithlo's chosen term for how Native image makers draw upon this reproducible medium to counter the abstract violence of erasure. (Mithlo, 2020: 52)[5] Joining other Indigenous photographers in countering empire's suppression of animistic and kinship-oriented world views, she enacts an ongoing "collaboration with land," in her words, inviting us to "reimagine the land and sky as Indigenous." (Dorame, 2023) The viewership that Dorame's photographs intimate for us—where we regard land as relation rather than resource to be extracted—suggests the importance of positionality in decolonizing landscape photography. Her work affirms the generous challenge that Indigenous epistemologies pose to the field of art history: Necessitated by works of art that place pressure upon or are even incompatible with western frameworks, they call upon us to question and rethink the very concepts, standards, and language we have inherited from dominant culture and, ultimately, to reconsider how we relate to land itself. (figs. 6–10)

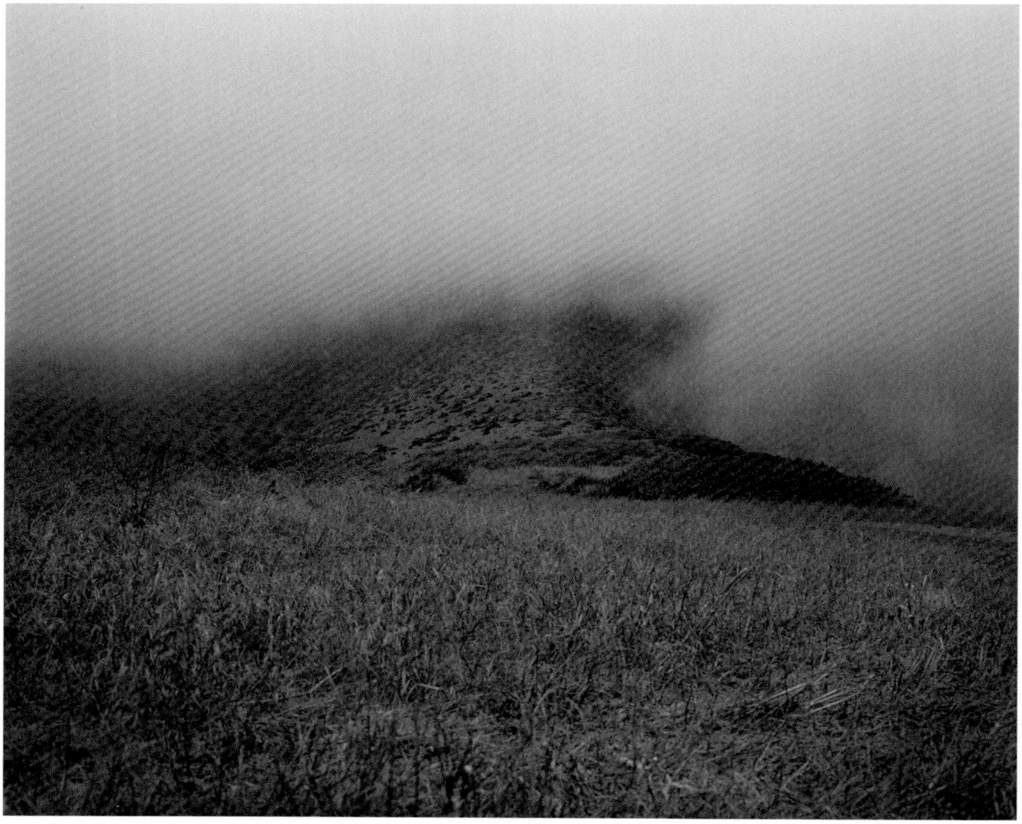

Fig. 5.

Mercedes Dorame, *Dissolving Mountainside – Kotiit 'Akuupyavet,* 2023. Inkjet print, dimensions variable.

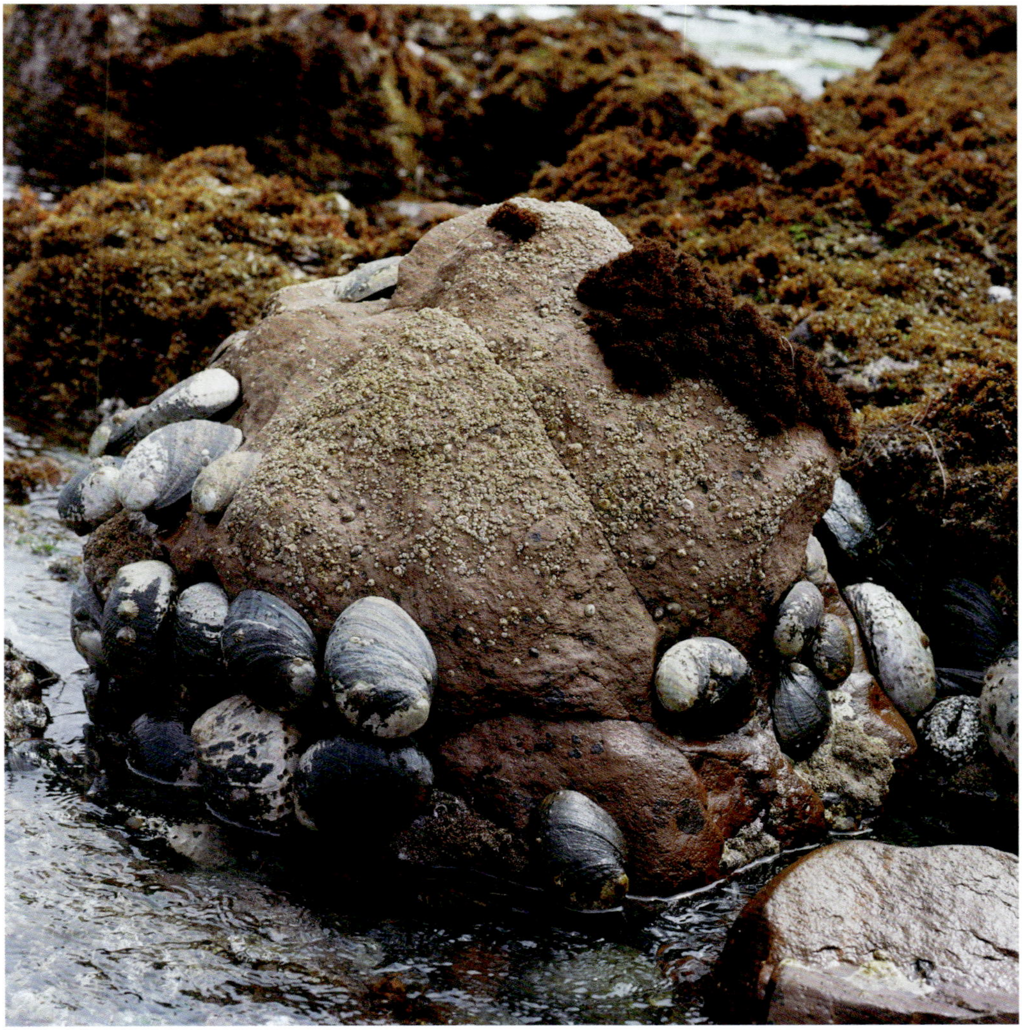

Fig. 6.
Mercedes Dorame, *I Will Come from the Ocean – Mooomvene Kimaaro,* 2022. Inkjet print, dimensions variable.

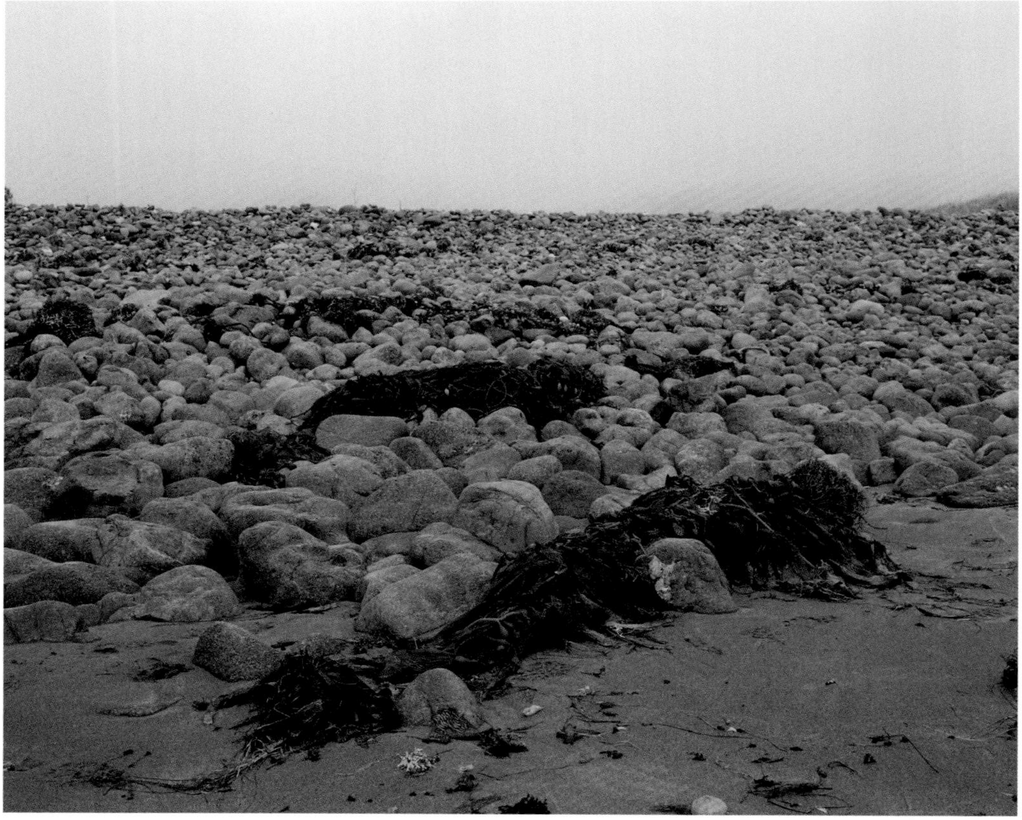

Fig. 7.

Mercedes Dorame, *Imprint – Xamee-evet III,* 2023. Inkjet print, dimensions variable.

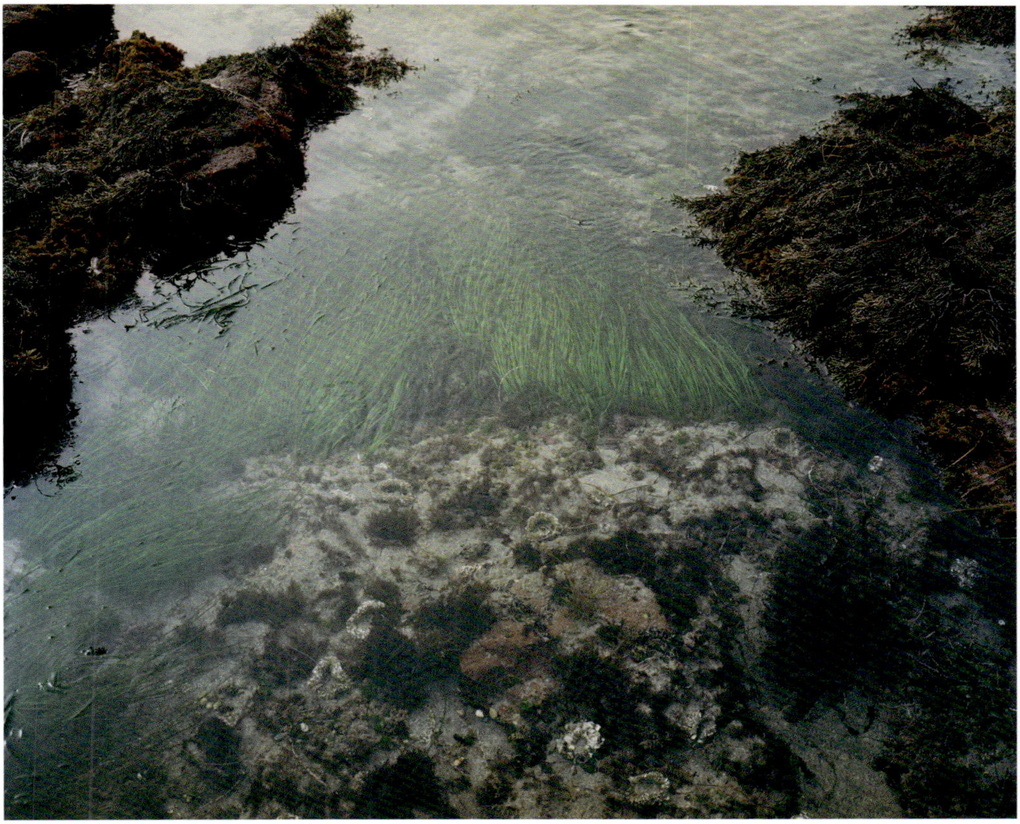

Fig. 8.
Mercedes Dorame, *Pathway Through – Paakyan,* 2023. Inkjet print, dimensions variable.

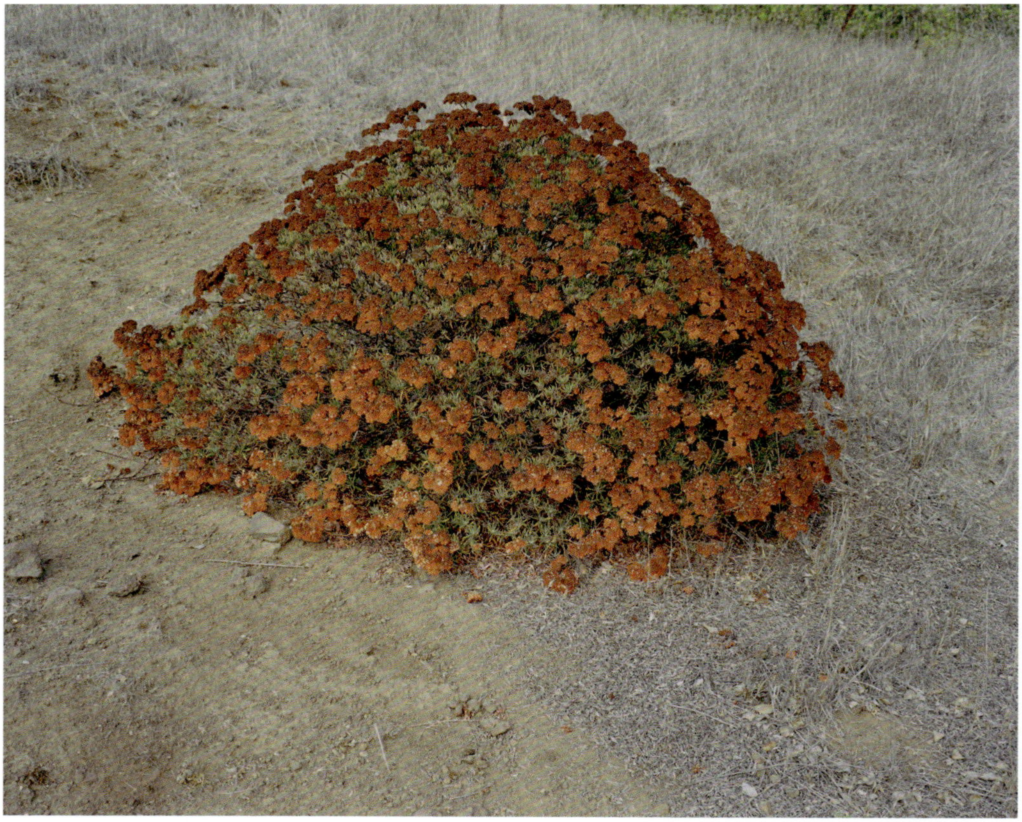

Fig. 9.

Mercedes Dorame, *Shifting Sun – 'Oroovevet Showookenax,* 2023. Inkjet print, dimensions variable.

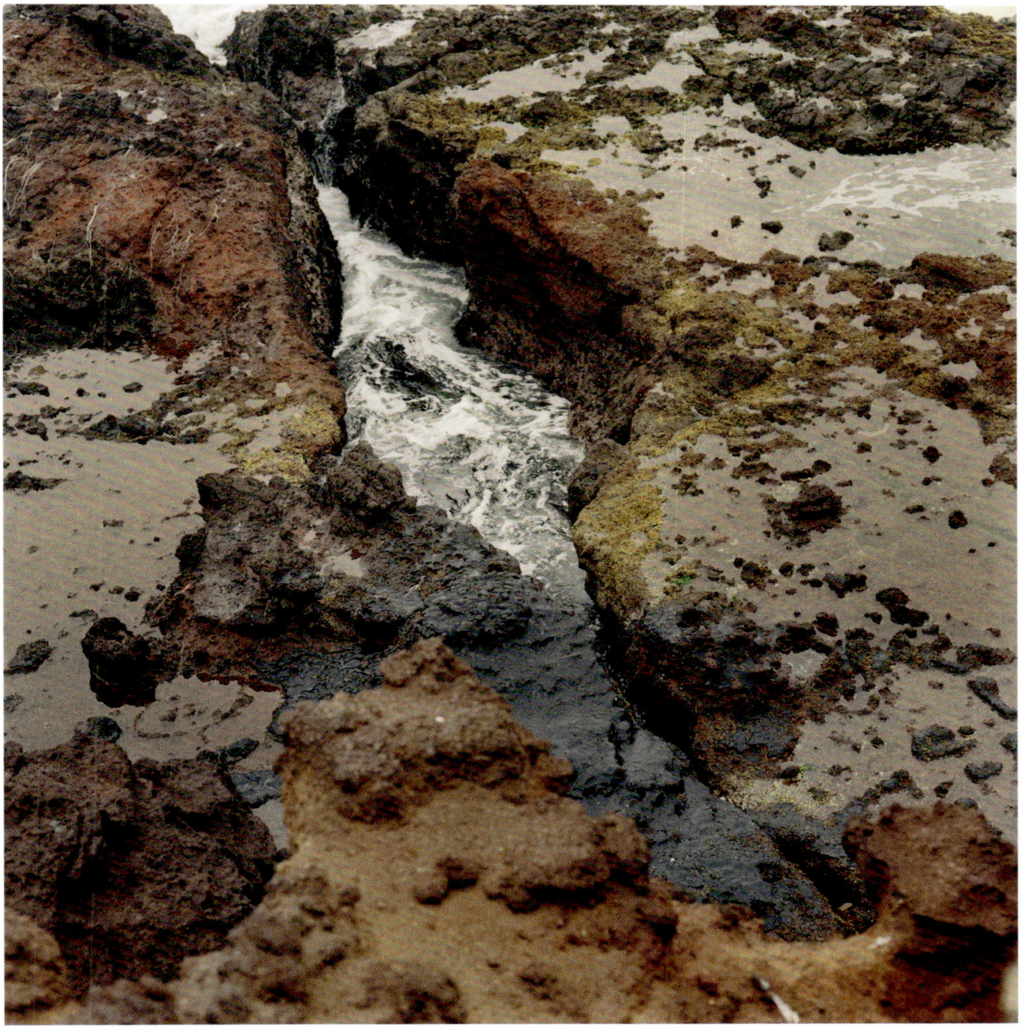

Fig. 10.

Mercedes Dorame, *Imprint – Xamee-evet I,* 2023. Inkjet print, dimensions variable.

Notes

1. Elaborated by Tohono O'odham/Chicano/Anglo ecologist and organizer Dennis Martinez in the context of land management, *kincentricity* refers to the Indigenous cosmology that eschews the binary between nature and humanity and is instead premised upon a model of interbeing, where people co-manage land with plants and animals. In her scholarship about interpretive frameworks for contemporary Indigenous art, Chickasaw curator heather ahtone has invoked kincentricity as the "vibration" that activates other immanent qualities of Native art, such as its materiality or symbolism (ahtone).

2. Photographer and educator Karolina Karlic launched Unseen California with cinematographer Joseph Angel as a multifaceted arts research initiative and two-year artist residency program in 2022. Dorame participated in the inaugural cohort with fellow artist researchers Karlic, Tarrah Krajnak, Dionne Lee, and Aspen Mays. Over the two-year period, each artist elected to work with a reserve in the University of California Natural Reserve System. Dorame made work at the Santa Cruz Island Reserve, managed by the University of California, Santa Barbara. See *Language Has No Weather*, 2024.

3. It should be noted that Santa Cruz Island is part of the northern Channel Islands, which are ancestral Chumash territory, and that the southern Channel Islands are ancestral Tongva territory, with the majority of their population living on Catalina Island. Dorame deliberately conceived of her work as exploring the "extensive exchange between the people of all the Channel Islands," emphasizing that the islands collectively sit near the Los Angeles basin. ("Unseen California Roundtable, Year 1," conversation recorded on 19 July 2022 and reproduced in *Language Has No Weather*, n.p.)

4. Mithlo identifies alternative temporalities as a defining feature of Native arts practice.

5. For Mithlo, reclamation can pertain to stories, places, assumptions, and photographs themselves.

Bibliography

heather ahtone, "Shifting the Paradigm: A Love Story," keynote address for *First American Art* Magazine symposium "We Have Words for Art," February 27, 2021, https://firstamerican.art/words/ (accessed April 28, 2026).

Mercedes Dorame, artist statement for *Everywhere is West*, 2023.

Language Has No Weather (Columbia, Missouri: theretherenow, 2024).

Dennis Martinez, "Native Perspectives on Sustainability: Dennis Martinez (O'odham/Chicano/Anglo)," interview by David E. Hall, January 3, 2018. Transcript available at https://www.nativeperspectives.net/Transcripts/Dennis_Martinez_interview.pdf (accessed April 28, 2026).

Nancy Marie Mithlo, "Native Arts' Visual Remix," in Nancy Marie Mithlo, *Knowing Native Arts* (Lincoln: University of Nebraska Press, 2020), 47–70.

Daniel R. Wildcat, "Indigenizing the Future: Why We Must Think Spatially in the Twenty-First Century," in *Destroying Dogma: Vine Deloria Jr. and His Influence on American Society*, eds Steve Palik and Daniel R. Wildcat (Golden, CO: Fulcrum Publishing, 2006).

PART 4
Against the Classical Landscape

CHAPTER 12
"Mightier than the Pen?" Amateur Photography and the Circulation of Environmental Narratives in California, 1880s–1920s

Yves Figueiredo

> It has been said that "the pen is mightier than the sword," but who shall not say that in *this* instance, at least, *the camera is mightier than the pen?*
> (Wildon, 1866: 106)

On January 7, 2025, around 10:30 a.m., a fire broke out in the hills of Pacific Palisades, north of Santa Monica, California. Within just a few hours, it had consumed more than 5,000 acres and seemed entirely out of control, destroying one of the most iconic neighborhoods of Los Angeles. By early evening, a second fire had ignited to the north of Altadena, near Pasadena, California, followed a few hours later by a third blaze in the northern San Fernando Valley. For a month, fires of unprecedented scale and ferocity ravaged the Los Angeles area, leaving devastation in their wake and shocking the world. International media outlets remained fixated on the damage and the startling spectacle of a metropolis seemingly on the brink of being engulfed by flames. Crucial context was missing, and commentary struggled to make sense of the situation. It was clear that something was deeply wrong with the planning and development of Los Angeles and Southern California, but pinpointing exactly what it was proved difficult.

But was it really so unpredictable? The encroaching wildland–urban interface, the advance of homes into the mountains and of trees into the city had long raised concerns that such a disaster could occur.

California's decades-long drought had already produced even larger fires elsewhere in the state, and many commentators pointed to the influence of global warming. Yet beyond these obvious explanations, it is crucial to question the historical and human causes of this situation: Where did the trees that burned come from, who planted them, when, and for what purpose? As political ecologist Diana Davis emphasizes:

> Just a couple of years ago California lost hundreds of millions of trees, in the 'mega-fires' that raged over more than four million acres in the fall of 2020, the largest in California history [...]. Most of the forest trees damaged were naturally occurring whereas the majority of the trees in urban and peri-urban areas had been planted over the last century and a half by well-meaning humans. Both, however, were strongly influenced by human actions driven by particular arboreal environmental narratives and imaginaries. (Davis, 2025: 90–91)

What went up in flames—the lives lost and the heartbreaking devastation—was not merely the result of natural forces, but of human decisions made over decades. These choices were shaped by dominant cultural narratives that were often seen as neutral or apolitical—innocent expressions of widely accepted beliefs, reassuring bits of conventional wisdom about what a desirable environment should look like, about the inherent value of trees or the deeply ingrained rejection of desert landscapes. Yet such narratives play a powerful role in shaping how we collectively perceive and engage with our environments. Far from being neutral or universal, they are deeply political and ideological. Over time—across decades or even centuries—they are circulated and solidify, passed down from one generation to the next through a host of different media: legends, classical tales, paintings and engravings, and, more recently and on a much larger scale, photography. Ultimately, they gain such authority that challenging or even questioning these narratives becomes difficult, sometimes unthinkable. Thus, to truly grasp today's environmental crises and land degradation, we must situate them within their historical context, recognizing that present-day land-use decisions are often the result of shifts and choices made in the past. (Gorostiza, 2015; Davis, 2025)

Making sense of the devastating Los Angeles fires requires us to unpack, deconstruct, and repoliticize the narratives that have long shaped our relationship with the environment. These stories have influenced everything from urban planning to personal ideals and tastes. Beneath these seemingly ordinary preferences lie powerful dynamics: struggles over land and resources, suppressed forms of knowledge, and dominant visions of nature and property that, in nineteenth-century California, were visually undergirded by quickly developing photographic practices. By examining the works of an increasingly organized and widespread community of amateur photographers, we can better understand how environmental imaginaries circulated and shaped Californians' perceptions of a landscape at once abundant and widely misunderstood in its aridity.

Indeed, amateur photographers did not simply imitate the iconic images of professionals, such as Carleton Watkins. Rather than passively reinforcing the established aesthetic canon, they often reinterpreted it, adapting its forms and, in doing so, exposing its limitations and internal tensions. While these images often failed to challenge dominant assumptions about race and class, they introduced a broader range of perspectives that subtly reframed the environmental message. Their value lies in their hybridity and fluidity: Neither fully private nor entirely public, these photographs circulated widely, were published, collected, or shared within social and civic networks. By the late nineteenth century, the scale of amateur photography was indeed remarkable, as amateur images moved through social clubs, local exhibitions, personal albums, and even commercial outlets, creating a vast, decentralized visual archive that reached audiences well beyond the elite circles in which professional photographs typically circulated. (Görgen, 2025) What defines them as "amateur" often has more to do with the photographer's profession or lack of formal artistic status than with the quality or reach of the work itself: Fame or technical expertise was not the primary factor. These images therefore occupy a unique space, both culturally and historically, in the broader visual economy of environmental representation.

This essay argues that amateur photography functioned as a key medium for the dissemination and legitimization of environmental practices and, more fundamentally, of the environmental narratives that underpinned them. After briefly examining the visual canon shaped by photographers such as Carleton Watkins and Eadweard Muybridge, I will explore how tourist photographs both drew upon and subtly reinterpreted this tradition and its environmental rhetoric. From there, I will focus on Theodore Parker Lukens, a Pasadena civic leader whose work exposes the inherent interdependence of environmental narratives and practices. Not a typical amateur, Lukens was a "forester-photographer" who both drove the reforestation of the San Gabriel Mountains and documented it photographically. In doing so, he circulated the key environmental narratives of his time while actively reshaping the landscapes they sought to define. His case shows that imaginaries and practices operated not separately but as mutually reinforcing forces, illustrating Karl Offen's argument that "as truth-narratives, environmental discourses [are] deeply implicated in affecting the processes they seek to describe." (Offen, 2004: 25)

Image Politics and the Birth of Environmentalism

"Carleton Watkins clicked, Abraham Lincoln signed, Yosemite was 'saved' and environmentalism born." (DeLuca and Demo, 2000: 241) Although the authors are careful to acknowledge that "it was not that simple," their claim remains essentially valid: The birth of American environmental politics, which took shape in the 1860s in California and laid the groundwork for the cultural and political movement that led to the establishment of the first national parks (what we now call "preservation"), was from the start closely associated with photographers and painters. (Huth, 1948; Palmquist, 1983; Ogden, 1990; Runte, 1990; Sellars, 1997) The men who were involved in the creation of the Yosemite park in 1864 were not concerned with ecosystems or endangered habitats (the concept of ecosystem would not be formulated until 1935, and from a strict ecological perspective, Yosemite Valley is not particularly significant), but they were deeply committed to the promotion of Californian scenery and landscapes, and the hope of endowing the nation with a new and original cultural heritage. (Schama, 1995)

Israel Ward Raymond, the San Francisco businessman who first approached California Senator John Conness in February 1864 with the proposal to set aside Yosemite Valley as a public park, took care to include several of Watkins's 1861 Yosemite stereographs with his letter supporting what would become the Yosemite bill. (fig. 1) (Huth, 1948; Green, 2018: 161–169) Although Raymond sent small-format photographs for obvious practical reasons, Watkins was already recognized and celebrated as the master of the "mammoth" print, the large-format images that most powerfully conveyed the valley's grandeur. (fig. 2)

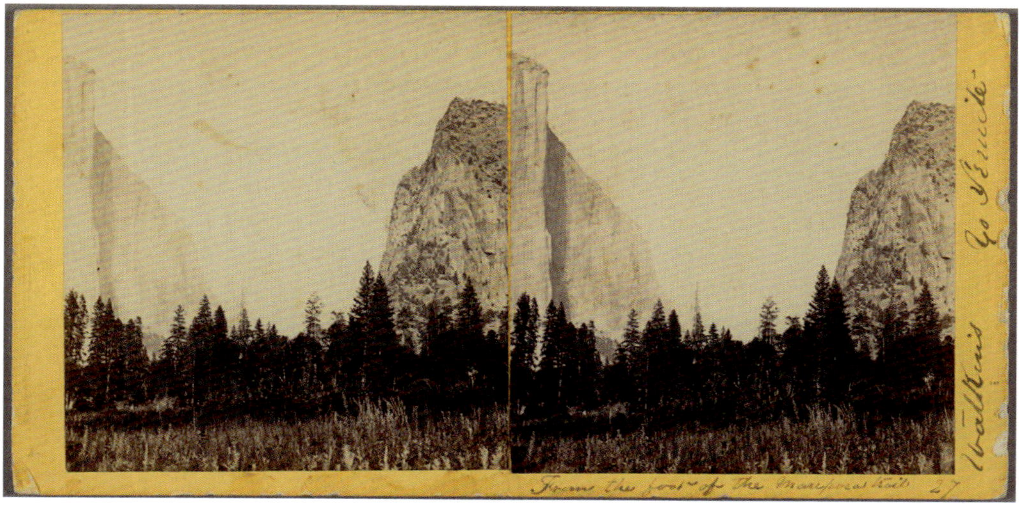

Fig. 1.

Carleton E. Watkins, *From the Foot of the Mariposa Trail. Watkins Yosemite*, 1861. Mounted stereographic print, 17.7×8.7 cm. Bancroft Library, University of California Berkeley.

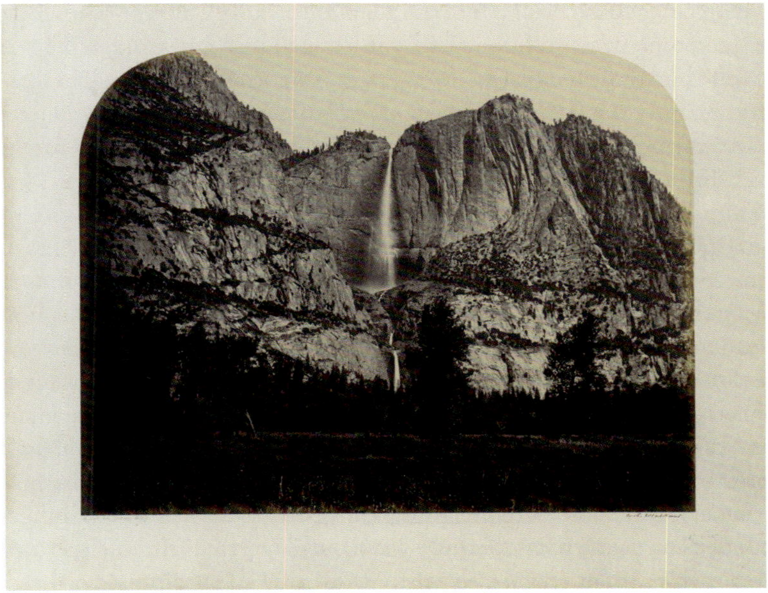

Fig. 2.

Carleton E. Watkins, *2637 Ft. Yosemite Fall, Front View,* 1861. Albumen silver print from glass negative, 40.9× 51.3 cm. Courtesy of the Huntington Library, San Marino, Calif.

Preservationist thought drew heavily on these monumental representations of the American wilderness, so much so that the aesthetic approach has since been described as "monumentalism."

Watkins's photos were circulated among Conness's colleagues and proved extremely useful in conveying the valley's overall character and helping to rapidly pass the bill. (Sanborn, 1981: 99; Fels, 1983: 34) Frederick Law Olmsted, the landscape architect of Central Park fame, was appointed the park's first commissioner and was tasked with devising a comprehensive site management framework from scratch, which would later serve as a guide for future National Park Service policy. His first initiative toward that goal was to write to Carleton Watkins, Thomas Hill, and Virgil Williams, whom he regarded, as artists (a photographer and two painters), as the leading experts on the subject: Aesthetic values evidently played a decisive role in shaping early preservation policy from the very beginning, profoundly influencing its direction and priorities. (Kelsey, 2013) This landscape-based and racialized interpretive framework provided a shared, intelligible ground for Euro-Americans to appreciate California's natural environments. (Berger, 2003) The aesthetics of monumentalism, as established by Watkins's photographs and the paintings of Albert Bierstadt or Thomas Moran, served as interpretive tools that made the landscape intelligible to many. As historian Alfred Runte noted, "monumentalism is easy to understand." (Runte, 1991: xiv)

Indeed, their works exemplify a political rhetoric that helped shape the contours of American environmentalism—a rhetoric that, to a remarkable extent, still resonates today. The paradigmatic function of Watkins's pristine views and their decisive influence in informing later public policies positioned the wilderness ideal, or what geographer William M. Denevan terms "the pristine myth," as the dominant environmental narrative. (Denevan, 1992) Americans' deep familiarity with the myth of a virgin, wild, and untouched continent, a fantasized *tabula rasa*, simultaneously awe-inspiring and fearsome, inviting both conquest and reverence, rendered the pristine images emerging from California all the more powerful and persuasive, not because they appeared to reveal something new, but because they seemed to affirm what was already imagined and believed. (Nash, 2001; Oelschlaeger, 1991; Sandweiss, 2002) The power of this rhetoric helps explain why it prevailed even though most 1860s depictions of Yosemite presented open, park-like settings, broad meadows and unobstructed vistas that contradicted the wilderness narrative of dark, impenetrable forest. (Reznick, 2025: 54) (fig. 3) Attempts to frame iconic views with trees (fig. 4) may suggest dense cover and align more closely with the wilderness ideal but should not mask the fact that the valley's landscape, at least its floor, simply did not match that model at the moment of the park's creation (the systematic erasure of the cultural landscapes of Yosemite and its political implications are addressed in the final section of this essay).

Amateur Photography, Sublime Landscapes and Personal Experience

In the 1860s, natural landscapes in the American West were primarily appreciated through secondhand media: paintings, lithographs, and literary descriptions. But by the 1870s and increasingly thereafter, thanks to advances in portable photography and, crucially, the introduction of Kodak cameras around 1890, these landscapes became immersive environments visited firsthand by tourists. (Sanborn, 1981; Brunet, 2019) As historian Anne Farrar Hyde has observed: "By the late nineteenth century, the art world's search for the sublime and the ideals of Romanticism and Transcendentalism had begun to move Americans from the contemplation of paintings of dramatic landscapes in art galleries and drawing rooms to the observation of real landscapes." (Hyde, 1987: 210)

This shift from mediated contemplation to direct observation introduced a form of relative personal agency in the appreciation of nature—relative because visitors were still guided, more or less subtly, by a scenic design that structured the valley through a sequence of viewpoints, whose names show just how visual the proposed framework for interpreting the valley was. Prominent examples include "Discovery View," "Prospect Point," "Valley View," "Contemplation Rock," and, of

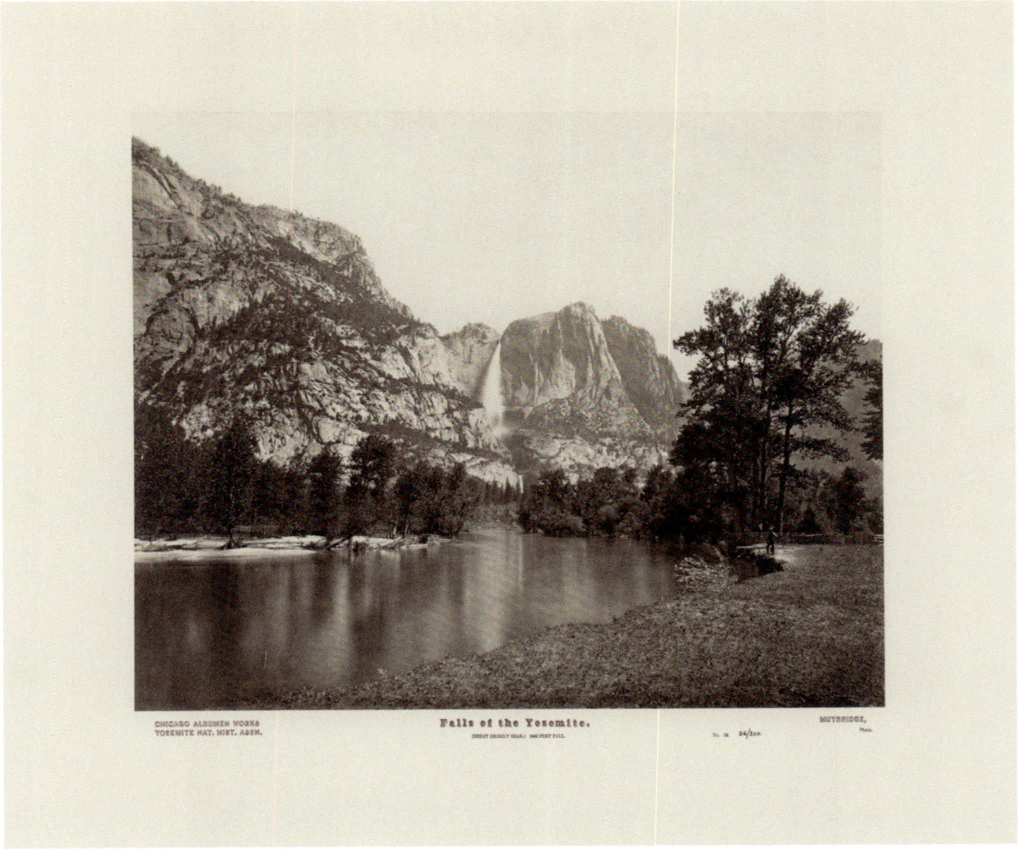

Fig. 3.

Eadweard Muybridge, *Falls of the Yosemite*, 1872. Albumen silver print from glass negative, 42.9×54 cm. Courtesy of the Huntington Library, San Marino, Calif.

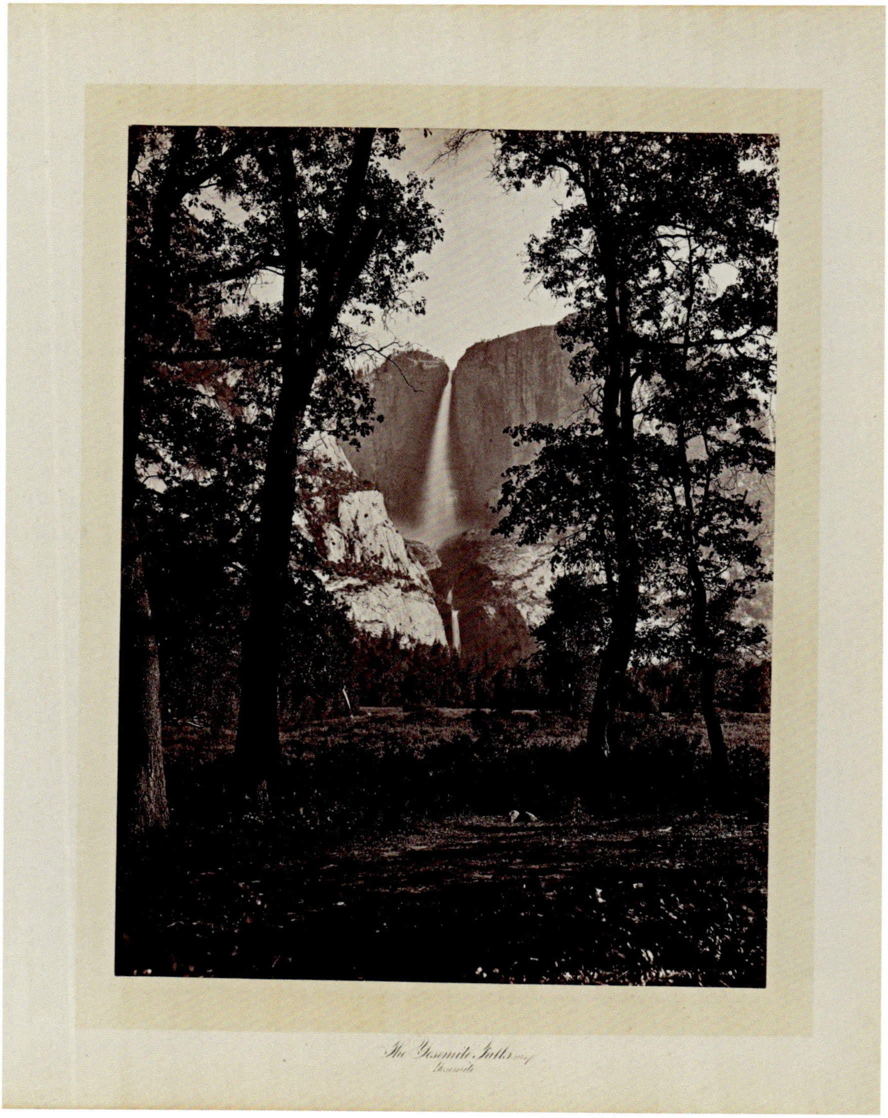

Fig. 4.

Carleton E. Watkins, *The Yosemite Falls, 2634 ft.*, 1865–1866. Albumen silver print from glass negative, 41.4×52.9 cm. Courtesy of the Department of Special Collections, Stanford University Libraries.

course, the two "inspiration points" ("Old" and "New"). Some, unfortunately lost, conveyed astonishment, such as "O! My! Point," or a sense of reverence, such as the "Stand Point of Silence." Despite these efforts to promote a uniform experience, engagement with the landscape gradually became more subjective and individualized. Local photographers offered their services, but tourists increasingly chose to document the experience themselves, a practice actively encouraged at the time. Galen Clark, the famed guardian of the valley, advised in the "Hints to Yosemite Visitors" section of his book: "Kodak films and plates can be obtained in the Valley. Developing and printing are done in the Valley. TAKE YOUR CAMERA." (Clark, 1904: n.p.) Although Bierstadt, Moran, and Watkins retained their iconic status, their images were now placed in dialogue with those of countless amateur photographers.

Unlike the grand masters, who often excluded human figures, tourists typically placed themselves, family, or friends within their shots. These photographs rarely depicted untouched wilderness; instead, they captured humans within nature, often treating it as a scenic backdrop rather than a sublime force. Heeding Olmsted's advice, they sought the "union of the deepest sublimity with the deepest beauty of nature," which, in his view, constituted "the Yo Semite the greatest glory of nature." (Olmsted, 1865: 8) As Kevin DeLuca and Anne Demo explain, "in the union of the sublime and the beautiful is born the tourist gaze. The beautiful foreground gives the tourist a pleasing place from which to view the spectacular spectacle of the sublime." (DeLuca and Demo, 2000: 248–249) By softening the strict wilderness rhetoric, these images helped reimagine the valley as a park—a place not only for preservation, but also for public enjoyment, exactly as the Yosemite Act of 1864 decreed: "the premises should be held for public use, resort, and recreation."

One site in particular offers valuable insight into the transition between photographic styles and helps to examine the role of amateur photography in the evolving perception of the Yosemite landscape. This site, the iconic Overhanging Rock, has been extensively photographed since the park's opening and continues to function as a visual landmark. George Fiske, a Yosemite photographer who at times collaborated with Watkins, is best known for his striking images of individuals seated or standing atop the rock, their feet suspended thousands of feet above the valley floor. (fig. 5) Yet, these photographs, despite placing human figures at the compositional center, remain closely aligned with the rhetoric of the sublime that characterizes Watkins's work. The presence of a calm, solitary figure on the precipice does not disrupt the scene's grandeur; rather, it amplifies the vastness, silence, and stillness of the wilderness. The human subject is introduced into the landscape, but discreetly, almost tentatively, as if subsumed within a sublime natural world that appears both awe-inspiring and overwhelming.

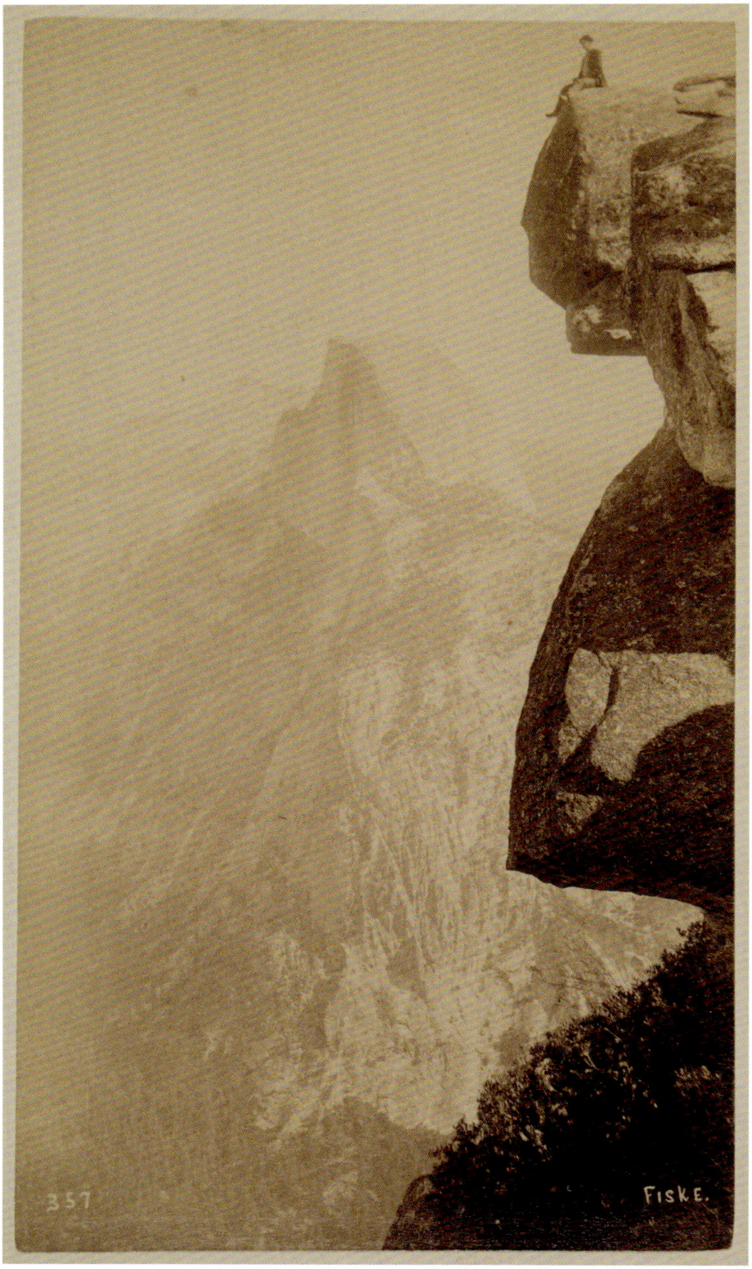

Fig. 5.

George Fiske, *Half Dome (5000 Feet) and Glacier Point (3200 Feet)*, ca. 1879–1881. Albumen silver print from glass negative, 13.2×21.4 cm. Amon Carter Museum of American Art, Fort Worth, Texas.

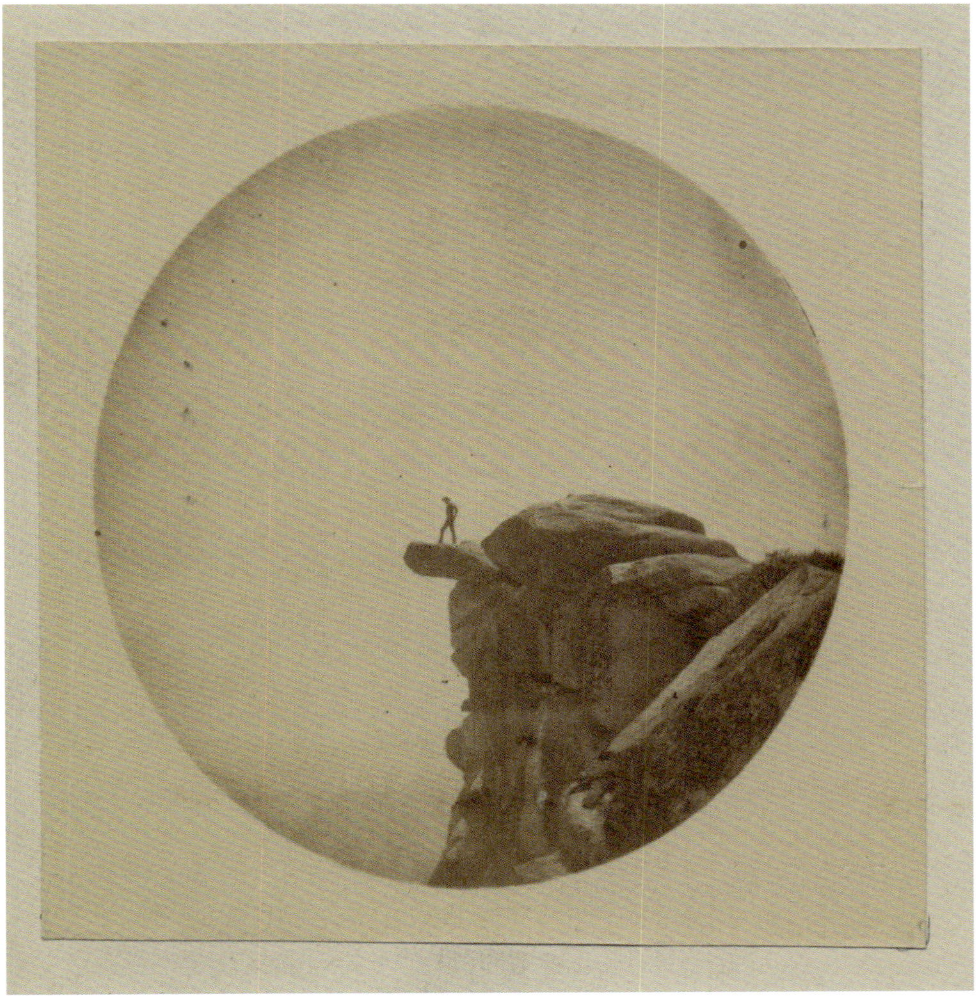

Fig. 6.

Photographs from Charles Palache's *Six Weeks in the Saddle: A trip to Stockton in July and August 1889*. Photographic print on paper, 6.7×6.7 cm. Bancroft Library, University of California Berkeley.

Compared with Fiske's professional work, tourist photographs showcase the human presence in a far more assertive and less contemplative way. (fig. 6) While the composition closely echoes Fiske's image, the effect is markedly different. Here, a man stands at the edge, tentatively stepping forward as if to approach the precipice. His apprehension is palpable, yet the image conveys a desire to master that fear; his courage (not the rock) becomes the central subject of the photograph. The small, round format of the image, typical of Kodak camera lenses, underscores its nature as a personal snapshot taken with a private camera, marks it as an object of popular consumption, and evokes the feeling of peering into an intimate scene, like glimpsing through a keyhole: What the image captures is not so much a landscape as a deeply personal experience. Over time, posing on Overhanging Rock evolved into a performative ritual, a staged demonstration of bravery and self-possession. The timeless stillness conjured by artists such as Bierstadt or Watkins gives way to scenes that are both personal; fleeting; and, at the same time, endlessly reproducible. As the amateur scene expanded in California and blended with the growing tourist industry, guided in part by active photography clubs, such as the California Camera Club, and fueled by the growing appeal of illustrated magazines, such as *Camera Craft* or *Sunset*, these new images became both commonplace and deeply desirable. (Görgen, 2025)

These new photographs reframed Yosemite: No longer a silent, monumental backdrop, the park became a site for human drama, with tourists cast as protagonists. (Demars, 1991) The cliffs, waterfalls, and dizzying heights ceased to inspire awe on their own; instead, it was the daring actions of visitors climbing, posing, and confronting fear that generated emotional impact. (Patin, 2012) As family albums and postcards suggest, the representational focus shifted, from overwhelming nature to scenes celebrating human conquest and domestication. Awe and fear now emanated less from the landscape's sheer scale than from the boldness of those who engaged with it. Tourists sought to inscribe themselves into a landscape already consecrated as a national icon. As art historian Kate Ogden notes: "Yosemite was already becoming somewhat hackneyed as a subject for art. What was important to the tourists of 1880 was proof of their own visits, rather than (or in addition to) documentation of the valley's glories." (Ogden, 1990: 147) Their photographs should not, however, be viewed as rejections of the visual legacy established by Bierstadt or Watkins, but, rather, as acts of appropriation: deliberate engagements with an established visual tradition. Intentionally or not, by repeatedly posing in front of iconic Yosemite landmarks, visitors were, one after the other, inscribing themselves into the broader visual mythology of the American wilderness. To some, these gestures may have appeared vulgar or irreverent, a form of disrespect toward a landscape once framed as dignified and untouchable. And yet, such images perform a significant symbolic function: They reintroduce the human figure into what had long been portrayed as an empty, silent, and desolate space. In doing so, they challenge the ideal of a pristine, unpeopled wilderness and begin to reshape the narrative of what it means to encounter nature.

This shift echoes the concerns raised by environmental historian William Cronon in his still-controversial 1995 essay, "The Trouble with Wilderness; or, Getting Back to the Wrong Nature." Cronon criticizes the wilderness ideal as the foundation of environmental thought, arguing that it erases human history and excludes more sustainable, lived relationships with the land. We cannot truly reconnect with nature, he writes, "if we hold up to ourselves as the mirror of nature a wilderness we ourselves cannot inhabit. To do so is merely to take to a logical extreme the paradox that was built into wilderness from the beginning: if nature dies because we enter it, then the only way to save nature is to kill ourselves." (Cronon, 1995: 83) By foregrounding human presence in the valley, the photographs discussed here push the landscape toward an environment rather than a fantasy of untouched nature. Far from being casual tourist snapshots, they anticipate these critiques, quietly challenging the pristine myth and asserting humans as part of the scene. In doing so, they helped soften the more radical, aporetical edges of early preservation—much as John Muir himself did when he committed the Sierra Club to organizing outings that opened the parks to ever-growing numbers of visitors.

Amateur Photography and Environmental Practice

Amateur photography, however, quickly expanded beyond its touristic and commemorative functions. As technological improvements—from glass plates to roll film—made photography increasingly accessible, it became a tool for documenting environmental change and mobilizing public concern. Amateur images gradually assumed a normative power, shaping what came to be seen as legitimate or desirable views of the Californian landscape. These photographs did more than echo established aesthetics—they participated in defining them. By highlighting certain features and ignoring others, they praised, condemned, or legitimized aspects of the landscape in line with the contemporary social and cultural expectations of the predominantly White leisure class. The environmental narratives they conveyed were actively constructed, filtered through the values and judgments of their makers. Through magazines, albums, letters, and postcards, amateur photography reached wide audiences and left a lasting cultural imprint. (Görgen, 2025)

One of the most persistent and ideologically charged themes in these visual narratives was the representation of forests and tree cover, which both reflected and reinforced broader environmental discourses of the period, especially those tied to reforestation. This vision of environmental health privileged trees as the primary markers of ecological integrity, exemplifying what Diana Davis calls "arborocentric thinking," or the long-standing cultural adoration of trees. (Davis, 2025: 91) Within this framework, forests were treated as the pinnacle of a "normal" environment, rendering the relative treelessness of many Californian landscapes not as natural variation but as evidence of degradation. Efforts to "restore" such landscapes therefore sought to reshape them according to inherited European ideals of temperate, forested environments, often blaming the absence of dense tree cover on Indigenous mismanagement rather than recognizing it as a product of sophisticated ecological stewardship. (Offen, 2004; Davis, 2018) Amateur photography played a central role in circulating and normalizing this arborocentric vision—from sequoias to redwoods to dense woodlands—even as the Yosemite Valley itself was undergoing a profound transformation. Once park-like and maintained as such through Indigenous burning practices, the valley documented in Watkins's 1861 photographs, of broad meadows and scattered groves, gradually gave way to one with thick forest cover. (Figueiredo, 2007; Reznick, 2025)

This transformation was not merely ecological; it was also deeply political, reflecting Indigenous dispossession and the suppression of "good fire" in favor of rigid fire-control policies. (Offen, 2004: 26; Anderson, 2005; Cermak, 2005; Everett, 2008: 1809; Deur and Bloom, 2020) Amateur images not only accompanied and recorded the erasure of Indigenous environmental knowledge and land management traditions, they helped legitimize it by visually naturalizing settler assumptions about what the landscape should be. (Solnit, 1994) Under settler-colonial governance, Indigenous ecological practices were systematically undermined; dismissed as unscientific or primitive; and, ultimately, suppressed. The photographic narratives that circulated during this period thus did more than depict a changing landscape; they shaped public perceptions while reinforcing the marginalization of Native voices and their stewardship of the land. Photographing a dense forest growing within a naturally arid landscape is far from a neutral or apolitical gesture. It constitutes (voluntarily or not) an engagement with environmental narratives that contributed to the erasure of Indigenous cultures. A situated analysis of these images reveals the racial dynamics embedded within the arborocentric vision championed by prominent "tree-lovers" such as John Muir. (Pomeroy, 2001; McNally, 2024)

While Yosemite Valley is, unsurprisingly, one of the best-known examples of a landscape reshaped to fit dominant environmental ideals, similar efforts unfolded across California. A striking case affected the San Gabriel Mountains above Pasadena, where Theodore Parker Lukens (one of Pasadena's earliest civic leaders) became a driving force in redefining the region's vernacular landscape. An Illinois transplant who arrived in 1880, Lukens quickly rose from newcomer to prominent booster, serving twice as mayor and helping steer Pasadena's rapid growth. (Sargent, 1969; Farmer, 2017: 145) A passionate student of trees, he viewed the stark, treeless slopes of the San Gabriels as signs of environmental decline, an interpretation which was fully aligned with prevailing notions about the necessary presence of trees in "healthy" landscapes. In his dual role as a civic promoter and advocate for tree planting, he recognized that the region's persistent water scarcity, by then widely acknowledged as Southern California's defining environmental limitation, posed a serious threat to both agricultural and urban development. With most of the water supply dependent on limited mountain run-off, demand for this scarce resource was fierce, setting farms, orchards and rapidly growing communities in direct competition.

The power struggle was intense and complex, and Lukens came to play a major role in it by undertaking the formidable task of reforesting the mountains. He began the project independently, later gaining support from the federal government when he was hired by the newly formed US Bureau of Forestry, led by Gifford Pinchot. He established a tree nursery at Henninger Flats, just above Pasadena, and, with the help of university students, set out to collect pine cones; gather the seeds; grow them in the nursery; and, finally, plant the seedlings in the mountains. It began in January 1899, and during the first week, they planted 12,000 seeds. During the second week, they planted a total of 53,000 seeds. They started by planting the ridges, so that the first trees would be visible from the city, as they wanted the work they were doing to be seen. (Lukens Papers, box 7) It would be fair to say that Lukens literally brought forestry, predominantly as a practice, to Southern California.

His position in this undertaking deserves closer scrutiny, if only because it was so unusual. He was, on the one hand, very closely connected to John Muir, with whom he exchanged a steady stream of extensive letters. Muir clearly saw him as both a friend and his voice in Southern California, pushing him to organize, to found a club, and to spread the message of preservation in a city, Los Angeles, that was growing at breakneck speed. On the other hand, Lukens was employed by the federal government to implement an experimental reforestation program in the region—likely thanks to Muir, who put him in touch with Pinchot to secure precisely this kind of state support. (Lukens Papers, box 2) This meant Lukens occupied two worlds at once: He was firmly linked to activist preservationist circles (dominant in Northern California but still emerging in Southern California) while simultaneously serving as an agent of US governmental policy. Straddling both the passionate realm of preservationist politics and the machinery of federal action, he blurred boundaries in ways rarely seen, even in this place and period.

In this ambiguous context, his photographs occupied a particularly intriguing place and status. Lukens was an avid photographer, producing hundreds of images every year. And here, again, ambiguity prevailed: The federal government typically financed the negatives—and therefore claimed them for its archives—while his prints circulated freely across Southern California and far beyond. His personal files brim with letters from friends, colleagues, and powerful regional figures, such as Harrison Gray Otis of the *Los Angeles Times* or attorney Henry W. O'Melveny, thanking him for the photographs he sent them.

He also transformed these images into lantern slides for the many lectures he delivered on reforestation, and into illustrations for articles in local publications or occasional essays. He used them to build networks of influence even before reforestation work formally began. In 1895 alone, he sent John Muir 150 photographs and in both 1896 and 1897, another 200. His efforts were rewarded: In 1896, Muir appointed him to the Sierra Club Standing Committee on Publications and Communications, making him a key actor of the club's communications efforts. In a letter dated November 27, 1896, Muir wrote that he hoped he should "be able to make good use of some of them as illustrations," presumably in the *Sierra Club Bulletin*. (Lukens Papers)

This corpus thus occupies a genuinely hybrid space, neither wholly personal nor fully professional, and that hybridity is precisely what makes it so remarkable. Lukens was no ordinary amateur, yet he was never a professional photographer; his images made no aesthetic claims and did not circulate within the usual artistic circuits. Instead, they possess extraordinary documentary value, revealing how environmental imaginaries and environmental practices spread together and intertwined, mutually reinforcing and in many ways inseparable. His photo albums indeed demonstrate Lukens's keen interest in trees (several albums are almost entirely composed of photographs of tree specimens) and his acute concern with aridity and its perceived threat to life in Southern California. (fig. 7) The caption "Desert in sight" echoed the alarm of Southern Californians at the time and evidently justified calls for widespread reforestation. However, these images and the work they documented merit closer scrutiny.

First, as historian Jared Farmer has pointed out, the San Gabriel Mountains were never densely forested. What Lukens and his contemporaries described as "reforestation" was, in fact, "afforestation"—the planting of trees in areas that had historically been treeless. As Farmer notes, Lukens actively worked to introduce tree cover in landscapes where it had not previously existed. (Farmer, 2017: 146)

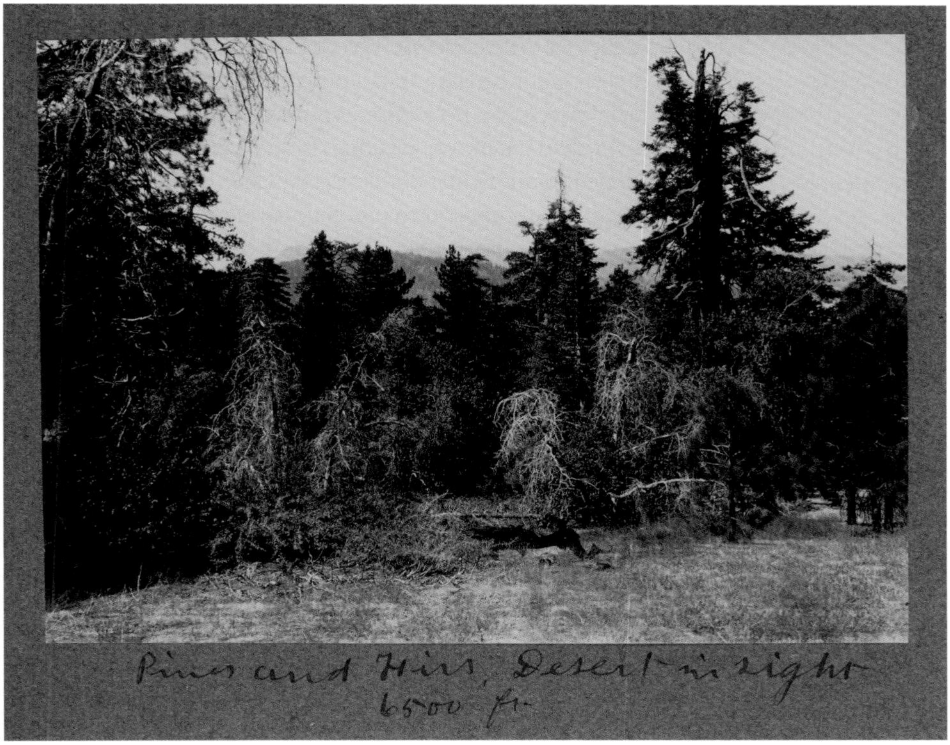

Fig. 7.
Theodore Parker Lukens, "Pines and Firs. Desert in Sight. 6500 ft.," date unknown. Photographic print on paper, 27×18.5 cm. Courtesy of the Huntington Library, San Marino, Calif.

Second, the environmental practices advocated by Lukens align with the suppression of fire and grazing—strategies widely endorsed by conservationists such as Lukens and Muir, whose aversion for "hoofed locusts" was famous. Sheep and wildfires were seen as destructive forces that devastated mountain ecosystems. Yet, more recent scholarship has shown that both fire and grazing were long used as tools by Indigenous communities for maintaining ecological balance and forest health (Everett, 2008; Davis, 2025) These practices, far from being purely destructive, were integral to sustainable land stewardship.

Third, the narrative presented in these images conforms to a now-debunked belief that reforestation could stimulate rainfall and positively alter local climates. This "rain follows the plow" theory, while once widespread (not just in California or the American West but internationally) had been scientifically discredited by the 1890s. (Figueiredo, 2013) Its persistence in Lukens's imagery reflects the enduring power of environmental myths and the challenges involved in dislodging deeply rooted narratives.

Lukens's photographs were also instrumental in building his authority within Southern California's reforestation movement. Designed to look like scientific documentation, they present a well-run, disciplined, and efficiently managed operation: men working in neat rows along the mountainside (fig. 8), or the carefully ordered tree nursery at Henninger Flats. (fig. 9) These are not casual snapshots but images deliberately constructed to project competence and institutional legitimacy, framing Lukens's work as both serious and reliable. Their concise, informative captions further heightened the strength of their message. One photograph used to illustrate his essay *Forestry in Relation to City Building* bears the caption "A portion of the Roots of one spruce tree. Hardly possible for the water to escape through such a network of root-dams." (fig. 10) This image, and the way Lukens used it, exemplifies his approach to photography and its uses: Photography was not merely a means of recording reforestation efforts, but a tool to promote his own understanding of ecological processes, communicate his sense of environmental urgency, and shape public perception. Through these images, Lukens advanced not only a scientific argument for reforestation but also a political one, establishing himself as a leading voice in the conversation around forest management and environmental planning in the region.

The photographs' compositions and didactic captions give their message a powerful sheen of legitimacy and authority. They project a sense of certainty that, despite resting on neither rigorous field analysis nor any serious consideration of the region's aridity, presents itself as indisputable. The result is a forced alignment of the landscape with a dominant ideology, each reinforcing the other in a continuous feedback loop. The environment itself is largely ignored—until the sheer incongruity of this colossal reforestation project finally reveals itself to all in the form of a devastating wildfire.

Just as William Cronon famously claimed in his 1995 wilderness essay that "the time has come to rethink wilderness," perhaps now, thirty years later, the time has come to rethink received wisdoms and the environmental narratives and imaginaries that sustain them. If we are to understand environmental catastrophes like the Los Angeles fires—let alone prevent them—we must urgently "set the historical record straight, de-simplify society-nature processes, critique our Western and self-righteous notions of what constitutes and generates 'development,' and reinsert the ingenuity and agency of subordinated peoples into the official histories that continue to overlook and subjugate them." (Offen, 2004: 31) As we watched Henninger Flats burn in the Eaton Fire, we were reminded of the enduring impact of environmental choices made in the past and passed down as neutral, rational, and apolitical expressions of common sense. The visual traces left by Californian amateurs at the turn of the twentieth century offer a useful entry point for analyzing and unraveling these narratives, exposing the political and ideological frameworks that underpin them.

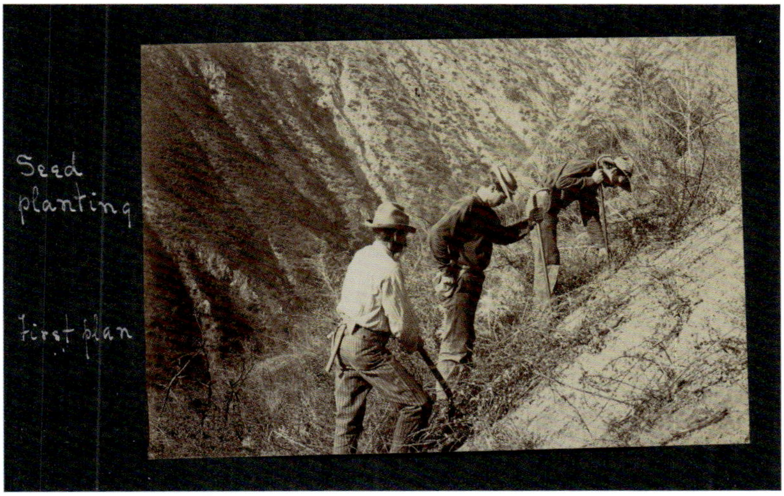

Fig. 8.

Theodore Parker Lukens, “Seed Planting. First Plan,” date unknown. Photographic print on paper, 27×18.5 cm. Courtesy of the Huntington Library, San Marino, Calif.

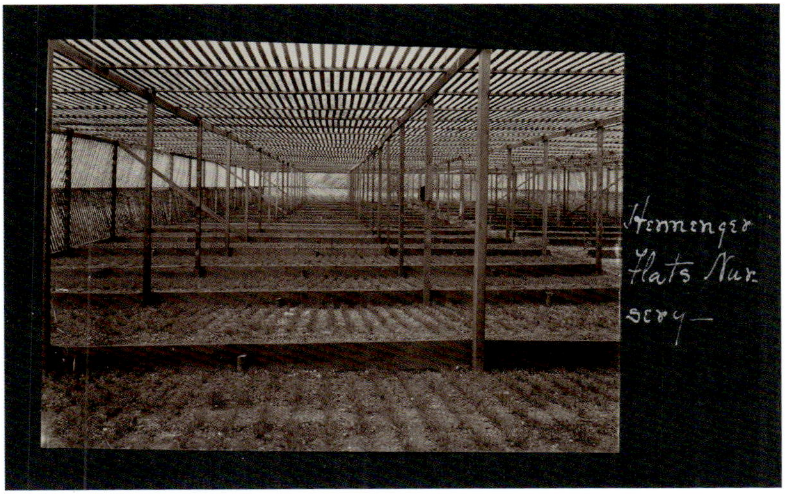

Fig. 9.

Theodore Parker Lukens, “Henninger Flats Nursery,” date unknown. Photographic print on paper, 27×18.5 cm. Courtesy of the Huntington Library, San Marino, Calif.

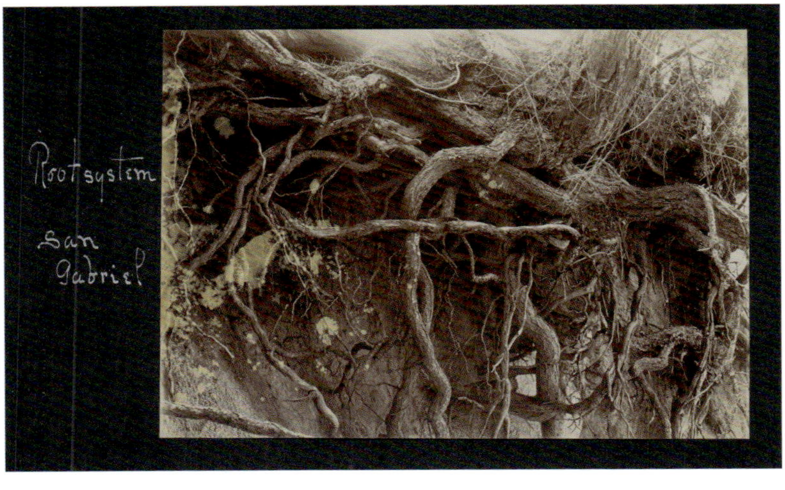

Fig. 10.

Theodore Parker Lukens, “Root System. San Gabriel,” date unknown. Photographic print on paper, 27×18.5 cm. Courtesy of the Huntington Library, San Marino, Calif.

Archival Collections

Theodore Parker Lukens papers, The Huntington Library, San Marino, California.

Bibliography

M. Kat Anderson, *Tending the Wild, Native American Knowledge and the Management of California's Natural Resources* (Berkeley: University of California Press, 2005).

Martin A. Berger, "Overexposed: Whiteness and the Landscape Photography of Carleton Watkins," *Oxford Art Journal* 26, 1 (2003): 1–23.

François Brunet, *The Birth of the Idea of Photography* (Cambridge, MA: MIT Press, 2019).

Robert W. Cermak, *Fire in the Forest: A History of Forest Fire Control in California, 1898–1956* (Washington, DC: USDA Forest Service, 2005).

Galen Clark, *Indians of the Yosemite and Vicinity: Their History, Customs and Traditions* (Yosemite Valley, CA: G. Clark, 1904), http://www.yosemite.ca.us/history/indians_of_the_yosemite/ (accessed April 20, 2025).

William Cronon, "*The Trouble with Wilderness; or, Getting Back to the Wrong Nature," in Uncommon Ground: Rethinking the Human Place in Nature, ed. William Cronon* (New York: W. W. Norton, 1995), 69–90.

Diana K. Davis, "Ecologies of the Colonial Present: Pathological Forestry from the *taux de boisement* to Civilized Plantations," *Environment and Planning E: Nature and Space* 1, no. 4 (2018): 447–469.

Diana K. Davis, "Narrating Nature: Decolonizing Socio-ecological Assemblages," in *Doing Political Ecology*, eds Gregory Simon and Kelly Kay (New York: Routledge, 2025), 90–103.

Kevin M. DeLuca and Anne T. Demo, "Imaging Nature: Watkins, Yosemite, and the Birth of Environmentalism," *Critical Studies in Media Communication* 17, no. 3 (2000): 241–260.

Stanford E. Demars, *The Tourist in Yosemite: 1855–1985* (Salt Lake City: University of Utah Press, 1991).

William M. Denevan, "The Pristine Myth: The Landscape of the Americas in 1492," *Annals of the Association of American Geographers* 82, no. 3 (1992): 369–385.

Douglas Deur and Rochelle Bloom, "Fire, Native Ecological Knowledge, and the Enduring Anthropogenic Landscapes of Yosemite Valley," in *The Routledge Handbook of Indigenous Environmental Knowledge*, eds Shonil A. Bhagwat and Thomas F. Thornton (New York: Routledge, 2020), 126–135.

Richard G. Everett, "Dendrochronology-based fire history of mixed-conifer forests in the San Jacinto Mountains, California," *Forest Ecology and Management* 256 (2008): 1805–1814.

Jared Farmer, *Trees in Paradise: A California History* (2013; Berkeley: Heyday Books, 2017).

Thomas W. Fels, *Carleton Watkins, Photographer: Yosemite and Mariposa Views from the Collection of the Park-McCullough House* (Williamstown, MA: Sterling and Francine Clark Art Institute, 1983).

Yves Figueiredo, "Inventing Yosemite Valley: National Parks and the Language of Preservation," *Historical Geography* 35 (2007): 12–37.

Yves Figueiredo, "Vers une approche environnementale de l'Ouest nord-américain: l'apport de John Wesley Powell," in *L'espace du nouveau monde, mythologies et ancrages territoriaux*, eds Nicolas Bernard, Gilles Chamerois, Arlette Gautier, Georges-Henri Laffont, Denis Martouzet (Rennes: Presses Universitaires de Rennes, 2013), 131–145.

Carolin Görgen, *The California Camera Club, Collective Visions in the Making of the American West* (Norman, OK: University of Oklahoma Press, 2025).

Santiago Gorostiza, "Liquid Power: An Interview with Erik Swyngedouw," *Undisciplined Environments* (2 July 2015), https://undisciplinedenvironments.org/2015/07/02/liquid-power-an-interview-with-erik-swyngedouw/ (accessed May 6, 2025).

Tyler Green, *Carleton Watkins: Making the West American* (Berkeley: University of California Press, 2018).

Hans Huth, "Yosemite: The Story of an Idea," *Sierra Club Bulletin* 33 (1948): 47–78.

Anne Farrar Hyde, "Temples and Playgrounds: The Sierra Club in the Wilderness, 1901–1922," *California History* 66, no. 3 (1987): 209–219.

Robin Kelsey, "Sierra Club Photography and the Exclusive Property of Vision," *RCC Perspectives* 1 (2013): 11–26.

Robert Aquinas McNally, *Cast Out of Eden: The Untold Story of John Muir, Indigenous Peoples, and the American Wilderness* (Lincoln: University of Nebraska Press, 2024).

Roderick F. Nash, *Wilderness and the American Mind* (1967; New Haven, CT: Yale University Press, 2001).

Max Oelschlaeger, *The Idea of Wilderness: From Prehistory to the Age of Ecology* (New Haven, CT: Yale University Press, 1991).

Karl H. Offen, "Historical Political Ecology: An Introduction," *Historical Geography* 32 (2004): 19–42.

Kate Nearpass Ogden, "Sublime Vistas and Scenic Backdrops: Nineteenth-Century Painters and Photographers at Yosemite," *California History* 69, no. 2 (1990): 134–153.

Frederick Law Olmsted, *Yosemite and the Mariposa Grove: A Preliminary Report* (1865; Yosemite National Park, CA: Yosemite Association, 1995).

Peter E. Palmquist, *Carleton E. Watkins: Photographer of the American West* (Albuquerque: University of New Mexico Press for the Amon Carter Museum, 1983).

Thomas Patin (ed.), *Observation Points: The Visual Poetics of National Parks* (Minneapolis: University of Minnesota Press, 2012).

Elizabeth Pomeroy, *John Muir, a Naturalist in Southern California* (Pasadena, CA: Many Moons Press, 2001).

Jordan Reznick, "The Warp and Weft of Landscape: Indigenous Ecological Knowledge and Colonial Photography in the Yosemite Valley," *California History* 102, no. 3 (2025): 34–72.

Alfred Runte, *Yosemite: The Embattled Wilderness* (Lincoln: University of Nebraska Press, 1990).

Alfred Runte, "Foreword to *Our National Parks*," in John Muir, *Our National Parks* (1901; San Francisco: Sierra Club Books, 1991), i-xiv.

Margaret Sanborn, *Yosemite: Its Discovery, Its Wonders and Its People* (New York: Random House, 1981).

Martha A. Sandweiss, *Print the Legend: Photography and the American West* (New Haven, CT: Yale University Press, 2002).

Shirley Sargent, *Theodore Parker Lukens, Father of Forestry* (Los Angeles: Dawson's Book Shop, 1969).

Simon Schama, *Landscape and Memory* (New York: A. A. Knopf, 1995).

Richard West Sellars, *Preserving Nature in the National Parks: A History* (New Haven, CT: Yale University Press, 1997).

Rebecca Solnit, *Savage Dreams: A Journey into the Hidden Wars of the American West* (San Francisco: Sierra Club Books, 1994).

Edward Wildon, "Views in the Yosemite Valley," *Philadelphia Photographer* 3, no. 28 (1866): 106–107.

CHAPTER 13
Black Gold: Petroleumscapes in Southern California Pictorialism and Beyond

Carolin Görgen

Sparkling bubbles gather on a thick, blackish surface. (fig. 1) Taking up almost two-thirds of the photographic print, they shape a foamy continent in a dark sea. Smaller islands drift off the mainland and glitter like pearls. If it were not for the shadow of a valve handle over the sparkle, the viewer would hardly suspect the industrial setting of the scene. Captured around 1930 in an oil ditch by Los Angeles–based Japanese American photographer T. K. Shindo, *Oil and Water* reflects the preciousness associated with both liquids. Also titled *California Oil*, the composition's bubbly geography vaguely echoes the Southern California coastline. Its gold toning rendered material the desirability of the Southland's "black gold" at a time when the region's budding film industry evoked glitter and fame—much like the lush substances in the abstract composition. As a member of a thriving camera-club network that expanded at the same time as the petroleum industry in the 1920s and 1930s, Shindo experienced firsthand the oil fever taking over the region—offshore drilling, beach contamination, and automobile frenzy included.

The blend of modern subject matter with ephemeral aesthetic pleasure, as seen in *Oil and Water*, was common in Japanese American photographer circles; yet Shindo's position as advertising director of the city's largest Japanese-language newspaper, *Rafu Shimpo*, also made him sensitive to the attractive aesthetics of the latest resource that Californians extracted from the earth. (Reed, 2016: 13–14, 22–27) In fact, the attraction to industrial scenes and everyday beauty was shared by a growing group of Southland Pictorialists, both Japanese and Euro-American. Many of these photographers continued the painterly style cultivated by fellow camera clubs on the West Coast since the late nineteenth century. (Martin and Bromberg, 2011; Görgen, 2025)

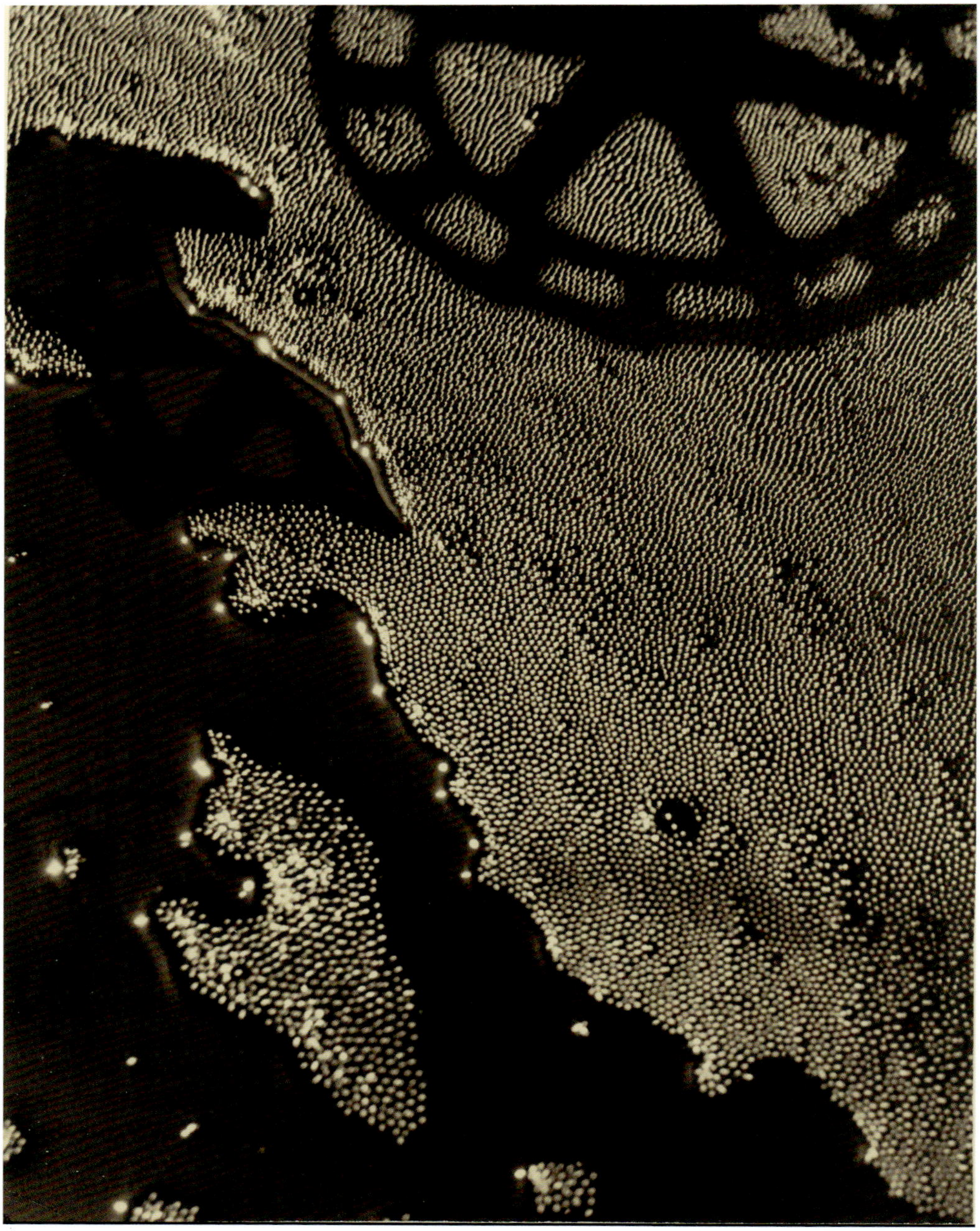

Fig. 1.

T. K. Shindo, *Oil and Water*, ca. 1930. Gelatin silver print, 34 × 27.2 cm. Los Angeles, J. Paul Getty Museum. © Estate of T.K. Shindo.

While industrial and abstract compositions such as Shindo's have been considered a departure from the classic landscape and genre scenes of Pictorialism, the petroleum-inspired photographs of the Southern California Pictorialists actually prolonged the deep roots of the state's photo networks in extractive infrastructures.

The organic shapes of oil, its industrial apparatus blending into the Southland, and its versatility for gold-toned prints turned oil into yet another resource to be exploited physically and aesthetically. After minerals, timber, and fruit, oil completed the iconography of abundance that regional photographers had developed since the nineteenth century. By the time Japanese American and Euro-American photographers embedded the aesthetics of oil production in their oeuvre, greater Los Angeles had essentially turned into a "petroleumscape"—a place where oil infrastructure and societal and cultural practices overlap and are enduringly shaped by petroleum, as Carola Hein conceptualized. (Hein, 2022: 15–18) Viewed through the petroleumscape, Southern California Pictorialist photographs of the 1920s and 1930s emerge as typically Californian visions of abundance, metaphorically and materially drenched in oil. They partake in the centuries-old fantasies of plenty associated with western American landscapes, lavishly depicted in photographs from the nineteenth century. Rather than constituting a break from previous aesthetics because of their industrial nature, I argue, the prints produced by Los Angeles–based photographers at the height of the oil boom are a smooth continuation of the "extractive gaze" (Gómez-Barris, 2017; Angus, 2024) that had long characterized western American photography. This way of seeing normalized seemingly infinite resource access as the promise of modernity, offered by mineral resources; coal; and, now, oil. Looking at oil and its regional ecosystems, at material promises and environmental conflicts, this chapter unearths and renders visible the ecologies of oil entangled in Southern California Pictorialism and explores alternative visions of the fossil fuel.

Oil Promises

Petroleum penetrated many layers of Southern Californian life by the 1920s, a reality perhaps best embodied in C. C. Pierce's panoramic view of a Long Beach residential neighborhood. (fig. 2) In this image, the orderly arrangement of single-family homes is framed by hundreds of oil derricks stretching across the horizon, making oil production the dominant field of vision just a year after its discovery on Signal Hill. Although the unpaved roads appear rudimentary, the infrastructure on the horizon and power lines connecting the homes promised efficient energy, all entwined with Southern California flora. A few automobiles parked in

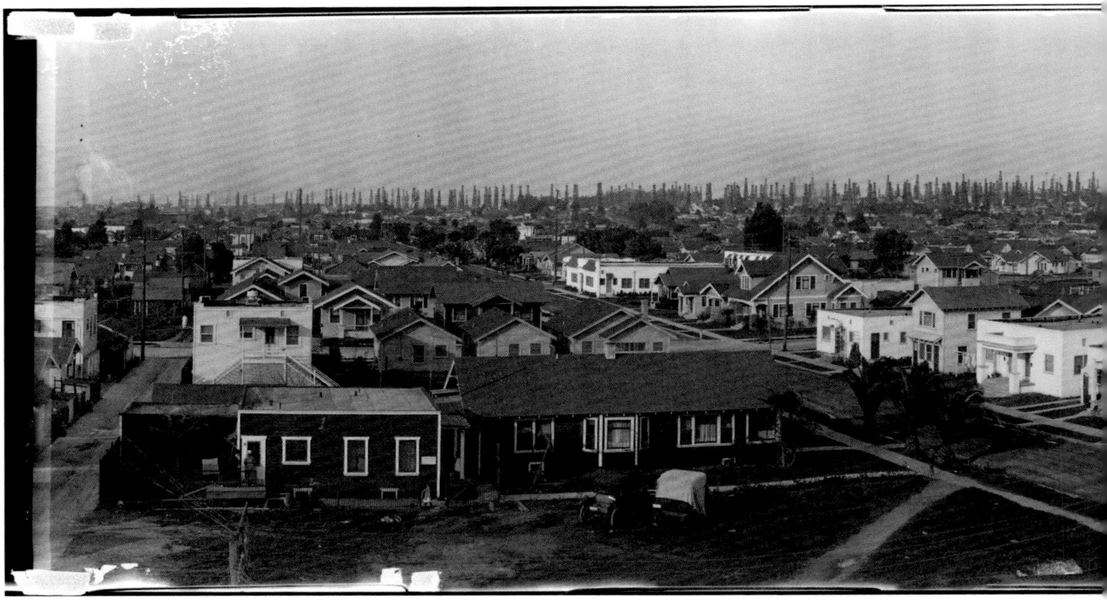

the foreground complete the scene, linking the petroleum infrastructure on the horizon to the fuel supply needed for the latest consumer gadgets. This neighborhood, which would become the city of Signal Hill by 1924, emerged from an unlikely alliance of oil companies and residents. The direct access to energy suggested in Pierce's view demonstrates how oil became both "domesticate[d]" and "disorientingly natural," blending into lands cleared for Euro-American settlement and amplified by the large format of the panorama. (Angus, 2024: 41–44) The view foreshadowed a culture of affluence sparked by oil discoveries, which eventually made oil an "inconspicuous" yet ubiquitous element of consumption. (Black, 2012: 43–49) It is exactly this vision of Californian abundance that fascinated camera enthusiasts such as Pierce, who had been a fixture in the city's camera-club scene from the early 1900s.

The Signal Hill oil field, then the world's richest per acre, was among the first in a series of oil discoveries, including Venice Beach in 1929 and Wilmington in 1932, that transformed the city and the wider region. The 1932 oil find notably reshaped San Pedro Bay, home to the port of Los Angeles and a key geostrategic location after the completion of the Panama Canal. (Dunbar-Hester, 2023: 11–13) In the same decade, Los Angeles experienced a demographic boom during which its population doubled, reaching 1.2 million by 1930. Fueled not only by petroleum discoveries but also by the growing real estate and motion picture industries, the city lost its bucolic reputation, as historian Nancy Quam-Wickham has shown. Inhabitants instead embraced the

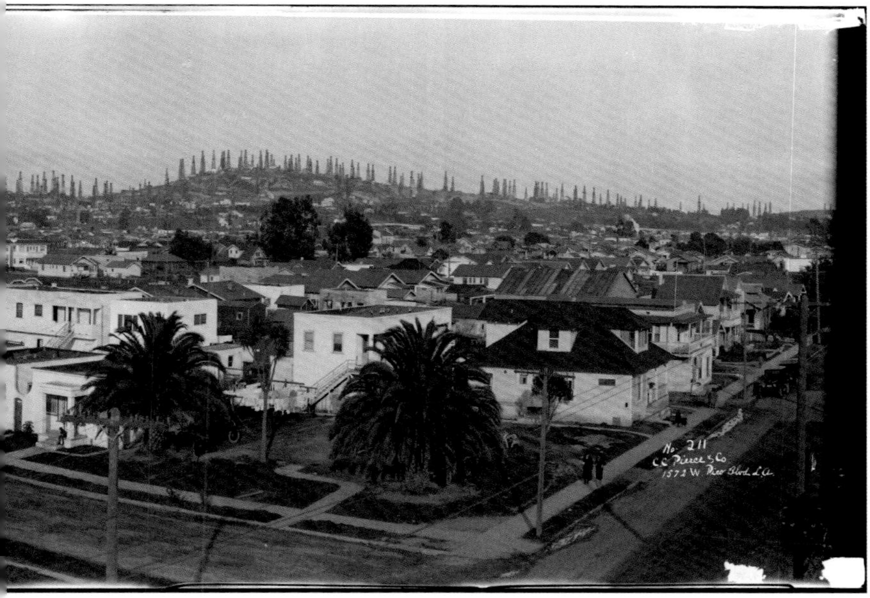

Fig. 2.

C. C. Pierce, *Long Beach Residential and Signal Hill Oil Field, Facing North and East*, 1922. Negative, 26 × 94 cm. San Marino, Calif., Huntington Library, Art Museum, and Botanical Gardens. photCL 470 (164).

many manufacturing industries—steel, railroad, trucking, lumber—that catered to petroleum's expansive production chain and that emerged alongside new, oil-dependent sectors, such as aviation. Within a few years, California oil made up one-fifth of global production and turned a once-remote corner of the American West into a top global exporter. (Quam-Wickham, 2001: 125–129) Next to the population boom and the increase in exploitative industries that resembled nineteenth-century mining frenzies, the 1920s petroleum discoveries also generated a new understanding of land ownership and resource access. (LeMenager, 2014: 15) At Signal Hill, Venice Beach, and elsewhere, tracts of land were sold alongside mineral rights. Following the "rule of capture," residents could extract oil on their properties, which incentivized rapid drilling and, in the process, multiplied extraction sites across residential neighborhoods. (Elkind, 2012: 82–85)

The quick access to petroleum that transformed the cityscape also had an undeniable cultural impact. Readily available energy derived from oil was a precondition for twentieth-century US mass culture, as environmental studies scholar Stephanie LeMenager has demonstrated. Oil furnished a new "energy infrastructure" that not only physically fueled but also materially shaped the media of modernity, including films, cars, and museums. (LeMenager, 2014: 5–6, 11) The growth of Los Angeles as an automobile city and home of the motion picture industry directly attested to the physical and material impacts of petroleum. Photography was also an oil-derived medium, as cameras—including the body and the roll film they used—relied on the industry. (Angus, 2024: 41) The very resurgence of a Southern California camera-club network during the oil boom (after short-lived efforts around 1900) testifies to the intricate links between energy systems and art making. As the region's economies centered around oil and its adjacent industries, the changing environment lured many a local photographer toward an "aesthetics of petroleum" (LeMenager, 2014: 67), whether abstracted as raw material, as in Shindo's composition, or depicted as energy source, as in Pierce's Signal Hill panorama.

The complexity and reach of the Southern California photo network become tangible in Florence Kemmler's *Oil Drums*. (fig. 3) Created at the height of the oil boom, the print echoes the shimmering aesthetics seen earlier, while engaging more directly with petroleum as a packaged commodity—depicting fifty identical barrels awaiting shipment. As a member of the Camera Enthusiasts of San Diego, Kemmler exhibited her work not only across Southern California, including at the Los Angeles Camera Club (founded 1915) and the Camera Pictorialists of Los Angeles (founded 1914), but also far beyond the region. Labels on the back of the print document its exhibition at camera clubs in Portland, Seattle, Omaha, Cleveland, Newark, and Providence, illustrating a much wider trajectory. While the iconography of the Los Angeles petroleumscape traveled across the country, it left its most lasting impact on Southern California, where both the variety of Euro-American clubs and the Japanese Camera Pictorialists of California (founded 1926) embraced it. Direct exchanges between both groups were largely one-sided, however, as Dennis Reed has demonstrated: Camera Pictorialists such as Edward Weston exhibited with the Little Tokyo photo community, yet Japanese photographers were never returned the favor. (Reed, 2016: 40–49) Still, elements of Japanese Pictorialism found their way into Euro-American practice. This is evident in *Oil Drums*, particularly in the absence of a horizon line and in the contrast between the barrels' shadows and the sea's glittery surface. Kemmler, like Weston and others, admired the Japanese Pictorialists. Their members, notably Shigemi Uyeda, had photographed petroleum landscapes since the early 1920s and circulated prints thereof across Europe and as far away as the Soviet Union. (Reed, 2016: 35–36, 51)

Visions of petroleum emerged in the dark glitter of oil ditches, in the forest of derricks populating the horizon, and in neatly packaged barrels—extending a distinctly Californian photographic tradition that framed the land through its extractive potential in a romantic vision of plenitude. Gold Rush–era photographs had pictured minerals as readily available, their surfaces integrating the very minerals used in photographic processes. (Kelsey, 2018: 394) By the turn of the century, Kodakers captured Californian sequoias as majestic yet imminently

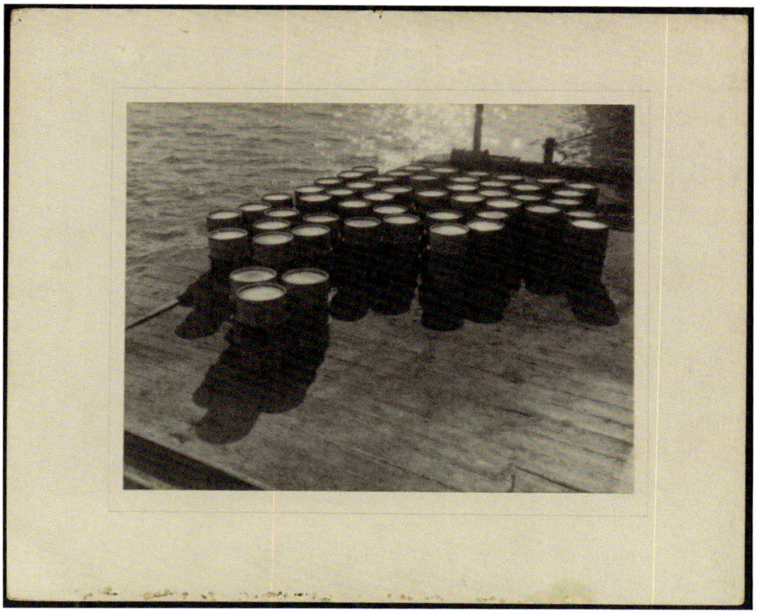

Fig. 3.
Florence B. Kemmler, *Oil Drums*, ca. 1929. Gelatin silver print, 26.3×35 cm. Los Angeles, J. Paul Getty Museum. © Estate of Florence B. Kemmler.

consumable, easily reduced to paper in little time. (Görgen, 2025: 173–175)[1] Pictorialists and Modernists continued this trend, notably Edward Weston, who framed the fruits and vegetables of Californian and Mexican orchards as aesthetically pleasant commodities, veiling any traces of the violent labor struggle of the 1920s and 1930s in the process. (Macfarlane, 2025) Building on this visual tradition, oil joined the long-standing iconography of abundance. Its portrayal as black gold conjured up memories of the 1850s gold rush; yet oil distinguished itself from gold through its supposed "everything-ness," that is, its inconspicuous yet extremely widespread presence as energy, commodity, and raw material—an omnipresence spurred on by the rule of capture, which encouraged rapid consumption. (LeMenager, 2014: 11) These evolving iconographies were persistently accompanied by debates about who had the right to access the wealth of Californian lands, often entangled with attempts to erase the stewardship and labor of Indigenous Californians and immigrants. (Klein, 2020: 41) Yet such conflicts were not visible in the smooth surfaces and warm tones of Pictorialist prints.

Unlike the framing of noble metals, timber, and fruit, the framing of oil as an abundant and easily accessible resource was reinforced by its common perception as being less labor-intensive. Oil extraction appeared to rely more on automated pumping than on visible physical work, unlike was the case for coal. This illusion of effortlessness was mirrored in photography, a medium that promised abundant images through the simple press of a button, thus perfectly matching an industry that yielded massive energy through automated extraction, as art historian Siobhan Angus has contended. (Angus, 2024: 7, 41–42) The labor involved in petroleum extraction remained largely invisible, subject to a "grid" infrastructure that streamlined production and hence seemed less susceptible to workers' demands. (Mitchell, 2011: 36–39) When workers did appear in photographs—such as in Florence Kemmler's 1930s series of oil tanks—they were overshadowed by the scale of production, their backs turned to the camera, and their attire blending into the machinery. While such depictions were a far cry from the spectacular news imagery of "industrial blackface" that showed workers blackened by petroleum, they had a similar dehumanizing effect that erased facial and other physical features, thus minimizing the dangers and physical toll of oil labor. (LeMenager, 2014: 93–96) A similar visual pattern can be seen in nineteenth-century mining photography, where the fraught labor relations that marked large-scale extraction did not fit the photographic frame. (Angus, 2024: 80–85) Not unlike other extractive industries, oil production was also shaped by racialized labor exclusion. In Los Angeles County's "Black Gold suburbs" of the 1910s, a racially restrictive climate maintained a white monopoly on oil production chains, notably excluding non-white workers from union membership. This segregation mirrored the broader racial divisions in California's early oil industry, where access to labor opportunity and visibility—both in economic participation and in photographs—was systematically restricted. (Viehe, 1981; Kurashige, 2007: 43)

Oil Conflicts

The promises of oil and abundance unraveled as petroleum infrastructure encroached upon Southern California's beaches. Hisao Kimura's 1930 rendering of gentle currents on a sandy shore initially evokes the golden light associated with the California coast and the Pictorialist search for fleeting beauty. (fig. 4) The reflection of an oil derrick in a puddle of water at the top only minimally disturbs the scene, appearing small and slightly out of focus in comparison to the glimmering beach. Yet despite its subdued presence, a slight upward tilt of the camera would have revealed the oil infrastructure that dominated the coastal landscape. While beaches had become a favorite destination of leisure-seeking Americans in the 1920s, the Los Angeles waterfront was also a site of friction, where real estate owners with ocean-facing homes clashed with oil companies eager to extract the resources beneath them. (Devienne, 2024: 69–101) By the time Kimura pointed his camera toward the shore in 1930, Los Angeles was embroiled in a battle over offshore drilling, outdoor enjoyment, and beach contamination. As

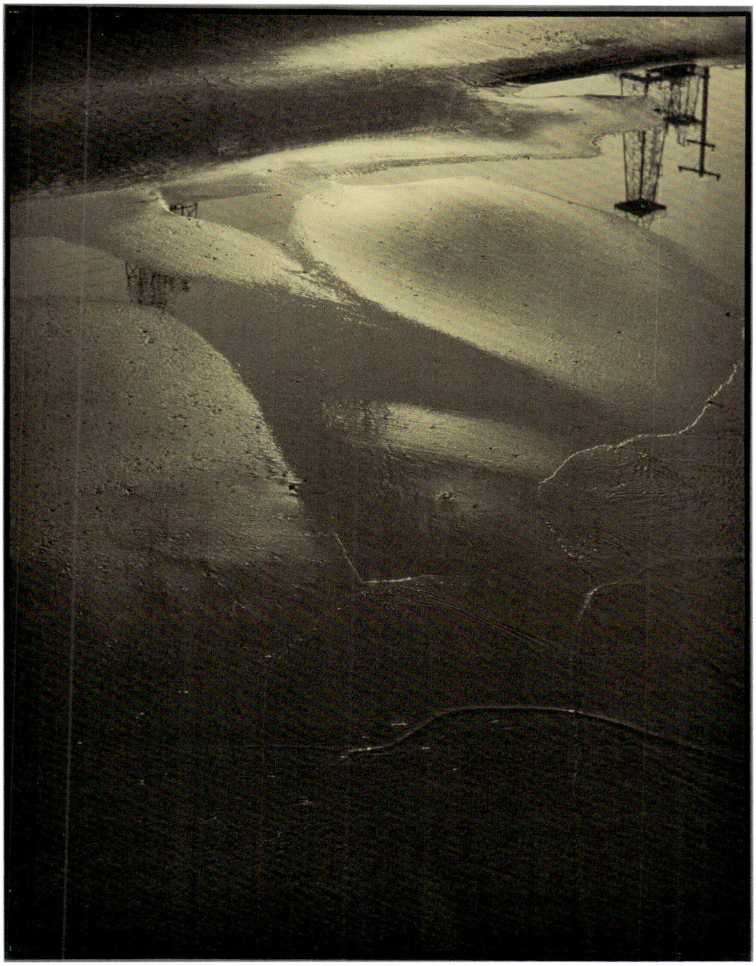

Fig. 4.
Hisao Kimura, *Oil Reflection*, ca. 1930. Gelatin silver print, 34.6 × 27.2 cm. Los Angeles, J. Paul Getty Museum.

Michael Dawson pointedly remarked, it seems "to have been a singular Southern California achievement to extract art from oil pollution," in a nod to the Los Angeles Pictorialists' attraction to the petroleumscape. (Dawson, 2005: 211) While Kimura's print only subtly hints at the disruptive presence of derricks in the sand, the virulent conflicts between petroleum wealth and landscape protection commenced on the beach.

As coastal petroleum—accessed through offshore drilling—was increasingly in demand in the late 1920s, Southern Californians were faced with the question of whether their shores were "a site for the extraction of raw materials, commercial harbors, and shipping or a serene place of relaxation, recreation, and play?" (Sabin, 2005: 97) Even though oil prices had dropped because of fierce competition during much of the 1920s and California was hardly in need of new petroleum resources, the effects of the Great Depression created a new dependency on revenues from the profitable oil fields, among others in Huntington Beach, as historian Paul Sabin has traced. As coastal petroleum became sought after, the means of extraction and their environmental consequences were debated, notably the construction of diagonal shafts that furnished access through slanted drilling. Though banned by the mid-1930s, slanted drilling had a significant impact on the Venice Del Rey oil field, where petroleum had been discovered in late 1929. (Sabin, 2005: 100–113) The process considerably increased the risk of leaks and fires, yet public concern over its methods stemmed less from environmental dangers than from economic motives. The local real estate market had advertised ocean homes as ideally situated, away from pollution—but environmental degradation now reached the coast and thus led to setbacks in the growing real estate and tourism industries. (Elkind, 2012: 86–87) At the height of the debate, the *Los Angeles Times* lamented that beachgoers should emerge from the ocean "glistening with drops of pure salt water instead of having their bodies smeared with oil," reflecting the era's emphasis on an attractive beach culture (quoted in Sabin, 2005: 104). Notably, the concern was with visible human consequences, rather than health effects or fragile ecosystems comprised of beachside flora and fauna. (see Devienne, 2024)

A glimpse at the infrastructure makes the density of the industrialized coast tangible. (fig. 5) By 1930, C. C. Pierce had extensively photographed the Venice Del Ray oilfield, oftentimes from a slightly elevated perspective, as in this case, thus bringing to the fore the many layers of drilling operations. Here, the central oil derrick shoots toward the sky like a cathedral spire, heightened in size by clouds of steam at its foot. Below, the streets, with power lines, cars, and sheds, suggest bustling activity. Except for the men casually seated on a porch below the derrick, human beings are almost absent—emphasizing once again the seeming effortlessness of petroleum access. The cloud of steam that originates from a pipe on the beach obstructs a broader view of the shore, leaving what is left of the beach to the imagination. Though the dreamy beach scenes of early Pictorialism had faded, Pierce still employed its characteristic elements—not surprising for a professional photographer who had frequented the region's Pictorialist clubs since the turn of the century. He specifically mobilized the plumes of smoke in the foreground to soothe and soften the industrial scene. While the photographer muted the confrontation with coastal drilling, the actual soundscape of this scene would have been deafening. Venice residents complained about the "clattering drills" and "incessant roar of diesel pumps" that now enveloped the coast. (Elkind, 2012: 86) The noises generated by engines, drills, whistles, and hammers could be heard around the clock in the city's working-class neighborhoods, where resident families faced the additional dangers of burning or drowning. (Quam-Wickham, 2001:

132–135) Laborers and residents of oil fields would have looked right through Pierce's thin veil of steam and likely linked it to the emerging patterns of "oil weather," that is, smog, that took over Southern California by the 1940s—a direct consequence of the cheap oil put into gas tanks. (LeMenager, 2014: 78–79)

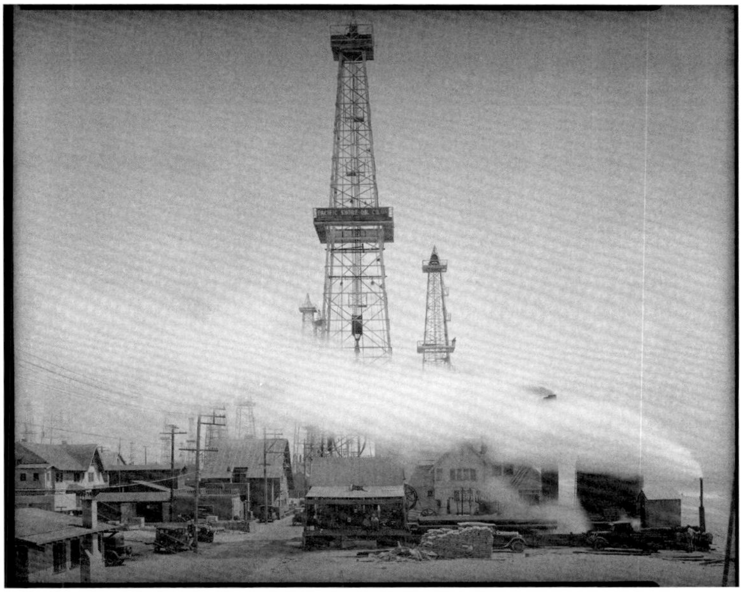

Fig. 5.
C. C. Pierce, [Oil Derricks among Small Houses Probably at the Venice Oil Field, Los Angeles], ca. 1930. Negative, variable dimensions. C. C. Pierce Photographic Collection, 1898–1934, Special Collections, University of California, Los Angeles.

The sandbags stacked in the foreground of Pierce's composition also hint at the fortification and protection necessary in case of leaks and fires. Los Angeles faced an exceptional fire risk, which was reinforced by high gas pressure in the region's deposits, as well as by rudimentary construction in densely populated fields, including the one at Venice. (Elkind, 2012: 84) News photographs of the time—such as those showing firefighters attempting to lay hoses against a sky set ablaze (fig. 6)—served as stark reminders of how quickly the seeming tranquility of residential neighborhoods could be upended. (fig. 2) The sheer scale of the fires oftentimes hindered attempts to contain them. The 1928 Santa Fe Springs fire pictured in fig. 6 is a case in point, as it burned for almost two months, rendering futile the somewhat helpless efforts of the four firefighters in the foreground. (Harrison, 2017) Such photographs also contain evidence of the slower, long-term damage wrought by extraction. Consider, for example, the house in the foreground: while the blast may have destabilized its structure, its slanted appearance also evokes homes shaken by an earthquake—an apt comparison, given that earthquakes were a frequent consequence of oil drilling, as in the case of the 1933 Long Beach earthquake. (Dunbar-Hester, 2023: 18)

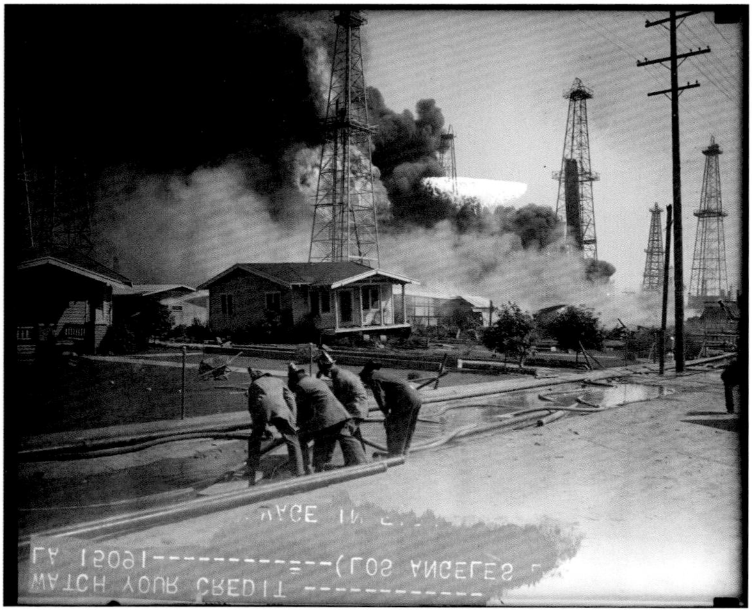

Fig. 6.

Unknown, [Firefighters Lay Hoses to Protect Homes from an Oil Well Fire in Santa Fe Springs], 1928. Negative, 10×12.5 cm. Los Angeles Times Collection, Special Collections, University of California, Los Angeles.

The dark, explosive nature of fires also resembled the gushers that relentlessly spewed oil toward the sky and blackened surroundings. At the time, oil companies went as far as leaving gushers open "for advertising purposes," thus unleashing thousands of barrels of petroleum into the soil and multiplying environmental risks. (Elkind, 2012: 84) Both phenomena—oil blasts and gushers—lent themselves to spectacular depictions that emphasized "the aesthetic properties of oil" as a boundless resource. (LeMenager, 2014: 93) Such large-scale events captivated the public, not least the 1910 spill at the Lakeview Gusher in Kern County, north of Los Angeles, which lasted for eighteen months and attracted hundreds of spectators eager to witness the energetic spewing of black gold. (Morse, 2012: 125–126) Spectacles like these may also be understood through the lens of "petro-masculinity," a concept articulated by Cara Daggett. From the spewing gusher to the drilling beneath the earth and the muscular bodies of workers, the physical aspects of oil infrastructure have long echoed sexual desires, as Stephanie LeMenager suggested with regard to Upton Sinclair's novel *Oil!* (1927) in which the 13-year-old narrator experiences the sight of an oil gusher as a kind of "arousal and premature ejaculation." (LeMenager, 2014: 93) But Daggett takes this further, arguing that petro-masculinity is also linked to a nostalgia for fossil fuels and a rejection of environmentalist protection, which is rejected as supposedly 'effeminate.' She posits that there are connections between oil nostalgia, climate change denial, and the desire for authoritarianism—a development that we have recently seen in the Trump administration's obsession with drilling. (Daggett, 2018: 25–44) Historically, the visual coverage of open gushers transformed real environmental destruction into artificial spectacle and obscured the aftermath of oil-soaked streets, soils, and waterways. (Quam-Wickham, 1998: 192–193). Such spectacular coverage thus suppressed meaningful engagement with ecological consequences—a phenomenon that may still be observed in news channels that rally around wildfire coverage. (Davis, 1998: 95–147; LeMenager, 2014: 93–97, 101)

The spectacle of the petroleumscape can, in no small part, be traced back to the emergence of Hollywood and its material entanglements with oil. Southern California's artistic photographic scene developed hand in hand with the motion picture industry, and many photographers smoothly transitioned between still and moving images, making theatrical compositions fashionable. (Reed, 1994: 71–73) Will Connell's study of curled film strips shows the entwined nature of still and moving images in the early twentieth century, both generated thanks to petroleum. (fig. 7) More than a mere technological connection, the extraction of petroleum literally fueled the growth of photo and film culture. Connell's additional inscription "in *Pictures*" on the back of the print provides a sly critique, as *Pictures* was his 1937 satirical photo book that mocked the Hollywood industry and its addiction to artifice and fame. Connell would have known about the growing number of stars partaking in the "oil well craze," acquiring derricks named after themselves and playing in studios whose entire production chain depended on oil. (Jacobson, 2025: 65–70) Albeit less extravagant than their film counterparts, photographers became just as entangled with the petroleumscape, from plastic film supports to the gasoline that powered their automobile excursions—including to the nearby beaches accessible via the recently opened Pacific Coast Highway. Think, for example, of Edward Weston's *Seeing California* series, which was sponsored by the Southern California automobile club. (Dawson, 2005: 215–216) Such collaborations not only reveal the direct link between artistic scenery and oil dependency, but also show how "the acceleration in carbon emissions and the turn away from the collective"—exemplified in automobile usage—literally fueled avant-gardist ideas of progress in the early twentieth century. (Ghosh, 2016: 79–80) The ecological footprint of Hollywood—ranking second in pollution in California, just behind the oil industry—underscores the inextricable connections between the energy consumption of art production and the environmental impact of oil. (Corbett and Turco, 2006; Castro, 2025)

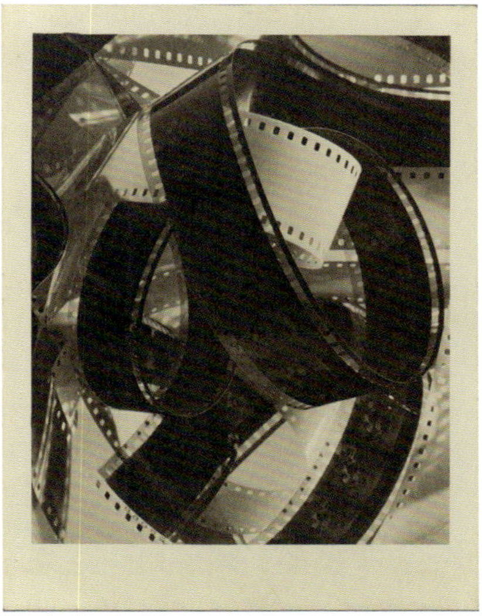

Fig. 7.
Will Connell, *Film*, ca. 1931. Gelatin silver print, 29.6 × 24.3 cm. Los Angeles, J. Paul Getty Museum. © Will Connell.

Hollywood's oil-driven film economy, which attracted many photographers, was tightly bound to what Scott Kurashige termed a "new economy of racism," and both of these intensified during World War II. The industry's rise—boosted by D. W. Griffith's movie *Birth of a Nation* (1915)—relied on circulating white-supremacist ideas, and by the 1920s, the Ku Klux Klan's resurgence, along with racially restrictive housing policies, made areas such as Hollywood inaccessible to many minorities, including Japanese American and Black residents. (Kurashige, 2007: 21–24) Japanese American photographers nonetheless built a close community in Little Tokyo's vibrant camera-club scene, but this was cut short after the Japanese attack on Pearl Harbor, in 1941. Under Executive Order 9066, some 120,000 Japanese Americans were incarcerated, devastating Los Angeles's photo community, as Reed has demonstrated in decades-long research. (Reed, 2016: 53–57) As the war escalated, so did Los Angeles's racial and economic structures. The oil industry—criticized in the 1930s for unchecked drilling—was revived under wartime patriotism, and petroleum became vital to the war effort, muting concerns over extraction. (Elkind, 2012: 88–90) Thus, World War II reinforced both the oil economy and the racial hierarchy, reshaping the Southland's industrial, social, and artistic landscape.

Oil Re-visions

As oil has come to dominate all aspects of modern life, historians have attempted to trace our dependence on the fossil fuel, highlighting the intertwined economic, political, and cultural stakes embedded in its consumption. (Sabin, 2004; LeMenager, 2014) As Southern California Pictorialists pointed their cameras toward the petroleumscape, their

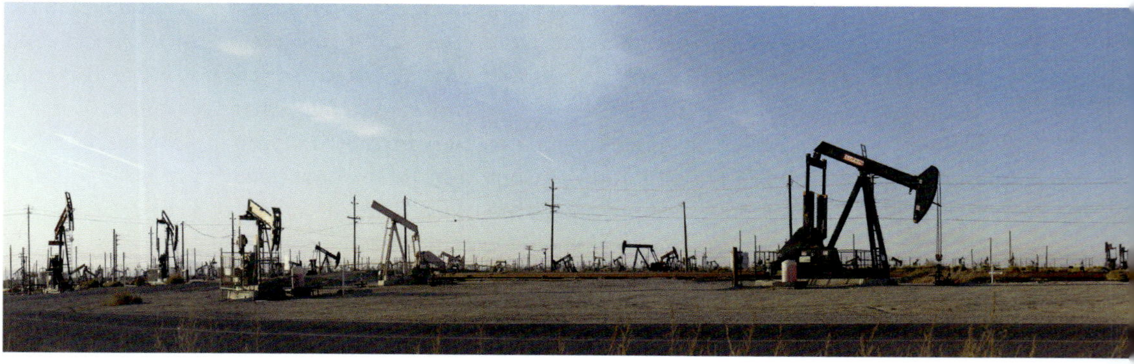

works—intentionally or not—benefited from oil and participated in the extractive fantasies fueled by oil as a commodity, motif, and creative matter. Yet the ubiquity of oil may also be envisioned from an alternative perspective—one that diverts our gaze from "settler ecologies" and the relentless calculations of resource extraction. (Dunbar-Hester, 2023: 21) Consider, once again, Shindo's *California Oil* (see fig. 1): While the gold-toned print evokes the wealth tied to oil as raw material, it may also invite us to engage with the substance itself. Geologically, petroleum has existed for millennia in the fossilized remains of plants and animals. Slumbering within the warm layers of the earth under immense pressure, the fossils became carbonized. Uses of petroleum date back to early human history and have not always been viewed as merely a financial gain; in the Los Angeles basin, the Gabrielino and Tongva applied it as sealing for baskets and canoes. The historical allure of oil was studied in the early twentieth century—including by companies such as Union and Salt Lake Oil. Union Oil geologist William Orcutt and University of California paleontologist John C. Merriam examined the fossil deposits at the La Brea Tar Pits in Los Angeles in the early 1900s as a site of geological interest, rather than an investment opportunity. As Métis scholar Zoe Todd reminds us, the matter itself is not inherently detrimental to our ecosystems. (Dunbar-Hester, 2023: 48–49) Rather, the extraction, processing, and industrial uses of oil turn it into a destructive, even "weaponized," substance:

> It is not the material drawn from deep in the earth that is violent. It is the machinations of human political-ideological entanglements that deem it appropriate to carry this oil through pipelines running along vital waterways, that make this oily progeny a weapon against fish, humans, water and more-than-human worlds. (Todd, 2017: 99)

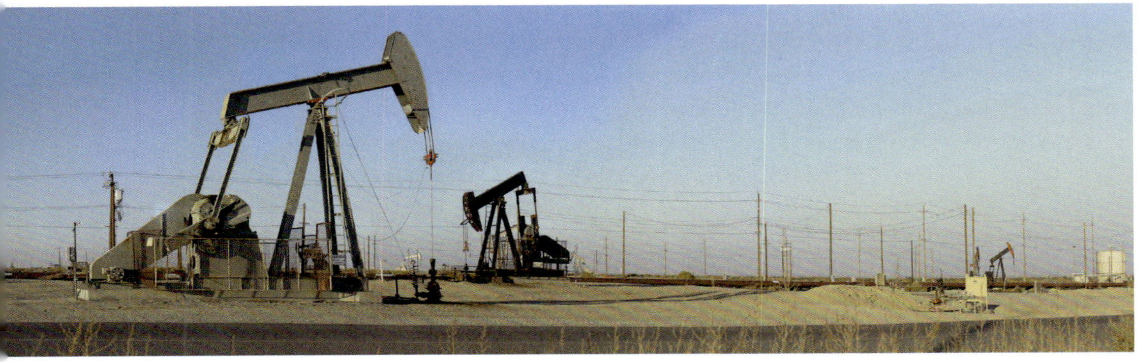

Fig. 8.
Lewis deSoto, *POWER, Lost Hills, California (Yokuts), 10.19.16*, 2016. Surecolor Epson inks on Epson archival semi-matte paper, 50.8 × 345.44 cm. © Lewis deSoto.
Academic Native and site research by Vivian Ngyuen, 2025.

Thinking of oil as geological "progeny," intimately linked to human evolution, allows us to reconsider our connection to the raw material rather than the extracted commodity. This way, Shindo's exploration of oil's material qualities and rich texture link us back to our fossilized origins, resuscitating its natural—rather than its economic—potential. Such deep connections highlight "the implication, or inherent interconnection" not only between art makers and their materials, but also between human and non-human beings and the earth, as Alan Braddock suggested. (Braddock, 2023: 2)

Undoubtedly, our historical kinship with oil is largely veiled by the legacy of its extractive regime, notably in Los Angeles County. Given the region's roughly twenty-thousand oil and gas wells—some active, some idle—oil infrastructure near schools and parks appears less like an ally and more like an environmental threat. It has led to soaring asthma rates, especially across Black and Hispanic communities, who tend to live closer to active wells. (Johnston and Shamasunder, 2021) The oil infrastructure dotting the Southland also reminds us of our continued dependency on petroleum, as consumption has only increased across California since the 2010s. (Rapier, 2019) The tentacles of the industry reach all the way into this essay, as half of the reproduced artworks are drawn from the Getty Museum, an institution built on the wealth of oil tycoon J. Paul Getty.

Yet, even the continued threat posed by idle and active oil infrastructure can be seen from a new angle. In his *Power Series* (fig. 8), Cahuilla artist Lewis deSoto uses panoramic photographs of rigs, gas pipelines, and other infrastructure. Shifting our understanding of power as spiritual, DeSoto invites viewers to engage with "the life energy of all things." Echoing the historical panoramas of oil infrastructure (see fig. 2), deSoto encourages a look beyond the seemingly endless extraction machinery, detaching the energy beneath the earth from its monetary value (deSoto, n.d.).

From this perspective, the notion of abundance itself may be reconsidered, as Cara Romero (Chemehuevi) alludes to in *Oil Boom*. (fig. 9) Bare-chested dancer Cannupa Hanska Luger (Three Affiliated Tribes of the Fort Berthold Reservation) floats beneath a landscape dotted with derricks, not unlike the supposed landscapes of plenitude we saw earlier. Luger, a leading activist during the Dakota Access Pipeline protests at the Standing Rock Reservation in 2016, embodies Indigenous resistance to extractive infrastructures and the defense of sovereign lands. (Fox, 2021: 81) Here, his body reflects the golden shimmer of the California sun and the earth beneath, while his dark dress evokes drops of petroleum, or black gold. The suspended position of the floating performer—tied up in between the surface and the underground—makes our connection to the earth, to the fossils in ourselves, remarkably clear. What the Pictorialists saw as a petroleumscape of abundance may thus be reimagined as a survival dance of energy.

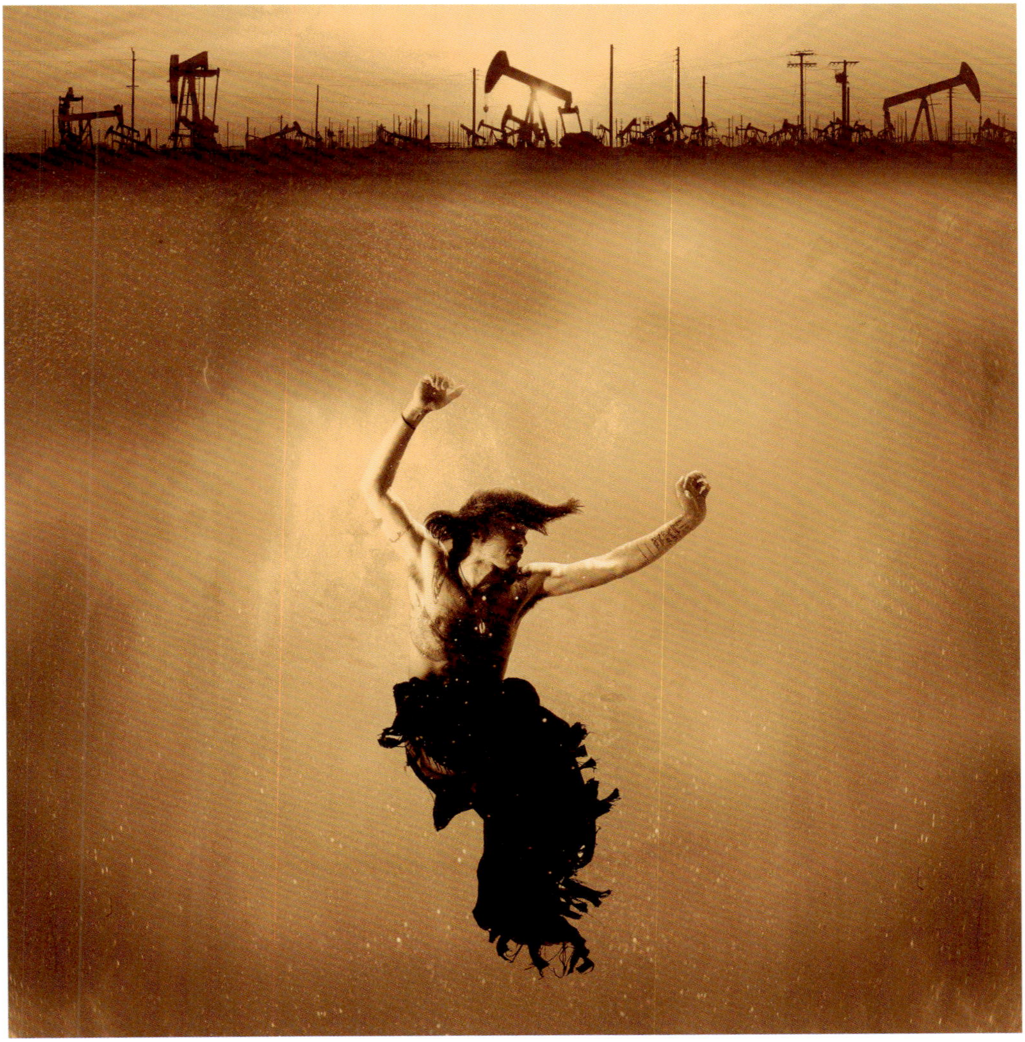

Fig. 9.

Cara Romero, *Oil Boom*, 2015. Archival pigment print, 102 × 100.6 cm. Hood Museum of Art, Dartmouth. Purchased through the Sondra and Charles Gilman, Jr. Foundation Fund. © Cara Romero.

Notes

1. On minerals and paper, see essays by Alice Cazenave and Katherine Mintie in this volume.

Bibliography

Siobhan Angus, *Camera Geologica: An Elemental History of Photography* (Durham, NC: Duke University Press, 2024).

Brian C. Black, "Oil for Living: Petroleum and American Conspicuous Consumption," *Journal of American History* 99, no. 1 (2012): 40–50.

Alan C. Braddock, *Implication: An Ecocritical Dictionary for Art History* (New Haven, CT: Yale University Press, 2023).

Teresa Castro, "Renouveler les récits: le cinéma au défi de l'environnement et de l'écologie," *Écrire l'histoire* 25 (2025). https://doi.org/10.4000/14mom (accessed April 28, 2026).

Charles J. Corbett and Richard P. Turco, "Sustainability in the Motion Picture Industry" (Los Angeles: University of California Los Angeles, Institute of the Environment, 2006), https://www.ioes.ucla.edu/wp-content/uploads/2016/10/mpisreport.pdf (accessed March 11, 2025).

Cara Daggett, "Petro-masculinity: Fossil Fuels and Authoritarian Desire," *Millennium: Journal of International Studies* 47, no. 1 (2018): 25–44.

Mike Davis, *Ecology of Fear: Los Angeles and the Imagination of Disaster* (New York: Verso, 1998).

Michael Dawson, "Transitions in Southern California Landscape Photography: 1900–1940," in *Land of Sunshine: An Environmental History of Metropolitan Los Angeles*, eds William Deverell and Greg Hise (Pittsburgh: University of Pittsburgh Press, 2005), 204–219.

Lewis DeSoto, "Power (Panoramas) 2012-Present," n.d., https://lewisdesoto.net/wallworks/power_panoramas/index.html (accessed February 10, 2025).

Elsa Devienne, *Sand Rush: The Revival of the Beach in Twentieth-Century Los Angeles* (New York: Oxford University Press, 2024).

Christina Dunbar-Hester, *Oil Beach: How Toxic Infrastructure Threatens Life in the Ports of Los Angeles and Beyond* (Chicago: University of Chicago Press, 2023).

Sarah S. Elkind, "Oil in the City: The Fall and Rise of Oil Drilling in Los Angeles," *Journal of American History* 99, no. 1 (2012): 82–90.

Amitav Ghosh, *The Great Derangement: Climate Change and the Unthinkable* (Chicago: University of Chicago Press, 2016).

Macarena Gómez-Barris, *The Extractive Zone: Social Ecologies and Decolonial Perspectives* (Durham, NC: Duke University Press, 2017).

Carolin Görgen, *The California Camera Club: Collective Visions in the Making of the American West* (Norman, OK: University of Oklahoma Press, 2025).

William L. Fox, "Overburden," in *Landscapes of Extraction: The Art of Mining in the American West*, ed. Betsy Fahlman (Munich: Hirmer Publishers, 2021), 71–81.

Scott Harrison, "From the Archives: Santa Fe Springs Oil Well Fires," *Los Angeles Times*, April 25, 2017.

Carola Hein, *Oil Spaces: Exploring the Global Petroleumscape* (New York: Routledge, 2022).

Brian Jacobson, *The Cinema of Extractions: Film Materials and Their Forms* (New York: Columbia University Press, 2025).

Jill Johnston and Bhavna Shamasunder, "Los Angeles' Long, Troubled History with Urban Oil Drilling is Nearing an End after Years of Health Concerns," *The Conversation*, June 2, 2021, https://theconversation.com/los-angeles-long-troubled-history-with-urban-oil-drilling-is-nearing-an-end-after-years-of-health-concerns-175983 (accessed March 13, 2025).

Robin Kelsey, "Photography and the Ecological Imagination," in *Nature's Nation: American Art and Environment*, eds Alan Braddock and Karl Kusserow (New Haven, CT: Yale University Press, 2018), 394–405.

Shana Klein, *The Fruits of Empire: Art, Food, and the Politics of Race in the Age of American Expansionism* (Oakland, CA: University of California Press, 2020).

Scott Kurashige, *The Shifting Grounds of Race: Black and Japanese Americans in the Making of Multiethnic Los Angeles* (Princeton, NJ: Princeton University Press, 2007).

Stephanie LeMenager, *Living Oil: Petroleum Culture in the American Century* (New York: Oxford University Press, 2014).

Ellen Macfarlane, *Politics Unseen: Group f.64 Photography and the Problem of Purity* (Oakland, CA: University of California Press, 2025).

David F. Martin and Nicolette Bromberg, *Shadows of a Fleeting World: Pictorial Photography and the Seattle Camera Club* (Seattle: University of Washington Press, 2011).

Timothy P. Mitchell, *Carbon Democracy: Political Power in the Age of Oil* (New York: Verso, 2011).

Kathryn Morse, "There will be Birds: Images of Oil Disasters in the Nineteenth and Twentieth Centuries," *Journal of American History* 99, no. 1 (2012): 124–134.

Nancy Quam-Wickham, "'Cities Sacrificed on the Altar of Oil': Popular Opposition to Oil Development in 1920s Los Angeles," *Environmental History* 3, no. 1 (1998): 189–209.

Nancy Quam-Wickham, "Another World: Work, Home, and Autonomy in Blue-Collar Suburbs," in *Metropolis in the Making: Los Angeles in the 1920s*, eds Tom Sitton and William Deverell (Berkeley: University of California Press, 2001), 123–142.

Robert Rapier, "California's Oil Hypocrisy Presents a National Security Risk," *Forbes*, June 21, 2019, https://www.forbes.com/sites/rrapier/2019/06/21/californias-oil-hypocrisy-presents-a-national-security-risk/ (accessed March 12, 2025).

Dennis Reed, "Southern California Pictorialism: Its Modern Aspects," in *Pictorialism in California: Photographs 1900–1940*, eds Michael G. Wilson and Dennis Reed (Los Angeles: J. Paul Getty Museum; San Marino: Henry E. Huntington Library and Art Collections, 1994), 67–88.

Dennis Reed, *Making Waves: Japanese American Photography, 1920–1940* (Los Angeles: Japanese American National Museum, 2016).

Paul Sabin, *Crude Politics: The California Oil Market, 1900–1940* (Berkeley: University of California Press, 2004).

Paul Sabin, "Beaches versus Oil in Greater Los Angeles," in *Land of Sunshine: An Environmental History of Metropolitan Los Angeles*, eds William Deverell and Greg Hise (Pittsburgh: University of Pittsburgh Press, 2005), 95–114.

Zoe Todd, "Fish, Kin and Hope: Tending to Water Violations in Amiskwaciwâskahikan and Treaty Six Territory," *Afterall: A Journal of Art, Context, and Enquiry* (2017): 95–99.

Fred W. Viehe, "Black Gold Suburbs: The Influence of the Extractive Industry on the Suburbanization of Los Angeles, 1890–1930," *Journal of Urban History* 8, no. 1 (1981): 3–26.

CHAPTER 14
Documenting "Restless Earth": John S. Shelton's Photographs for *Geology Illustrated* (1966)

Jennifer Tucker

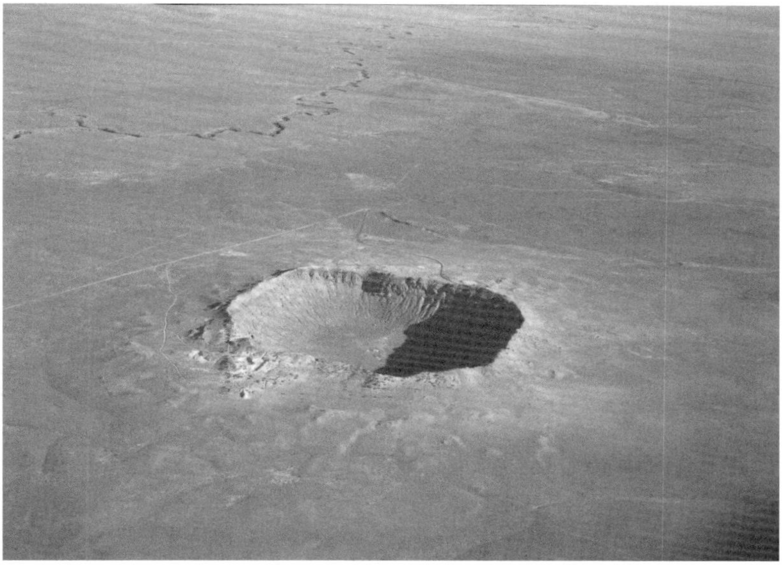

Fig. 1.
John S. Shelton, [Aerial view of meteor crater near Winslow, Arizona, looking NW from 4,500 feet above the plateau], 1958. From *Geology Illustrated* (1966): Frontispiece. Glaciers and Landforms Collection, University of Washington. PH. Coll. 859.1525.

In 1966, geologist, professor, and photographer John S. Shelton published his book *Geology Illustrated*, which featured a striking photograph of a meteor crater as its frontispiece. (Shelton, 1966) (fig. 1) The image, taken from an airplane above Arizona on July 12, 1958, draws the eye to the dark depths of a depression in what is otherwise a flat, planar landscape, and is a visual reminder of another famous scientific photograph, from exactly fifty years earlier: the high-speed photograph of the "permanent splash" where a projectile entered a plate of metal armor, which was the frontispiece of physicist Arthur M. Worthington's notable book

A Study of Splashes. (Worthington, 1908) Both photographs, which reached a wide viewership in scientific and popular culture, were evidence of new technologies of modern photographic vision, enabled in one case by faster shutters and lenses and more sensitive photographic materials and, in the other, by flight and the technical capabilities of aerial reconnaissance cameras. Yet the photograph of the meteor crater represented a different magnitude of time: a representation of the prehistoric past deciphered from the geological formations observed in the landscape of the American Southwest.

This essay offers what is to my knowledge the first detailed account of a significant photographic project in the history of American geology: the nearly 400 photographs and line drawings that appeared in John S. Shelton's *Geology Illustrated*. A professor of geology at Pomona College, Claremont, California, from 1945 to 1960, John Sewall Shelton (not to be confused with Oklahoma-based geologist John W. Shelton) took aerial photographs of the earth's surface while piloting his small plane during trips out of a local airport. In the early 1960s, he was an active contributor to the American Geological Institute's influential Earth Science Curriculum Project (ESCP), a major National Science Foundation–backed initiative aimed at modernizing secondary school earth science education with an inquiry-based approach. (Romey, 1971) He won numerous awards for his contributions to public understanding of earth science, including (in 1993) the American Geological Institute Legendary Geologist Award for "Outstanding Contribution to Public Understanding of Geology."

Photographs that were published in *Geology Illustrated* became integral to twentieth-century geological science and environmental education. They also added novel images and scientific ideas to an already existing and rich historical tradition of photographing the American Western landscape. (Sandweiss 2002; Kelsey 2007; Watts 2015; Meyers 1985) *Geology Illustrated* includes nearly 400 photographs and line drawings, the majority of them aerial photos of geological formations in the American West and Southwest. Reviewers of *Geology Illustrated* agreed that a fundamental contribution of the book was its remarkable photographs, which were arranged in sequences that developed some of the main principles of geology. The textbook was used widely in classrooms well into the 1990s. In 1999, it was recognized as one of the top 100 science books of the twentieth century by *American Scientist* magazine. (Morrison, 1999) Further, the photographs created for the book reached tens of thousands of viewers through their circulation in slide sets, videos, and educational films. In 1994, the American Geological Institute produced a videocassette with more than 3,500 images, including 100 photographs by John S. Shelton. In 2004, Crystal Productions, in collaboration with Shelton, created classic earth science slide sets, available digitally.

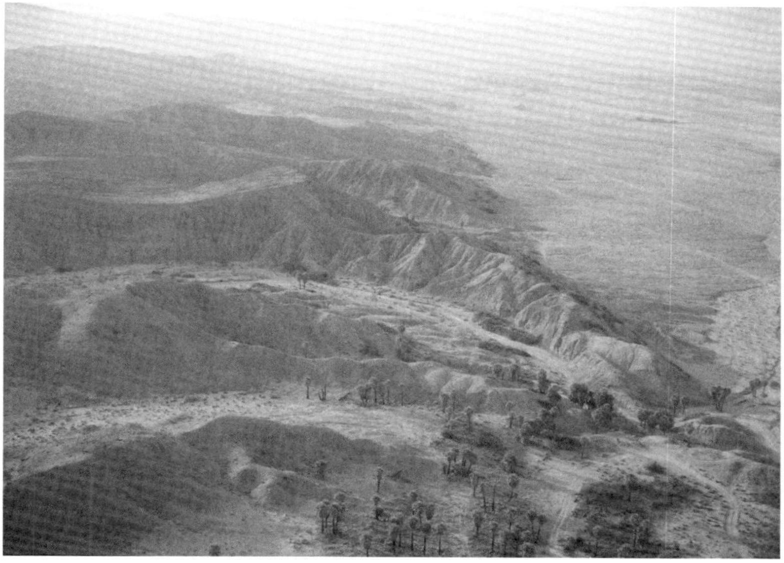

Fig. 2.
John S. Shelton, *Branch of the San Andreas fault system north of Salton Sea,* July 20, 1962. University of Washington. PH Coll 859.4585.

I suggest that during the 1960s and after, the "American West" itself became defined in no small part through Shelton's geological photographs and the artwork of his brother Hal, who introduced new ways of making natural-color maps of Western landscapes. *Geology Illustrated* shaped new, mid-twentieth century understandings of the region's ecology and formations, from its ancient origins to the significance of sedimentary and metamorphic rocks. It also helped establish geological landmarks, such as the San Andreas fault (fig. 2), as points of historic and scientific interest. While describing the distant past, the book addressed itself to the present and the future, from concerns about groundwater supply and irrigation to the value of public scientific education.

Geology in the Aviation Age

John S. Shelton was born September 2, 1913, in Hanover, New Hampshire, and his family moved to La Jolla, California, in 1923. He graduated from Pomona College, where he majored in math and music, and after a few years teaching geography at Francis Parker School in San Diego, he pursued graduate studies in geology at Yale University. Geologist Alan Bateman led the department, and another influential figure was Adolph Knopf, a pioneer of using new radiometric methods of measuring geologic time. By the 1960s, it was possible to determine the approximate age in years of a growing number of mineral occurrences in a range from a few hundred thousand to more than three billion years. (Shelton, 1966: 246)

Yale was also a national leader in visual education, promoting the use of slides and film in the teaching of history and science. Carl Owen Dunbar, who was director of the Peabody Museum, at Yale University, initiated the construction of the museum's first dioramas, such as the 110 foot-long, Pulitzer Prize–winning mural "The Age of Reptiles," painted by Rudolf Zallinger between 1942 and 1947. (Skinner and Narendra, 1985) After graduating from Yale, Shelton moved west in 1941, where he joined the geology department at Pomona College. California, home to the aeronautical industry, was an important center of the evolving relationship between scientific research and university education. Shelton received his pilot's license in 1939, and he began taking aerial landscape photos to combine his three passions: flying, photography, and geology. Figure 3 is a photograph taken of him at Cable airport, near Pomona College, around this time. His contemporaries included Sherman Mills Fairchild, whose company, Fairchild Aerial Survey, produced thousands of aerial photographs of the United States and México.

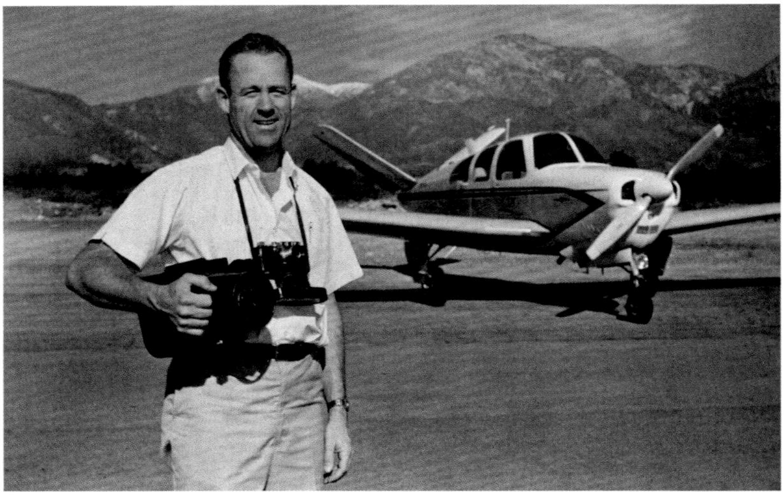

Fig. 3.
[Photographer unknown], Portrait of John S. Shelton at Cable Airport with cameras, ca. 1950. From *Geology Illustrated* (1966): Back cover.

The expansion of aviation combined with photography ushered in new paradigms of vision in science and art. (Amad, 2010; Weems, 2011, 2015; Schwartz, 2020; Angus, 2024) The airplane, no less than cinema, was media in motion. (Schwartz, 2020) The arid nature of much of the American West meant that there was less foliage to obscure terrestrial formations. An aerial perspective enabled Shelton to capture the vast scale of the dramatic topography of the West: "scaling up our vision," to borrow a term from historian Naomi Oreskes. (Oreskes, 2014: 379; Oreskes, 2001) The US West came to be seen as especially ancient, conveying a powerful sense of what geologist Marcia Bjornerud describes as "timefulness," or a "feeling for distances and proximities in the geography of deep time." (Bjornerud, 2018: 17) One begins to understand, she writes, "that rocks are not nouns but verbs." (Bjornerud, 2018: 9)

Shelton wrote that there were "many different kinds of deforming movements in the crust, ranging from such obvious things as displaced shorelines, deformed sedimentary layers, and volcanism, to the more subtle but often more impressive implications of unconformities, metasediments, exposed plutonic rocks, tectonic landscapes and the very existence of mountains." "Each of these, in its own way," he continued, "embodies evidence of a *restless earth*" (my emphasis). But he added: "only a few contain even a hint of the *time* involved in their creation." (Shelton, 1966: 246) Photographic views of geological objects were crystallizations of time. The American West was a dynamic landscape whose appearance revealed otherwise hidden clues to underlying changes and movements: the weathering of soils; erosion; deposition, debris flows, and historic landslides. Shelton was fascinated by the work of streams and rivers, and by their "infinite complexity" of interrelated forces ("discharge, velocity, turbulence, shape and roughness of channel, quantity of load and the sizes and shapes and proportions of the different kinds of particles"). (Shelton, 1966: 146)

Similar Landforms, Contrasting Origin Stories

From 1960 to 1974, Shelton was the Chief Scientific Advisor to the Elementary Earth Science Program of the American Geological Institute and the chief science adviser for Encyclopedia Britannica Films. In addition to writing books, such as *Geology Illustrated,* he worked on films, including titles such as *Why Do We Still Have Mountains?*, *How Solid is Rock?*, and *The Beach: A River of Sand,* which included discussion of the effects of urban development on natural processes.

Shelton's book *Geology Illustrated* included photographs of glaciers and other landforms throughout the world, some by other photographers, including Bradford Washburn, Ted Nichols, Carl Fries, Jr., and Berlin Moneymaker. It followed a long tradition of educational geology textbooks, guidebooks, and specimen slides, as explained in Martin J. S. Rudwick's path-breaking work, *The Emergence of a Visual Language for Geological Science, 1760–1840*. (Rudwick, 1976) In the nineteenth century, the spread of railways profoundly altered understandings and aesthetics of space and time, as the excavation of cuttings exposed new sediments for geologists. (Tucker, 2014) The British scientist T. H. Huxley's book *Physiography: An Introduction to the Study of Nature*, first published in 1877, was a landmark text that introduced a holistic, observational approach to understanding the earth's physical processes.

Another precedent is a book titled *Down to Earth: An Introduction to Geology* (1936), by two geologists at the University of Chicago, Carey Croneis (1901–1972) and William C. Krumbein (1902–1979), which included sixty-four plates of rotogravure pictures and numerous diagrams and drawings.

What the creation of rail travel had done for nineteenth-century geology, the plane—and also drilling, which unearthed rock samples at thousands of feet beneath the surface—would do for twentieth-century geology. (Shelton, 1966: 160; Tucker, 2014) *Geology Illustrated* was a landmark textbook in the quality, number, and novelty of its geological views. Shelton urged readers to go into the field and see the rocks for themselves. He added that to stop at reading a book "would be like learning to fly an airplane and then never going anywhere." Shelton explained that "a major goal of geology is knowledge of conditions, relations, processes, and events that in large part cannot be directly observed, either because they are out of sight beneath the earth's surface or because they belong to times that predate human history." (Shelton, 1966: xi) The exercise enabled viewers to travel back in planetary time, to "reconstruct conditions of the distant past." (Shelton, 1966: xi)

While the majority of the illustrations in *Geology Illustrated* were photographic views of the landscape from the air, there were also many photomicrographs showing the results of geological processes, such as progressive mylonitization (deformation through shearing) of granitic rock in the southeastern San Gabriel Mountains. The photographs provided novel views of landslides, sand dunes, dust storms, erosion, dams, and drainage patterns. (Shelton, 1966: 165) According to Shelton, "One of the simplest and most graceful forms in which windblown sand accumulates" was the curved or crescent shaped dune, or barkhan, an example of which is illustrated in Shelton's photograph of barkhan fields

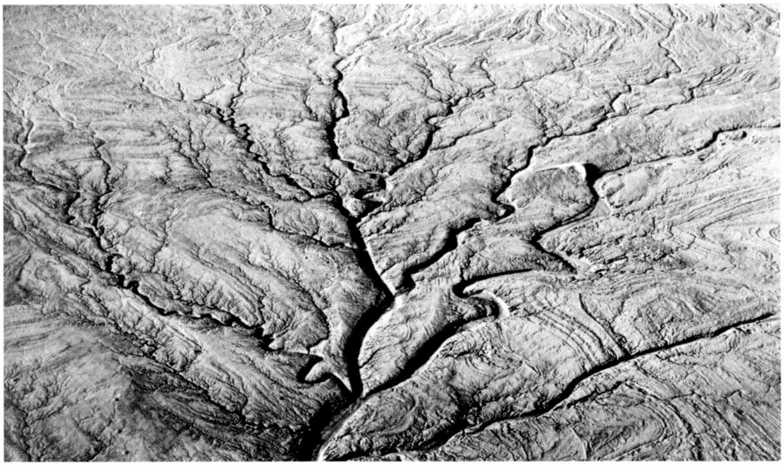

Fig. 4.
John S. Shelton, *Upper Tule Wash, and Its Barchan, Looking approx SW from about 5,500 feet*, November 7, 1959. From *Geology Illustrated* (1966): 165. University of Washington, PH Coll 859.3086.

in Tule Wash. A view upstream over upper Tule Wash, west of Salton Sea, California's largest inland lake at over 340 square miles (fig. 4), allowed readers to examine the sinewy lines that marked the pattern of "weak and resistant" rocks on a draining pattern on an ancient lake floor. To describe the visual effect that tributaries produced when they worked in weak rocks, Shelton used an artist's term: etching. (Shelton, 1966: 166) The main gullies in the branching drainage system have a dendrite pattern (like nerves), with some of the newer ones following the weak layers in the strata, as seen in the left foreground. (Shelton, 1966: 166)

Multimedia Documentary

Shelton drew readers' attention to the "ways the exposed parts of Earth's rocky structures have been carved and modified by surface processes to give the landscapes the shapes we see," such as lakes, shorelines, glaciation, and even groundwater. One photograph captured a dust storm over a vineyard district of Southern California. (Shelton, 1966: 196) The aerial perspective at different altitudes (normally below 6,000 feet) approximated, in a limited sense, the experience of viewing a geological specimen under different magnifications. Photographs offered both general and close-up views and were in a black-and-white format.

A photograph looking north at Ship Rock in northwestern New Mexico (fig. 5) represents an example of rock that is a volcanic neck: an igneous rock form that once fed a volcano. The inset photograph in the photograph shows the granite inclusion in tuff at the base of Ship Rock. Shelton's use of a pen for scale in the inset photograph recalls an earlier photograph by the photographer Timothy O'Sullivan (1840–1882) for Lt. George M. Wheeler's survey of the 100th Meridian west, which shows a ruler against El Morro, also known as Inscription Rock, a sandstone bluff in western México with a pictograph showing Spanish conquistadors. (Kelsey, 2007; Kelsey, 2003; Tucker, 2023: 128; Sandweiss, 2002). With its yardstick ("mensurational sight"), it signifies a new moment of geological exploration and historical conquest. While measuring scales (such as rulers) occasionally were used in *Geology Illustrated*, ordinary objects (pens, paper clips, a cowboy hat, sometimes another human) were more commonly used as scale markers.

The photograph of Ship Rock served to illustrate the way that similarly looking landforms often had completely different origins. The present height of Ship Rock was due to its "passive resistance" to the erosion that removed the surrounding elements, rather than the result of the upward thrusting of a column of hard rock. Unlike the depression created by a meteor crater near Winslow, Arizona, which was a "huge bowl," 4,000 feet across and 600 feet deep (see fig. 1), Ship Rock was probably

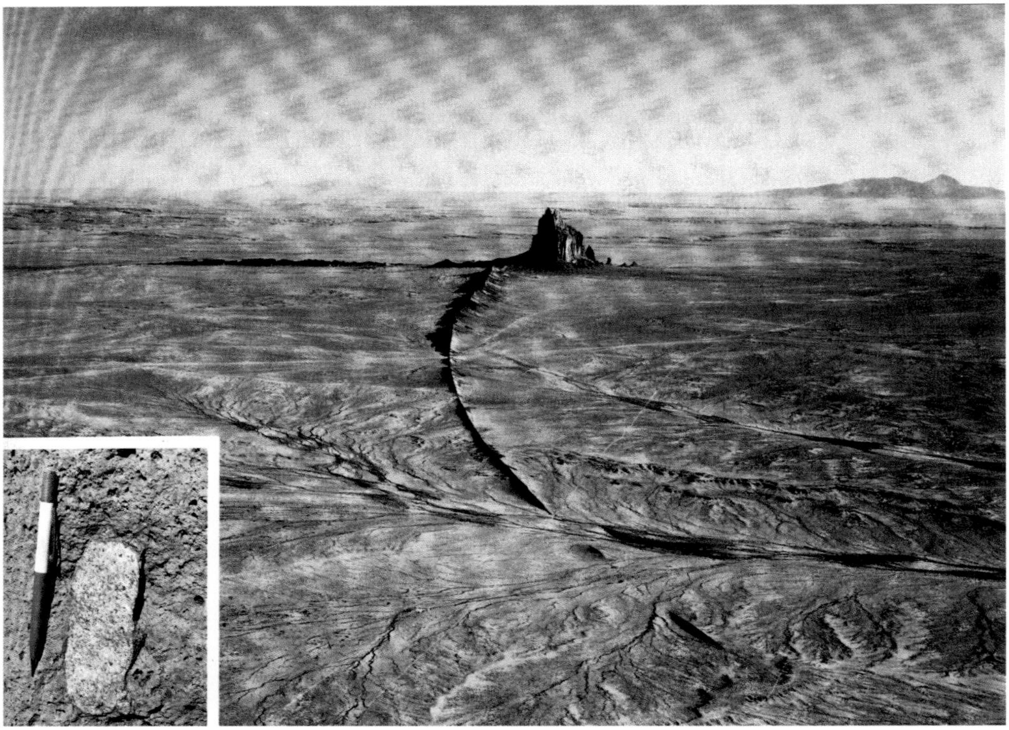

a "neck" ("the body of igneous rock that solidified in the throat, or pipe, that once was a volcano.") (Shelton, 1966: 14) A crater in Zuni Salt Lake in western New Mexico was also the result of a volcanic eruption.

Communicating the dynamic effects on rocks of time, water, wind, and other natural forces called for experiments in relationships of words, drawings, and photographs. The graphic practice of mixing media was a common feature of scientific image creation long before the coining of the term "intermedia" in art criticism in the 1960s, and the use of inset drawings and photographs for magnified details, cutaway views, and comparative examples was a standard pedagogical tool in science textbooks by the nineteenth century.

The photographs in *Geology Illustrated* were supplemented with explanatory line drawings by John's brother, Hal Shelton. We can see this in Hal's cutaway drawing (fig. 6) that accompanied the photograph above, which shows "the reconstruction of probable earlier conditions" at Ship Rock. (Shelton, 1966: 15) Hal Shelton earned a degree in scientific illustration from Pomona College before joining the United States Geological Survey during World War II. Hal Shelton's first attempt at natural-color mapping occurred while on temporary duty with the US Air Force. He used color and shading to replicate what the pilot would see from the cockpit. After the war, Hal settled in Golden, Colorado.

Fig. 5.
John S. Shelton, *Looking North at Ship Rock, New Mexico,* July 10, 1958. Inset: Photograph of fragment of sedimentary rock (granite) showing that small pieces of older rock in volcanic layers. By studying such remnants, geologists acquired knowledge about processes that resisted volcanic erosion. In *Geology Illustrated*, 1966, 15. University of Washington, PH Coll. 859.2375, PPH Coll. 859.1472.

An avid skier, he pioneered the use of the shaded relief map in depicting ski trails. His freelance recreational maps of Colorado caught the attention of United Airline pilot Elrey Jeppesen, who started a company that published aeronautical maps. (Patterson and Kelso, 2004) In the early 1950s, Hal made aerial maps that catered to the ever-increasing numbers of air travelers who were seeing Earth from above for the first time. The illustration in Figure 7 shows one of the maps from a booklet titled *Your Air Maps of United Air Lines* (1951). (fig. 7) John and Hal's

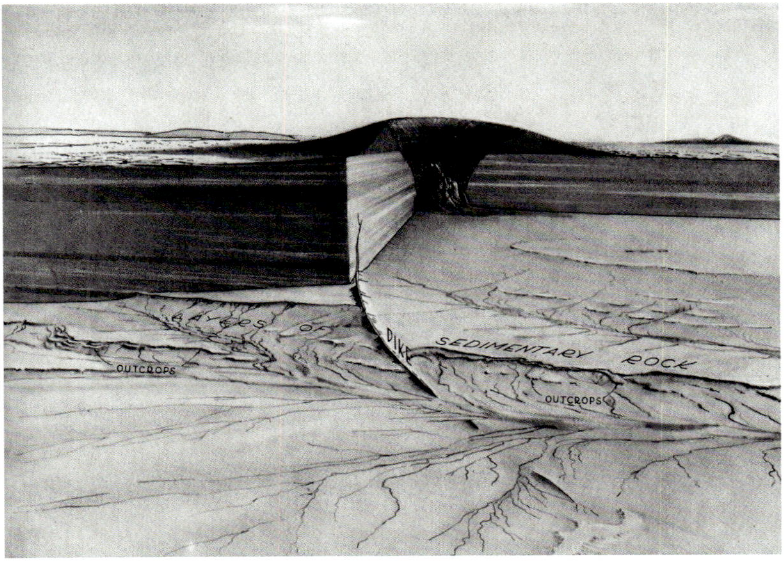

Fig. 6.

Hal Shelton, drawing. *Cutaway Restoration of Ship Rock, Showing Probable Earlier Condition*. From *Geology Illustrated* (1966): 15.

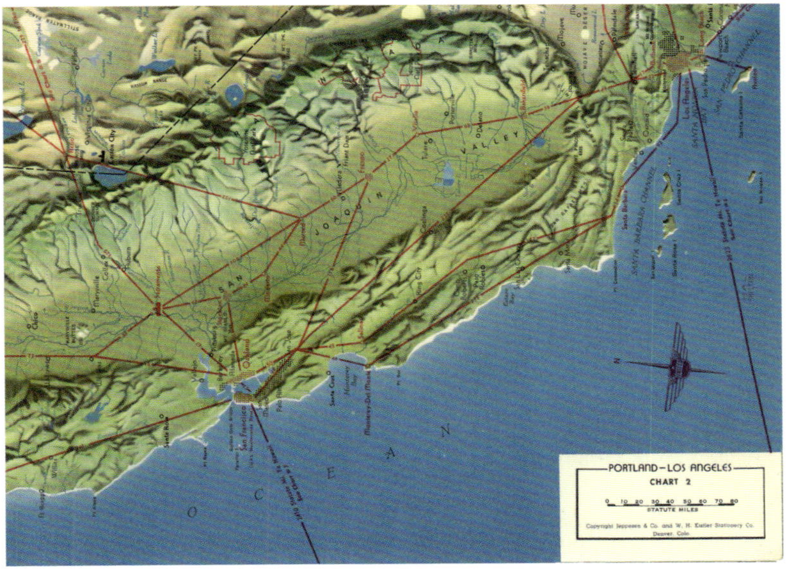

Fig. 7.

Hal Shelton, *Chart 2: Portland–Los Angeles*. Colored natural contour map (8–1/4" x 11") of the western coastline of the US. Reproduced in the 14-pp booklet, *Your Air Maps of United Air Lines: Main Line Airway* (Denver: United Airlines, 1951): 3–4.

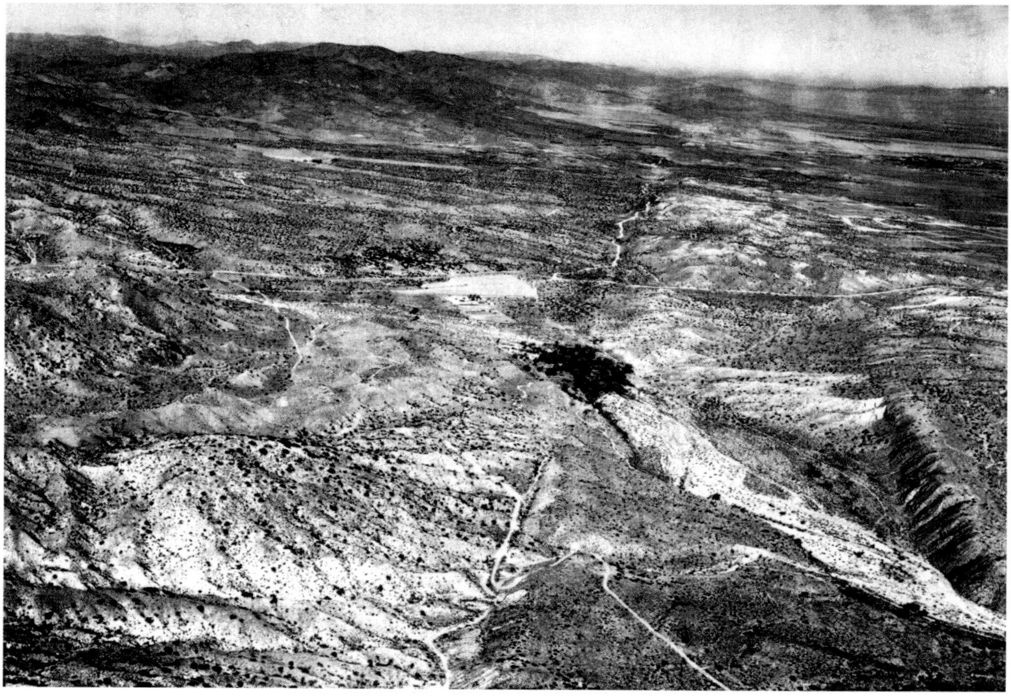

Fig. 8.
John S. Shelton, *San Andreas Fault, southeast of Palmdale, CA*, Sept. 9, 1959. From *Geology Illustrated* (1966): 243. University of Washington, PH Coll. 859.2783.

business association spanned two decades and yielded more than thirty titles in what became the Jeppesen Natural-Color Map series. NASA later employed this collection of maps by Shelton to index photos of Earth taken on early space missions.

In the 1960s, plate tectonics was emerging as the unifying theory of geology: an explanation of how major landforms, from the tallest mountains to the deepest oceans, are created as a result of Earth's subterranean movements. Shelton's photographs helped to establish the San Andreas fault and other landforms as recognizable geological features in the American West. Figure 8 is a photograph of the San Andreas fault that shows the view northwestward along the fault, between Valyermo and Palmdale, California, with a clump of trees in the distance marking the place where the fault crossed a dry bed of a creek, causing the groundwater to rise. (Shelton, 1966: 243) He also included aerial views of the Garlock fault, looking westward in the Mojave Desert of Southern California (once a hub of borax and gold mining). The photo in Figure 8 was included in a section titled "Man's Use of Water" and illustrates his wider interest in connecting the earth sciences to knowledge of human uses of the landscape. Shelton joined scientists and others in expressing concern about the pollution of groundwater in the western states due to agricultural and industrial usage.

Time Travel and "Relationships in Time"

John Shelton described his work in *Geology Illustrated* as a form of time travel. Writing before art critic John Berger's 1972 book popularized the term "ways of seeing" in art, Shelton proposed "looking beneath the surface" to understand how and where rocks were originally produced, and how they arrived where they are today. Variations in the process of formation and metamorphosis, from volcanoes to sedimentary rocks, are reflected in the "texture" of the material rock itself, which can be revealed through polarizing microscopy and x-ray spectrometry. (Keller, 2008: 23) As he described it, the challenge was to try to arrange geologic events in chronological order and to estimate the duration of each step in the sequence, in order to gain "some insight into the magnitude of geological time." "The most distinctive feature of any rock is more likely to be the imprint of the last process it went through." (Shelton, 1966: 55) Wind and water erosion were constantly changing the landscape, and were themselves profoundly transformed by human activity.

Photographic views of barkhans (or cresent-shaped dunes) at Tule Wash, west of Salton Sea, Southern California, captured another dimension of time travel in geology: the way that sand was moved by wind. (fig. 9) The numbered positions on the photograph document the advance of the sand dune (about fifty feet per year) as recorded in

Fig. 9.
John S. Shelton, Aerial view of Tule Wash, west of Salton Sea, southern California, in 1958, showing drainage patterns of tributaries. Retouched to show positions on photographs (not shown) taken in 1932 and 1950. Inset: Photograph by Shelton of the slipface of the same dune (33 feet high) to illustrate the ongoing advance of crescentic sand dunes. (Note the white cowboy hat, right hand corner, used for scale). From *Geology Illustrated* (1966): 198.

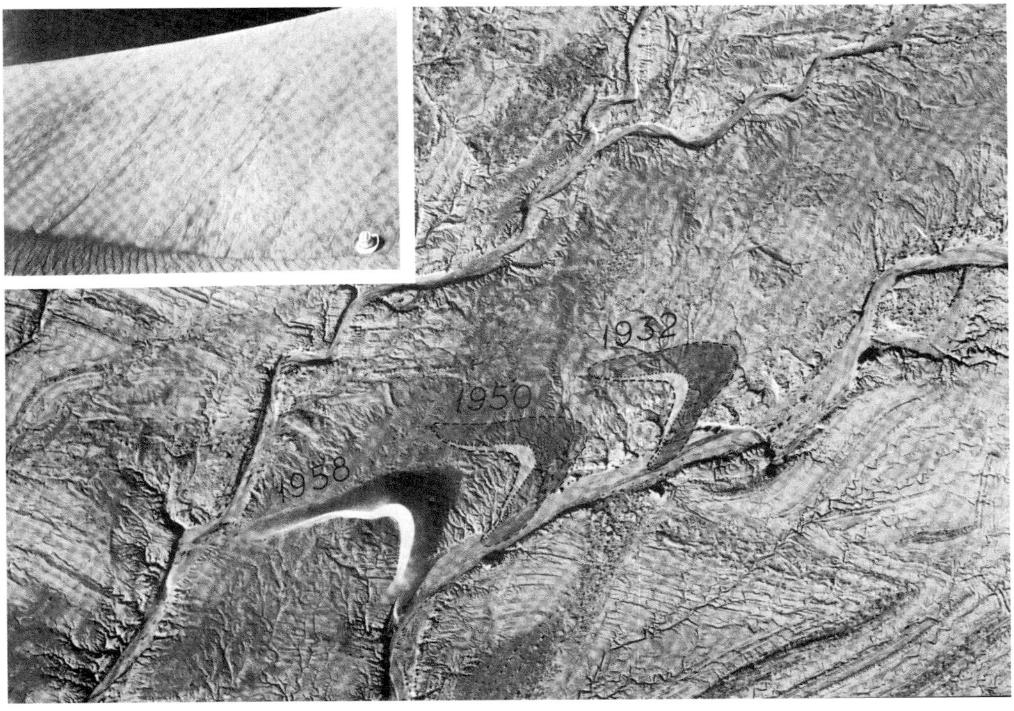

previous photographs, taken in 1932, 1950, and 1958. (Shelton, 1966: 198) The inset photograph (top left) shows a nearly vertical view (slipface) of the same barkhan, which measured thirty-three feet high. The white cowboy hat that Shelton used for scale recalls Timothy O'Sullivan's similar inclusion of a hat to mark scale in in his stereographic image of a "Melon Cactus, Colorado River." (1871) Although in the photo shown in Figure 9, he superimposed numbers to make chronology legible, Shelton believed that photographs were fully as important as the words that supplied the interpretation and, for that reason, he placed very few identifying marks directly on the photographs.

The geologist was a type of journalist or historian. Shelton assured readers that although no eyewitness could attest to what exactly happened in the earth's history, through field observation and scientific research, "the story can be pieced together." (Shelton, 1966: 122) Like a historic map, photographs helped sharpen the eye to identify what might not otherwise be seen in a landscape. For Shelton, "[e]very landscape is a geological battlefield"—a struggle, an interplay between two processes (fluvial and glacial vs. tectonic). (Shelton, 1966: 148)

The tension between surface and depth in the photography of geological landscape was ever-present. Only a few landscapes contained "even a hint of the time involved in their creation," wrote Shelton. Geologists made at least two rather different uses of time. (Shelton, 1966: 246) The first was to establish the sequence of events by determining their relative positions in time; for example, by recognizing older and younger rocks. By means of studying the "relations" of rocks (mostly with the aid of fossils), it was also possible to establish their contemporaneity, even on different continents. In untangling the "incomprehensibly long history" of the earth's crust, he wrote, "contemporaneity" and "sequence" were the only "relationships in time" that could be firmly established. (Shelton, 1966: 246) A second "use of time" was a measure of duration—or, the length of a geological event. For example, the deposition of fine-grained sediments whose bedding was rhythmic and seasonal could be used to count years in the same way as growth rings in trees. (Shelton, 1966: 246, 304–305)

Geology, a historical discipline, saw rocks as its primary sources: a way of reading into the past that historians, in another context, claim for photographs. (Meyers, 1985; Trachtenberg, 1989; Tucker, 2005; Daston and Galison, 2007; Mitman and Wilder, 2016) Shelton also asserted that "the past is the key to the present, for a sense of history is part of the fabric of geology." (Shelton, 1966: 247) With regard to human life, he said, all of recorded history was so short compared with the age of the earth that "almost every geological event or story we study is *pre*historic by a wide margin." The key to studying the geological past—the "geologist's maxim"—was first by studying the present. (Shelton, 1966: 247) Just as examination of modern glaciers led to recognition of a prehistoric Ice Age, therefore, studying rocks helped reconstruct their "geological history." (Shelton, 1966: 247) The photographs worked as more than illustrations, for they provided data that helped advance knowledge of changing geological conditions. Many geological processes could not be directly observed, either because they were "out of sight" of the earth's surface or because they belonged to times that predated human history. As an example, Shelton noted that three sources of information informed the study of the geological history of the Grand Canyon. The first was the structural relationship of the major rock bodies to each other. The second was the nature of the rocks themselves (their compositions and textures): "these tell about where they came from, how they arrived, and what has happened to them since." Finally, the fossils contained in the rocks helped create "the record of some of the life that witnessed and participated in these events." (Shelton, 1966: 252–253)

Shelton believed that history was itself a geological tool, explaining that "awareness of geologic history gives us a running start toward understanding the modern scene." (Shelton, 1966: 313) An awareness of geologic history is more relevant than ever, including to a new generation of environmentalists. (Prilliman, 2022) Geologists, for example, are currently debating which spot on Earth marks the first clear evidence of the Anthropocene—which many of them think is a new geological epoch that began when humans started altering the planet with various forms of industrial and radioactive materials in the 1950s. (Waters and Turner, 2022) A list of candidates today includes nine sites worldwide, each being considered for how reliably its layers of mud, ice, or other matter told the story of people's influence on a timeline that extends billions of years into the past. One of the new sites identified as marking new geological time is in California, Searsville Lake, chosen because it shows how human engineering (dam building) led to discoveries of rapid sedimentation from logging and industrial pollution, seeming to link human action to geological change over the past century. (Stegner, 2023)

At the time that Shelton was photographing the craters of the earth's crust, new space exploration efforts were finding new applications of geology. In February 1966, the Soviet Luna 9 spacecraft made the first successful robotic soft landing on the moon and provided the first close-up picture of the moon taken from its surface. The US responded with the Lunar Orbiter 1 robotic craft, the first to orbit the moon, in August 1966, taking photographs of the moon's surface. Shelton was pleased that the human exploration of space "has led to growing interest in the geology of the moon's surface. This interest has, in turn, stimulated more intensive study of craters on earth whose origins may shed some light on the causes of those on the moon." (Shelton, 1966: 368) Shelton was active at the same time as the creation of the Golden Record, a collection of images, diagrams, and sounds created for the Voyager II Mission in 1977. Carl Sagan and his team sought to assemble as accurate a representation of the evolution of the human and natural environment on Earth as possible in 118 images. Some of the artists and scientists who created to the Golden Record worked at the NASA Jet Propulsion Laboratory in Southern California. (Tucker, 2018: 129–130)

Photographic Aesthetic of a "Restless Earth"

Geologist and writer Marcia Bjornerud has suggested that the study of geology "is the closest we may get to time travel" and that rock formations are "the subtle, syntactically complex language of the Earth." (Bjornerud, 2018: 17) She cautions that natural geological processes that have unfolded over millions of years are now being accelerated, sometimes over mere decades, through the strip mining of coal, oil field fracking, and the unfettered combustion of these and other fossil fuels. (Bjornerud, 2018)

By the late 1960s, *Geology Illustrated* was a widely referenced source, and it helped establish the American West as an iconic star of the new geology. Through the medium of photography, Shelton taught students the elementary principles of geology, and he also hoped to stir a love of the outdoors. Reviewers universally praised the striking aesthetics and utility of Shelton's photographs as a teaching tool. The geologist Samuel S. Goldich, reviewing *Geology Illustrated* for the *Quarterly Review of Biology* in 1967, observed: "The very essence of this book is its approach through photographs and through questions, such as, what does this mean and how did it come about?" (Goldich, 1967: 416)

In the years after the publication of *Geology Illustrated*, Shelton worked as a scientific educational films adviser, including collaborations with Encyclopaedia Britannica Films, using his extensive aerial photography to teach, particularly around California. In his later writings and films, after 1970, he addressed the impact of human factors on geology: concerns that were traced to even earlier observations. For example, Shelton wrote in *Geology Illustrated* about the accelerating problem of polluted surface waters that was caused by irrigation and industry: the two factors (more than the growth in population) that he thought created the most acute problems: "The basic problems connected with man's use of water arise from sharp upward trends in the needed quantity and in demands for water of greater purity." He added: "Industry depends mostly on surface waters… Even when these industrial centers supplement their supply with groundwater, the waste is released to the streams." (Shelton, 1966: 243–243). Today, time-lapse satellite images of California's Salton Sea from 1984 to 2022 show the rapid expansion of exposed lake bed as the water has receded. (Johnston and Farzan 2026) Winds blowing across the receding lake bed is kicking up toxic dust left by years of agriculture chemicals – fertilizers, pesticides, salt and toxic metals – that has washed into Salton Sea. Scientists are finding that the effects on lung function of people living close to the Salton Sea are even greater than what studies find in urban California communities near busy roadways. (Johnston and Farzan 2026)

Shelton's aerial photographs of the American West not only continue to circulate in earth slide sets, available digitally; they are a valuable and neglected source of evidence of mid-century scientific vision and the interlocking of art and science, and they raise fundamental questions about the relationship of photography and history. The photographs in *Geology Illustrated* and their associated media outlets, such as educational slides and films, offer a record of the changes in landscapes that were being altered by industrialization (e.g. irrigation, fertilization, mining) and natural events, such as volcanoes and earthquakes. (Harris, 2012) Long before the coining of the word "Anthropocene," the photographs and text in *Geology Ilustrated* invited viewers to consider their place in deep geological time. The "view from above" was a mode of travel through space and time: a lens through which to ask questions that were driven, he said, by "plain curiosity": "Has it *always* looked like this?" The answer, he said, "will always be no," and the "quest for knowledge about what is out of sight underfoot and backward in time begins with what we can see now." (Shelton, 1966: xii).

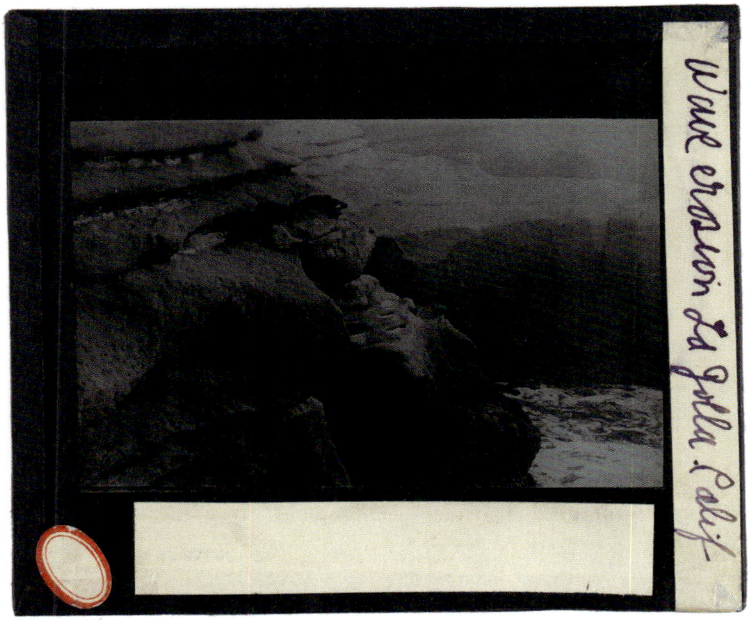

Fig. 10.

John S. Shelton. *Wave Erosion, La Jolla California*, glass lantern slide, c. 1960. Courtesy: John S. Shelton Educational Lantern Slide Collection, Wesleyan University's Special Collections and Archive.

Shelton died in 2008 in California, and no book-length biography of him or his work yet exists. However, hundreds of his photographs, slides, and educational films survive, and testimonies to the importance to his influence can be found in scientific and public libraries and archives. For example, this teaching slide showing prehistoric life (fig. 10) is one of hundreds that Shelton used in his classroom in the mid-twentieth century, and which students today can continue to view and research. (Shelton, 1974, 2004) *Geology Illustrated* is an enduring legacy of photographs and writings that profoundly shaped twentieth-century public understandings of earth science and the American West's unique geological history. The photographs and line drawings in *Geology Illustrated* brought the distant world of the American West up close to generations of students and stimulated the writing of new kinds of natural histories. Widely circulated in classrooms and science educational films by the 1970s, the geological photographs that Shelton made from the air had "social lives" beyond their reproduction in the textbook. Today, they serve to remind us of the many ways that geological photography continues to shape the imagery and layered meanings of the American West.

Bibliography

Paula Ahmad, *Counter-Archive: Film, the Everyday, and Albert Kahn's Archives de la Planète* (New York: Columbia University Press, 2010).

Siobhan Angus, *Camera Geologica: An Elemental History of Photography* (Durham, NC: Duke University Press, 2024).

Marcia Bjornerud, *Timefulness: How Thinking Like a Geologist Can Help Save the World* (Princeton, NJ: Princeton University Press, 2018).

Carey Croneis and William C. Krumbein, *Down to Earth: An Introduction to Geology* (Chicago: University of Chicago Press, 1936).

Lorraine Daston and Peter L. Galison, *Objectivity* (Princeton, NJ: Princeton University Press, 2007).

Samuel S. Goldich, "Review: *Geology Illustrated* (1966)," *Quarterly Review of Biology* 42 (September 1967), 416.

Melissa Harris, "Richard Misrach and Kate Orff Discuss Petrochemical America," *Aperture* (September 17, 2012), https://aperture.org/editorial/richard-misrach-and-kate-orff-in-conversation/ (accessed April 28, 2026).

Jill Johnston and Shohrech Farzan, "Toxic dust from California's Shrinking Salton Sea is Harming Children's Lung Growth – Our Study Tracked the Impact in 700 Kids," *The Conversation* (April 3, 2026), https://theconversation.com/toxic-dust-from-californias-shrinking-salton-sea-is-harming-childrens-lung-growth-our-study-tracked-the-impact-in-700-kids-279211 (accessed April 28, 2026).

Corey Keller, *Brought to Light: Photography and the Invisible, 1840–1900* (New Haven, CT: Yale University Press, 2008).

Robin E. Kelsey, *Archive Style: Photographs and Illustrations for U.S. Surveys, 1850–1890* (Berkeley: University of California Press, 2007).

Robin E. Kelsey, "Viewing the Archive: Timothy O'Sullivan's Photographs for the Wheeler Survey, 1871–1874," *Art Bulletin* 85 (2003): 702–723.

Amy R. W. Meyers, *Sketches from the Wilderness: Changing Conceptions of Nature in American Natural History Illustration, 1680–1880* (PhD diss, Yale University, 1985).

Gregg Mitman and Kelly Wilder, eds, *Documenting the World: Film, Photography, and the Scientific Record* (Chicago: University of Chicago Press, 2016).

Phyllis Morrison, "100 or so Books That Shaped a Century of Science," *American Scientist* (November 1999).

Naomi Oreskes, "Scaling Up Our Vision," *Isis* 105, no. 2 (June 2014): 379–391.

Naomi Oreskes (ed.), *Plate Tectonics: An Insider's History of the Modern Theory of the Earth* (New York: Westview Press, 2001).

Tom Patterson and Nathaniel Vaughn Kelso, "Hal Shelton Revisited: Designing and Producing Natural-Color Maps with Satellite Land Cover Data," *Cartographic Perspectives* 47 (Winter 2004).

McKenzie Prilliman, "Are We in the New Anthropocene? Geologists Could Define a New Epoch for Earth," *Nature* 613 (December 13, 2022).

William D. Romey, "The Earth Science Educational Program," *Journal of Geclogical Education* 19, no. 3 (1971): 119–123.

Martin J. S. Rudwick "The Emergence of a Visual Language for Geological Science, 1760–1840," *History of Science* 14 (1976): 149–195.

Martha A. Sandweiss, *Print the Legend: Photography and the American West* (New Haven, CT: Yale University Press, 2002).

Vanessa R. Schwartz, *Jet Age Aesthetic: The Glamour of Media in Motion* (New Haven, CT: Yale University Press, 2020).

John S. Shelton, *Geology Illustrated* (New York: W. H. Freeman, 1966).

John S. Shelton, *Guidebook to Earth Science Slides* (San Francisco: W. H. Freeman, 1974).

John S. Shelton and Crystal Productions, *Earth Science Slides* (electronic resource). (Crystal Productions, 2004).

Brian J. Skinner and Barbara L. Narendra, "Rummaging Through the Attic: Or, a Brief History of the Geological Sciences at Yale," *Geological Society of America Centennial Special*, Vol. 1 (1985): 355–376.

M. A. Stegner, E.A. Hadly, A.D. Barnovksy, S. La Selle, B. Sherrod, R.S. Anderson, S.A. Redondo, M.C. Viteri, K.L. Weaver, A.B. Cundy, P. Gaca, N.L. Rose, H. Yang, S.L. Roberts, I. Hajdas, B.A. Black, and T.L. Spanbauer, "The Searsville Lake Site (California, USA) as a Candidate Global boundary Stratotype Section and Point for the Anthropocene series," *The Anthropocene Review* 10 (2023): 116–145.

Alan Trachtenberg, *Reading American Photographs: Images as History* (New York: Hill and Wang, 1989).

Jennifer Tucker, "Voyages of Discovery on Oceans of Air: The Image of Science in an Age of 'Balloonacy.'" *Osiris* 11 (1996): 144–176.

Jennifer Tucker, *Nature Exposed: Photography as Eyewitness in Victorian Science* (Baltimore: Johns Hopkins University Press, 2005).

Jennifer Tucker, "Close Ties: The Railway Station and Photographic Networks," *Photoworks 21* (2014): 168–173.

Jennifer Tucker, "Visual and Material Culture," in S. Handley, R. McWilliam, L. Noakes, eds, *New Directions in Social and Cultural History* (London: Bloomsbury Academic Press, 2018): 129–142.

Jennifer Tucker, "Chemical Affinities: Photography, Extraction, and Industrial Heritage in Nineteenth-Century Northern England," *Nineteenth-Century Contexts* 44 (2022).

Jennifer Tucker, "Objectivity," in A. Byrd and E. Siegel. (eds.), *The Art Institute of Chicago Field Guide to Photography and Media* (New Haven: Yale University Press, 2023): 128.

Colin N. Waters and Simon D. Turner, "Defining the Onset of the Anthropocene Geological Epoch," *Science* 378 (November 17, 2022): 706–708.

Jennifer Watts, "Ansel Adams, William Current, and the American West," *The Huntington* (April 23, 2015).

Jason Weems, "Interpreting a 1930s Aerial Survey Photograph: The Artfulness of Technological Images," *History and Technology* (2011): 223–231.

Jason Weems, *Barnstorming the Prairies: Flight, Aerial Views, and the Idea of the Midwest, 1920–1940* (Minneapolis: University of Minnesota Press, 2015).

Arthur M. Worthington, *A Study of Splashes* (London: Longmans, Green, 1908).

[United Air Lines], *Your Air Maps of United Air Lines: Main Line Airway* (Denver: United Airlines, 1951).

CHAPTER 15
Arctic Time: Ecology and Photography in the Ice Fields

Elizabeth Hutchinson

> We were in a Summer of continuous brightness. For days and weeks together lamps were not thought of. It was a new heaven and a new earth; no morn, no noon, no night, no time for turning in or turning out. We lived and moved by the clock, not by the sun or the sky. (Bradford, 1885: 104)

The *Bear*, a three-masted ship equipped with steam power, occupies the majority of a roughly 8 × 10 inch print mounted on a larger album page. (fig. 1) Even without the handwritten caption that says the ship is caught in ice, we can observe that the ship's mobility has been arrested by something more than the camera. Chunks of ice surround the ship, rising from the surface of the sea in uneven blocks. Although there appear to be a couple of men on deck, the sails are down, no one occupies the crow's nest to advise on steering, and the steam engine funnel emits no smoke. Closer attention reveals ropes tethering the ship to the ice. The sailors appear to have been in this situation for a while, and to have no expectation of imminent departure. Like the ship's movement, time itself seems to have stopped in this picture. All photographs freeze a moment in time, of course, and nineteenth-century photographs, which required longer exposures, convey an enhanced sense of stillness. But the *Bear's* quandary occurred during Arctic summer, weeks when the sun never left the sky, making it a picture that destabilizes the experience of time's regular unfolding. As painter and Arctic traveler William Bradford reminds us, this is a time of "no morn, no noon, no night."

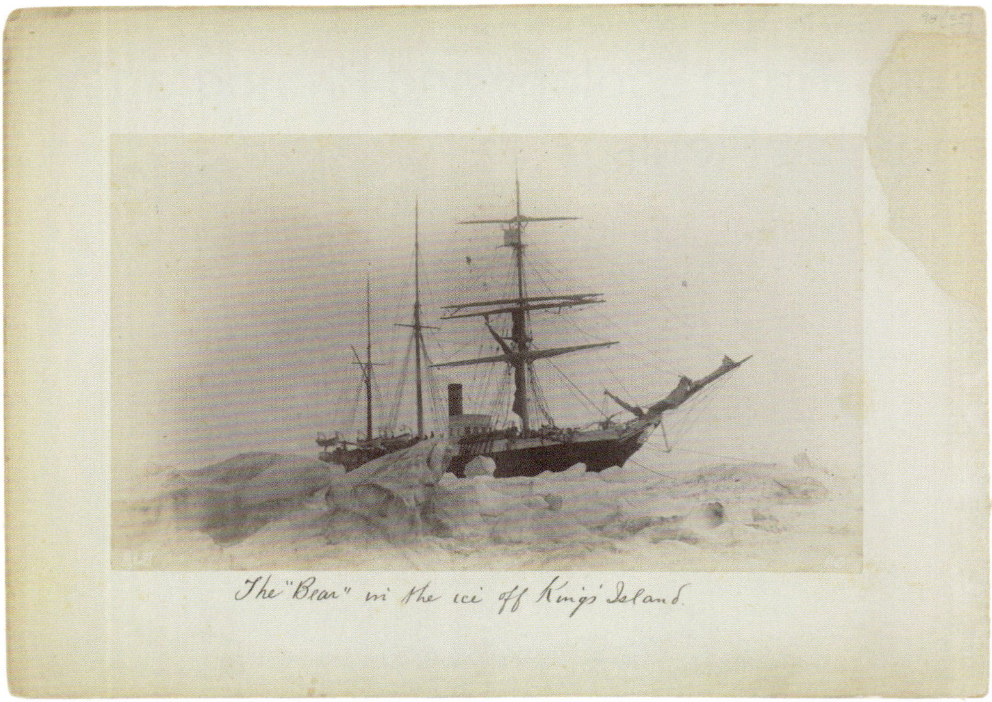

Fig. 1.
Alfred L. Broadbent, *The "Bear" in the ice off King's Island,* 1890s. Albumen print from plate glass negative, 18×27 cm. Carl S. Dentzel Photograph Collection of the American West, 1870s–1890s: Alaska Views. San Marino, Calif., Huntington Library. photCL 98.

This essay documents the photographic practices of marine travelers in the western Arctic. I argue that Arctic pictures belong in the corpus of western photography. Like images of the Rockies and the Pacific coast, pictures of Alaska and its environs resulted from journeys undertaken for artistic; scientific; and, as is the case for the *Bear*, governmental purposes. As with depictions of other parts of the West, such work gave audiences a visual introduction to newly acquired American territory and its physical features, visually incorporating it into the national imagination. Reproduced widely, photographs of glaciers and ice fields of the Alaskan coast integrated "the Final Frontier" into national imaginaries of expansion, resource extraction, and "civilization." While the Alaska purchase was questioned by many when it was made in 1867, early supporters were eager to exploit Alaska's resources—including fish, furs, timber, coal, and even ice—just like expansionists in the continental United States. Annexed at a time when the federal government was shifting its policies toward Native Americans from removal to unoccupied land to assimilation through education and missionization, Alaska was also an important site of American settler colonialism. We can thus see the region as one of the "alternative Wests" to which this volume is dedicated.

This essay will focus on a significant subset of photographs: those produced by employees of the United States Revenue Cutter Service (hereafter RCS), a fleet assigned to patrol Alaskan waters. Photographers traveled on the revenue cutters *Corwin* and *Bear* beginning in 1880, eventually creating an archive of around 1000 photographs. RCS pictures were used to illustrate government reports and newspaper and magazine articles, and they were also valued by individuals associated with the service and its work. Yale University, the Library of Congress, the Huntington Library, and the New Bedford Whaling Museum all hold albums that appear to have been assembled by Alfred Lee Broadbent, who made photographs during his service as an engineer on the *Bear* in the 1890s, presumably to present to passengers and friends. Other pictures connected to the revenue cutters were produced by government employees, rescued sailors, and tourists who traveled on the ships.

As I will explore, these photographers worked under challenging conditions, including changeable weather, temperatures ill suited to darkroom practices, and rolling seas. Because of this, photographs associated with the RCS's Bering Sea Patrol in the 1880s and 1890s offer a privileged window into how the environment helps shape its own representation. At every turn, the pictures manifest the necessity of engagement with the structuring limitations of an environmental context. After offering important contextualization, this essay will explore what the photographs reveal about the intertwined ecological concerns of climate and temporality.

Visual Cultures of Ice

The creation of so many pictures in Alaskan waters demonstrates the broad interest in images of the far north. Painted, printed, and projected representations of frozen Arctic seas were popular in the nineteenth century. Expeditions seeking the Northwest Passage had captivated international imaginations since Sir John Franklin's ships *Erebus* and *Terror* were abandoned in what is now Nunavut in 1845. At the urging of Franklin's widow, British and American rescue expeditions soon followed, establishing the careers of famed American explorers Elisha Kent Kane and Adolphus Greely, both United States Army officers. As Russell Potter (2007) has documented, such expeditions generated prodigious popular visual culture. In the 1850s and 1860s, there were no fewer than nine traveling panoramic displays related to Kane's voyages being shown in venues ranging from church halls to P. T. Barnum's American Museum. (fig. 2) The federal sponsorship of photography in the Arctic included Sergeant George W. Rice's pictures for the 1881–1884 Lady Franklin Bay Expedition, led by United States Army

Lieutenant Adolphus Greely, which were used as the basis for engraved illustrations of its reports. (Potter, 2007: 201) This development was undoubtedly related to the impact of photographers' work for the United States Geological Survey in the American West in the preceding decade, a fact that reinforces the association of the Arctic with other sites of US expansion in the nineteenth century.

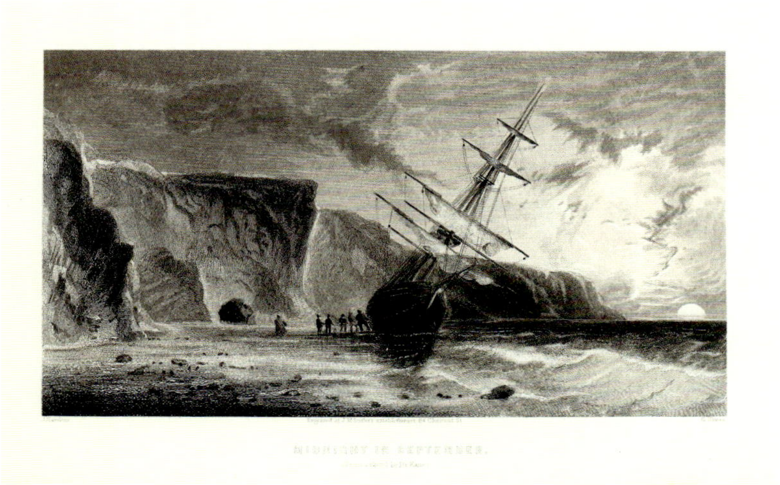

Fig. 2.

James Hamilton, from a sketch by Elisha Kane, "Midnight in September," from Elisha Kent Kane, *Arctic Explorations: The second Grinnell Expedition in Search of Sir John Franklin, 1853, '54, '55, Volume I* (Childs & Peterson, 1856), opp. 140.

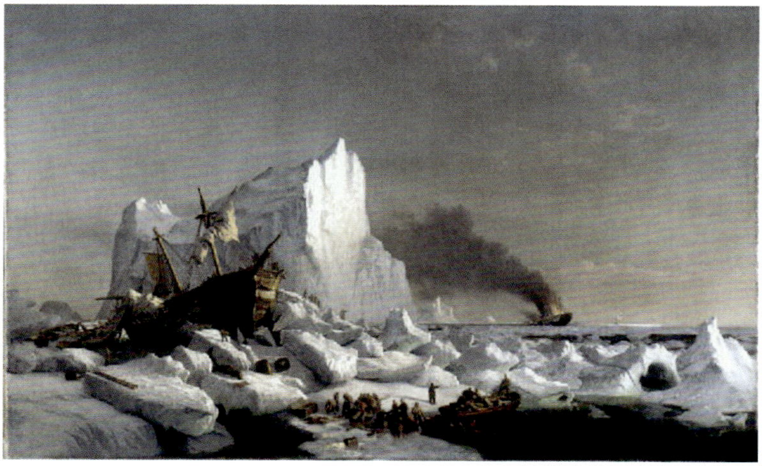

Fig. 3.

William Bradford, *Sealers Crushed by Icebergs*, 1866. Oil on canvas. 182.9 × 304.8 cm. Courtesy of the New Bedford Whaling Museum. 1972.33.

Nearly twenty years prior to the Lady Franklin Bay Expedition, painter William Bradford brought a camera on an Arctic cruise, likely making him the first to do so. Bradford used the photographs made during his 1860s expeditions as aids in the creation of paintings such as *Sealers Crushed by Icebergs* (1866). (fig. 3) Photographs made by John Dunmore and George Critcherson serve as tipped-in illustrations for the book

The Arctic Regions, which recounts his 1869 voyage to Greenland, and as lantern slides projected during public lectures about that expedition. (Bradford, 1872) Bradford praised Dunmore and Critcherson's "indefatigable [...] efforts to overcome the obstacles which were constantly presented, and which appeared really to have no end." (Bradford, 1872: viii) Even so, the photographs in *The Arctic Regions* bear traces of those obstacles. Many of the images taken at sea are in focus only in the center and are darker and blurrier in the corners, reflecting the challenges of making photographs while under said conditions and the limitations of their round lenses in the polar sunlight. (fig. 4) This haloed effect, which is visible in Figure 1 as well, may also be due to traces of the build-up of ice crystals on the lens. Although the artists brought a camera that held 14 × 18 inch glass plates on the expedition, many of these poorly exposed pictures came from 3 × 4 inch negatives, which the photographers would have chosen to allow for faster exposures. The fact that many of these small views still exhibit problems of exposure and focus reinforces the difficulty of making photographs under these conditions. Indeed, the larger photographs of icebergs included in *The Arctic Regions* were made when the ship was not moving. For example, one caption explains that the ship remained in place for three days. (Bradford, 1872: 74)

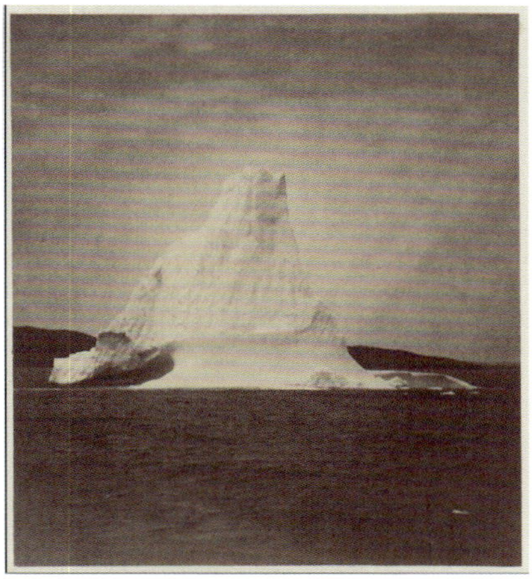

Fig. 4.
John Dunmore and George Critcherson, "Instantaneous View of an Iceberg" from William Bradford, *The Arctic Regions* (Sampson and Low, 1872), 52. The Miriam and Ira D. Wallach Division of Art, Prints and Photographs: Photography Collection, The New York Public Library. https://digitalcollections.nypl.org/items/7781a7d0-f814-0136-c573-0c7cf06c7f3d

While most early Arctic photographs relate to expeditions, others are connected to whaling. Thousands of whaling expeditions entered the dangerous Arctic seas in the nineteenth century, impacting the families of sailors, investors, and consumers of whale products. (Bockstoce, 1986) Some pictures of ships in the ice may have served as records that could return to home ports when the vessel had been abandoned or destroyed. For example, one photograph in the New Bedford Whaling Museum features a photograph by Broadbent of the *John Carver*, a 319 ton whaling bark that was wrecked in ice during a storm in the Bering Sea in June 1886. (fig. 5) The ship's logbook recounts this event, documenting the crew gathering up "what Bread we could [...] and taking to the boats." (John Carver Bark, 1885: 191) The crew was picked up by the bark *Atlantic* two days later, and the cargo was later salvaged. Just two weeks prior, the *John Carver* had made contact with the Revenue Cutter *Bear*, picking up mail and other papers. (John Carver Bark, 1885: 188) Because the ships were at anchor during this exchange, this encounter provided the opportunity to make a photograph of the *John Carver*. Because it happened rarely and was an exceptional event, logbooks

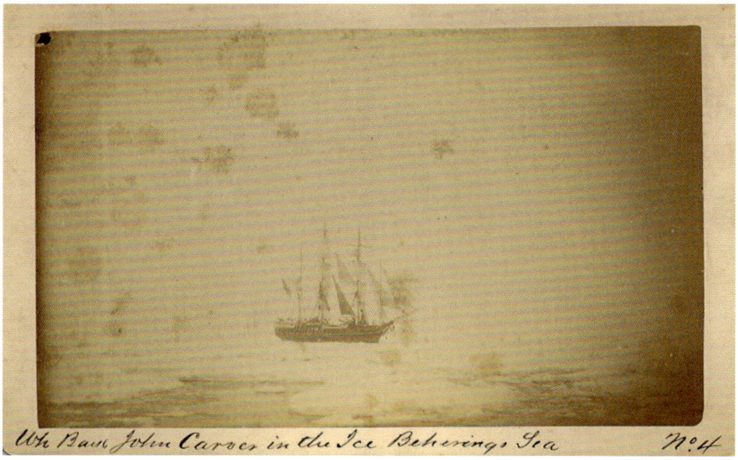

Fig. 5.

Photographer not known (possibly Albert L. Broadbent), *Wh Bark John Carver in the Ice Behring [sic] Sea. No. 4,"* 1886. Albumen print, 11.4×19.1 cm. Courtesy of the New Bedford Whaling Museum. 1982.40.9.

always noted when another ship was sighted, and from a similar impulse, Revenue Cutter photographers frequently depicted the other ships they encountered, both at sea and in port.

In each of these cases, photographers working for the government, for small commercial enterprises, and to contribute to personal and popular visual culture undertook their work as a means of providing American viewers with a glimpse of a region to which they were connected. The quest for a Northwest Passage related to their interest in trade goods from Asia. Whalers provided them with oil used to power lamps, lubricate machinery, and produce soap and other commodities. And the area more generally fueled a taste for stories of adventure and the human struggle against nature. As both the subject—ships foundering in icy seas—and the formal flaws in the pictures themselves demonstrate, this struggle was quite real.

Photography and Extraction in the Western Arctic

While the eastern Arctic created the market for photographs of icy seas, the largest corpus of these pictures was created on revenue cutter patrol boats in the waters surrounding Alaska. These ships were part of the Revenue Marine Service created within the Treasury Department in 1790 to enforce maritime law and, beginning in 1837 to service distressed vessels, a practice later taken over by the Coast Guard upon its founding in 1915. (Strobridge, 1999) Cutters were stationed in the Pacific during the Civil War, and immediately after the purchase of Alaska from Russia in 1867, they sailed north to extend the enforcement of customs law up the Pacific coast, soon after supporting the whaling industry there. American whalers had come to Beringia in 1848 in pursuit of bowhead, a species of baleen whale that spends its entire life in Arctic waters. This was extremely dangerous work, but the ongoing demand for whale oil and baleen and the diminishing populations of sperm whales and other species due to overhunting was irresistible. Whalers followed the bowheads into the so-called Arctic Grounds, stretching from the Bering Strait to Point Barrow, but could only do so during a few summer months when the ice was in retreat. When Autumn came early, ships could be caught in rapidly advancing pack ice; this halted their progress or, in the most extreme cases, ground them between floes. This is exactly what happened in September 1871, when whalers were forced to abandon thirty-three ships trapped in the ice, the majority of the Arctic fleet, knowing that rescue was not coming until the following summer. Many of the whale men were saved by ships that were able to break loose, but only after the rescuers had dumped their valuable catch to make room. The tragedy captured the American imagination and was illustrated in *Harper's Weekly* and other publications, creating a new chapter of ships-trapped-in-ice visual culture, perhaps the most luxurious being a five-plate series of lithographs by John Perry Newell. (fig. 6)[1] ("Crushed", 1871: 1128–1129)

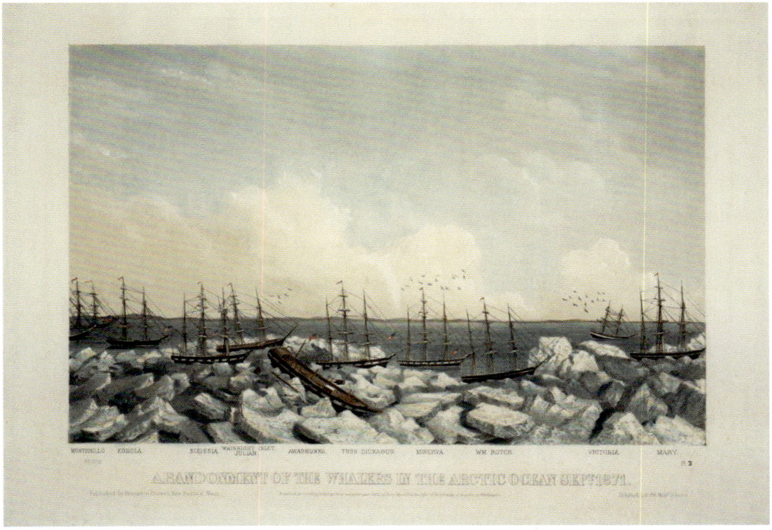

Fig. 6.
John Perry Newell, *Abandonment of the Whalers in the Arctic Ocean*, September 1871. Hand-colored lithograph, 36.5 × 53 cm. Printed by Benjamin Russell, New Bedford, Mass. J. H. Bufford's Lith, 490 Washington St, Boston, 1872. Theodore G. and Eleanor S. Congdon Maritime Collection, The Huntington Library, San Marino, Calif. priTEC 0054.

The permanent assignment of revenue cutters outfitted with iron ice-cutters on their bows occurred after the steam and sail ship *Jeanette* went missing in 1880 on the U.S. Arctic Expedition seeking the North Pole. Three ships—the *Richard Rush*, the *Corwin*, and the *Bear*—eventually made up a "Bering Sea Patrol" that reinforced an American presence in territorial waters around the Aleutian Islands and north up through the Chukchi Sea each summer, with the *Bear* continuing on into the

Beaufort Sea, along Alaska's North Slope, when the ice pack permitted. They also patrolled the Pribiloff Islands, the only breeding grounds for the northern fur seal. While policing illegal traffic in furs and alcohol was their priority, the ships also undertook search and rescue missions to recover vessels, men, and cargo lost due to the unpredictable movement of the ice.

RCS photography began almost as soon as the Bering Sea Patrol was created, perhaps in response to the inclusion of photographs in the annual reports of other projects, such as the US geological surveys of the trans-Mississippi West, including the United States Geological Survey. Robin Kelsey has argued that one reason survey employees made photographs in the late nineteenth century was to create documents that justified their budgets. (Kelsey, 2003) Prior to 1881, RCS officers had occasionally submitted sketches with their annual reports, which were transformed into illustrations in the final publications. These frequently took the form of depictions of coastlines useful for navigation but could include more sensational imagery. For example, the *Corwin* report for 1880 included an engraving of the ship caught in a "nip" (frozen ice field) off Siberia. (Hooper, 1884: opp. p. 6)

The following year, the ship's report included photographs made by naturalist Edward W. Nelson, who had traveled on the *Corwin* while serving as weather observer for the United States Army Signal Corps at St. Michael (at the mouth of the Yukon River) in 1881. Spencer Baird, the secretary (director) of the Smithsonian Institution, had arranged for this appointment so he could collect natural history specimens and document Inuit life for the National Museum. Most of the published photographs feature Alaska native Iñupiaq and Siberian native Chukchi people and settlements, not images of the sea. The report notes the difficulties Nelson faced taking pictures, which may have prevented him from creating other kinds of views. (Hooper, 1884: 23) Photographs appear in many subsequent publications about the RCS's work in the Arctic, such as *The Cruise of the U.S Revenue Cutter Bear and the Overland Expedition for the Relief of the Whalers in the Arctic Ocean,* indicating the support of the Secretary of the Treasury for this endeavor. (United States Treasury Department, 1898)

Nelson's pictures are claimed to be the first made in western Alaska. (Fitzhugh, cited in King and Lidchi, 1998:126). Within a year, other passengers and crewmembers on RCS ships were taking photographs, too. In addition to showing ships being rescued, pictures demonstrate the cutter's service to the Alaska Commercial Company (ACC), a San Francisco–based firm that had a twenty-year exclusive contract with the federal government to take seals from the Pribilof Islands in exchange for a lump sum and a tariff of $2 per pelt. The cutter's patrols were important, as the fur industry could yield more than $300,000 for the federal government a year. (Committee on the Library, 1885: 1) The cutters brought mail, supplies, and, sometimes, employees to company settlements. They also transported guests: ACC staff, federal employees, and Indian Affairs officials, such as Sheldon Jackson, the missionary who served as General Agent for Education for Alaska. Jackson oversaw the establishment of schools that were overseen by the Presbyterian church on behalf of the federal government.

All of these tasks undertaken by the RCS crews helped bring modernity—including American attitudes toward nature, religion, and work—to Beringia, advancing US settler colonialism in the region. Work for the ACC lured Indigenous people away from traditional seasonal activities. As historian Bathsheba Demuth has discussed, the establishment of commercial enterprises headquartered in the south and other disruptions of Yup'ik, Iñupiaq, and Chukchi ways of life unsettled long-standing practices developed in response of the rhythms of the migrating sea mammals and other cycles related to subsistence. (Demuth, 2019) Jackson also invested in economic development projects designed to bring Alaska Natives into a cash economy, something understood as essential to their assimilation to American culture. His major project was the development of commercial reindeer herding. Beginning in 1891, the *Bear* transported both reindeer and Chukchi reindeer herders across the waters to Iñupiaq communities on the Seward Peninsula. Pictures from these rescue and reindeer operations are among those reproduced in multiple albums and in the press, indicating the interest that the public in the south had in these activities. (fig. 7)

Beyond providing records for the government, the photographers' work may have served to reinforce ties between the RCS, ACC employees, and other settlers from the south through the visualization of their social interactions. An album in the Huntington Library Collections

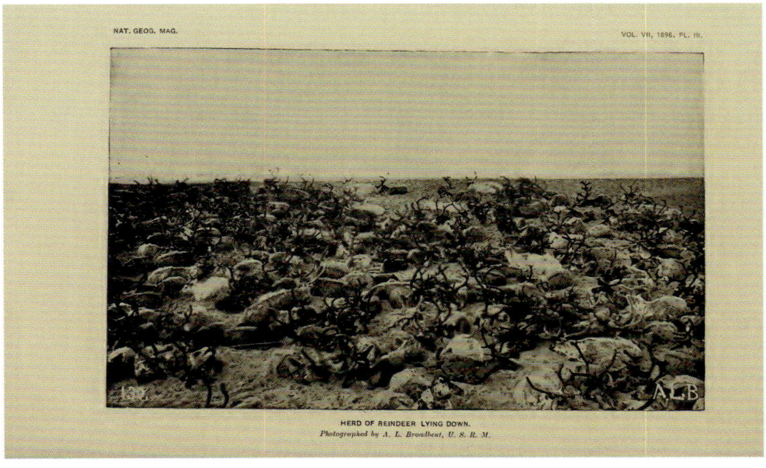

Fig. 7.

Albert L. Broadbent, *Herd of Reindeer Lying Down*, reproduced in Sheldon Jackson, "The Arctic Cruise of the U.S. Revenue Cutter Bear," *The National Geographic Magazine* (January 1896): opp. 29.

inscribed by Broadbent to "Mr. Baldwin of Dutch Harbor" demonstrates this. Baldwin was likely an ACC employee, as the ACC operated an office and coaling station at this location. Numerous images capture the socializing that broke up long periods of isolation for both sailors and those on land. One photo captures the performance of a civilian brass band on the *Bear* during a stop in Dutch Harbor. (fig. 8) Others capture visits on board from Iñupiaq locals hoping to trade with sailors, settlers standing smiling in front of their houses or places of work, and tourists both on ship and in port.

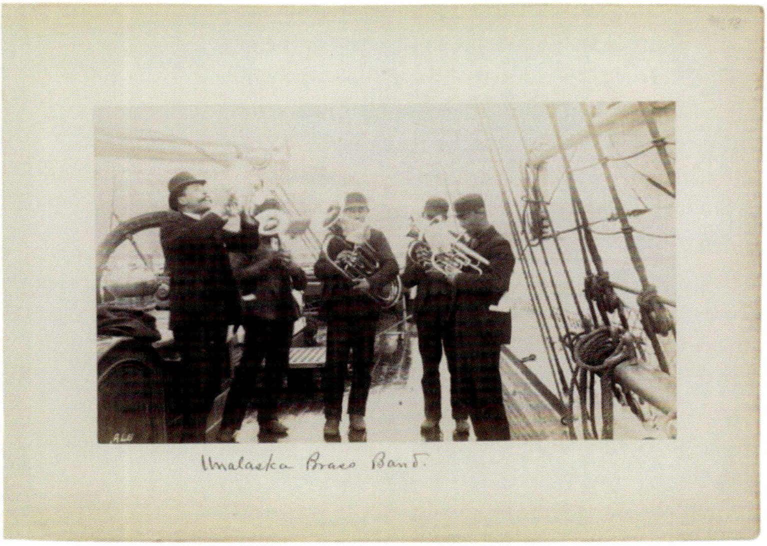

Fig. 8.
L. Broadbent, *Unalaska Brass Band*, 1890s. Albumen print from plate glass negative. 7×11 in. (18×27 cm). Carl S. Dentzel Photograph Collection of the American West, 1870s–1890s: Alaska Views, Huntington Library, San Marino, Calif. photCL 98.

The pleasure demonstrated in these pictures may reflect the fact that arriving in a port never happened on a schedule or in a specific order. Cutters could not predict how long a passage would take, and if they arrived to find a harbor blocked in with ice, they would leave, sometimes trying again later in the season and sometimes never arriving at all. Ellen Kittredge Lopp, a teacher and missionary who worked at the Iñupiaq village of Kingigin (Cape Prince of Wales) in the 1890s, worked on letters to be sent to family all year long, only finishing them when the *Bear* pulled into the harbor. On occasion, these letters recorded two full years of activity undertaken while waiting for its arrival. (Lopp, 2001) Lopp's letters demonstrate how time was experienced differently in the Arctic than in the continental United States, where transportation and communication was more routinized.

On some days, storms, fog, and other conditions changed the tasks sailors could undertake. The logbook's rhythm is occasionally interrupted by longer entries noting the need to drop anchor to make repairs or wait for clearer skies that allowed seeing what lay ahead. In late May, the log records being in "close ice" along with several other ships and slowly making for St. Lawrence Island.

As I will argue, the RCS photographs offer insights into how what I call "Arctic temporality" was experienced.

Arctic Temporality

For anthropologist Tim Ingold, meaning derives not from ideas imposed upon the world but through activities undertaken in and with it. This concept is well articulated in his use of Heidegger's concept of "dwelling," which he offers as a means of rejecting the binarism of conceiving of an "outer world" through constructs of an "inner mind." (Ingold, 2021: 237) Dwelling calls attention to how we spend time in the world, or in Ingold's terms, how we perform tasks. He points out that all tasks are contextual: "by watching, listening, perhaps even touching, we continually feel each other's presence in the social environment, at every moment adjusting our movements in response to this ongoing perceptual monitoring." (Ingold, 2021: 243) But this context is more than purely social; human activity is also shaped by what a location affords. This includes what he identifies as "rhythmic phenomena—the cycles of day and night and of the seasons, the winds, the tides, and so on." (Ingold, 2021: 248) In the Arctic, these cycles influence the movement of whales and other sea life and the growth and retreat of the ice shelf and the calving of ice bergs, conditions that structured the human experience. Seasonal changes limited ships' passage to the far north to the summer not only because of the reappearance of open water, but also because of the return of sunlight.

Adjusting activity to the environmental context is a particular necessity in the Arctic seas, given that conditions there are constantly shifting. As Demuth has written about Beringia, "human ways of seeing time mixed with the times of the place itself, the lives of animals and the cycles of the seasons." (Demuth, 2019: 8) This has been true for centuries for the Yup'ik, Iñupiaq, and Chukchi communities dependent on marine life for sustenance, but it was equally true for ships' crews turning sea mammals into commodities. She notes that change is a constant in the region, "from the earthquakes that jumble the Pacific Rim to the genetic drift of species over generations to the seasons, rhythmic in their succession but abrupt in lived experience." (Demuth, 2019: 10) Some of these changes occur in a matter of days; others unfold over a lifetime or, in the case of the formation of glaciers, tens of thousands of years.

The dependence of the cutters' itineraries on weather reminds us that these cycles are not purely regular. RCS crews experienced other uncertainties: the timing of ice receding and advancing, chance encounters between ships, and the need to respond to ships in distress ensured the unpredictability of the movement of the cutters and the work being conducted by their crews. This was exacerbated by the fact that the RCS worked during the long days of summer. A fascination with the midnight sun was part of the popular interest in images of the Arctic, as Figure 2 attests. The experience of Arctic temporality related not just to a shift in how days were experienced, but also to disorientation regarding seasons. For example, the naturalist and conservationist John Muir, who sailed with the *Corwin*'s 1881 rescue mission, noted that when he left San Francisco in May it was fifty-five degrees, but it was freezing snow in Unalaska, on the Aleutian Islands. He went on to describe the need to shelter in a harbor waiting for a change in weather: "I have never seen the sea in so hearty an uproar [...] it was all tossed into one white mass of hissing, roaring foam, blending in wild harmony with the wind-driven scud and sleet that thickened the air." (Muir, 1881: 1)

In addition to the unpredictability of the weather, changes in the ice could also suddenly slow or even halt ships' progress. The memoir of Hartson Bodfish, captain of the whaler *Beluga,* explains this:

> Arctic ice isn't like any other kind [...] the pack moves in summer and winter. At times there is no ice in sight, and then suddenly it will appear and pile up terribly. I recall one winter when I was lying behind a sandspit, waiting for spring, when the ice all disappeared. I left orders to be called at four o'clock, and to have steam up, ready to get under way. At midnight they called me to look things over, and the pack was back again. In fifteen minutes it had piled fifty feet high in the sandspit. (Bodfish, 1936: 41–42)

The book traces the *Beluga's* crossing and recrossing the sea for five months before making its first strike in late September, demonstrating the monotony of whaling and the simultaneous need to be ready for sudden environmental changes. Ships' logs similarly reflect these cycles of inactivity and preparation in response to changing conditions. Such calendar keeping suggests twenty-four-hour cycles were not meaningful ways to track events. The year itself unfolded unevenly, as ships could only hope to move through Arctic waters between May and September. Some would sail elsewhere in the winter, but others passed months with little activity waiting for the ice to break. During the months when whalers and other ships were in motion, the days were nineteen to twenty-four hours long and the sun's movement thus had little influence over the watches and other daily marine activities. These factors impacted work with cameras, too. RCS photography was facilitated by the popularization of the dry-plate process, which allowed negatives to be prepared and stored until use and also for a delay between exposuring and developing a negative. These conveniences were better suited to the unpredictable schedules aboard the revenue cutters. Nevertheless, storms, ice, and even the powerful glare of the sun could make their work more difficult. For example, equipment needed to remain undamaged, as light leaking into the camera interior could ruin a shot, and, of course, cracked plates or lenses were useless. Swells and storms thus threatened the work of photographers working on the *Bear* and the *Corwin*. In addition, photographers could not have their cameras out while undertaking their regular tasks. They generally made pictures when the ships' progress was halted by ice or while in port. Thus, the visual record of the RCS illustrates the acceleration and deceleration of tasks that occurred unpredictably in the Arctic, through what is pictured and what is absent. The predominance of images of ice testifies to the abundance of time available for image making when the ship's progress was impeded. Similarly, ships frequently spent lengthy periods in certain harbors to facilitate repairs, leaving photographers free to capture images of the residents and buildings in the settlements they visited.

Of course, events could also happen when ships were under sail. However, the movement of the sea made photography difficult. For example, when the *Corwin*'s crew sailed through a shower of volcanic ash from a new volcano in the summer of 1884, Lieutenant George H. Doty got out his camera. As the captain reported, one of the photographs was "somewhat blurred, unfortunately, owing to a sudden lurch of the ship." Later, Doty's work was stymied by an "immense volume of aqueous vapor." Nevertheless, the officer praised Doty's "(under the circumstances) brilliant success," writing, "Without the accompanying photographs an adequate description of this recent and most interesting addition of the Aleutian group of islands could hardly be given, while with them one can avoid creating false impressions, and can also remove erroneous conceptions." (Hooper, 1884: 42, 45)

Arctic conditions posed technical challenges to photographers beyond the moment of making an exposure. The difficulties followed them into the darkroom, where negatives were transported for developing and fixing. As I have explored elsewhere, these tasks were anything but straightforward when undertaken in the field, where a small tent replaced the darkroom and where chemicals were threatened by changing temperatures as well as dust and other hazards. (Hutchinson, 2020) The chemicals needed for these processes could flow poorly in cold temperatures. Western landscape photographer William Henry complained about this fact when photographing in frozen temperatures in the Tetons. He used a shovel heated over a fire to warm up his portable darkroom to ensure a smooth operation. (Jackson, 2005: 82) Arctic photographers no doubt improvised as well. Another problem was how to source the fresh, clean water that was essential for several steps of making photographs, including mixing solutions and rinsing the exposed print during development. Revenue cutter artists must have carried pure water with them, protecting it from freezing, when they left the ship to take pictures from the land or on the ice. Finally, finding an appropriate space to develop photographs could be tough. In 1894, *The American Annual of Photography and Photographic Times Almanac* included an article about the importance of photography for naval reports and those of the Coast Survey, noting, "on the new ships the facilities for photography as a rule, are very good. A store-room below the water line makes and [*sic*] ideal dark-room, for it is well ventilated by the ship's blowers." But this room had no windows and had only sea water available for use for the first rinsing during the development of a negative. (Gleaves, 1894: 48) Of course, these rooms were only functional as long as the sea was calm and the ship avoided surging ice.

Photographic prints carry traces of these challenges. Most notable is an unevenness in the exposure across many of them, perhaps an effect of the difficulty in managing the bright reflections of an icy landscape or of the poor flow of emulsion during the coating of the plate. (see figs 1 and 4) Other pictures appear faded all-over, perhaps indexing the light leak of a damaged camera. Finally, in one photograph, an iceberg appears to be viewed through a mottled overlay that is likely the effect of damaged emulsion. (fig. 9) The presence of these and many other flawed

Fig. 9.
Photographer unknown, *Ice-Berg off Point Barrow,* 1884. Photograph. 8×9 in. (19.5×24 cm). Photograph collection of the Revenue Cutters "Bear" and "Corwin" in Alaska, 1880s–1890s, Huntington Library, San Marino, Calif. photCL 131 (76).

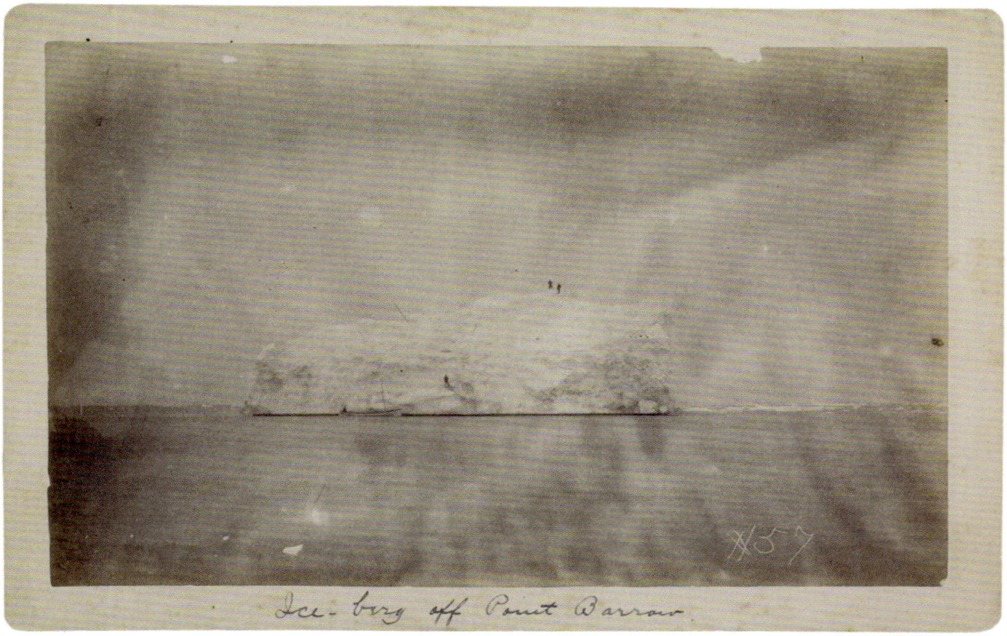

prints in collections of RCS photographs, even on the pages of the albums Broadbent prepared as gifts, demonstrate that their value was not grounded in the quality of the photograph, but in the record they offered of how time was spent in the Arctic. When Figure 9 was reproduced in the report of the *Corwin*'s 1884 cruise, the Government Printing Office cleaned the image up, removing the streaks, making a stronger distinction between the iceberg and the sky, and making the figures on the iceberg easier to see. (Healy, 1889: opp. p. 10) But the original offers the greater illustration of Arctic temporality. The laborious, imperfect nature of sight is constitutive of dwelling in this ecological zone.

Photography and Environmental Change

Whaling ships in the nineteenth century killed about 20,000 bowhead whales, decimating a population that was only 30,000 at its height, which was in the 1840s. Others were killed by shore-based kill. (Bockstoce, 1986: 15) Fur seals experienced similar devastation during the early period of the US occupation of Alaska. While the federal government intended to manage the population by sparing females and allowing enough males to survive to continue breeding, a reported 225,000–300,000 seals were killed each year in 1868 and 1869, although the ACC contract limited them to 100,000 seals taken a year. (Roppel and Davey, 1965: 451) As Demuth has explained, commercialized extraction brought to Beringia by the United States depleted resources at rates that could not be matched by regeneration, given the suppression of birth rates caused by indiscriminate hunting and the slow rate at which some species, such as bowheads, mature. (Demuth, 2019: 44–72)

Today the greatest threat to the icy habitat of the Western Arctic is climate change and the fossil fuel industries that drive it. The US National Oceanic and Atmospheric Association's 2024 *Arctic Report Card* indicates the dramatic reduction of sea ice over the preceding eighteen years and dangerously rising permafrost temperatures. Among the many resulting shifts in Arctic ecosystems, these changes have increased plankton blooms, which impact the oceanic food web, and dramatically reduced the population of migratory tundra caribou. (Moon et al., 2024: 2–3) These changes are likely irreversible.

As in the nineteenth century, contemporary photographers are documenting American practices of resource extraction and their impact on the environment. One example is Subhankar Banerjee, who began taking pictures in the Arctic National Wildlife Refuge in Alaska in 2000. As he spent time in the region, he came to understand that this coastal plain is "a biological nursery of global significance." (Banerjee, n.d.) Finis Dunaway has shown how Banerjee's work became central to legislative debates around the spread of extraction in the Refuge in 2003, when Senator Barbara Boxer showed one of his photographs on the Senate Floor and exhorted her colleagues to vote for a measure to prevent drilling. (Dunaway, 2009: 254–256) The region was opened for gas exploration in September 2019, though the political battle continues.

Banerjee's photographs illustrate the ongoing disruption of ecological cycles through their subject matter in pictures that are technically perfect. A century earlier, RCS photographers captured the evolution of Arctic temporality in the very form of their images. Through what they pictured, when they were able to shoot, and the visual appearance of the pictures themselves, they remind us how the Arctic environment shaped the experiences of the people who move through it. Studying images like these is important, as it reminds us that, while western American photography helped facilitate expansion and extraction, these projects did not unfold without challenges and that even today, ecological forces can unsettle human enterprises.

Notes

1. For more on the whaling disaster, see Nichols, 2010: 195–204, 227–256.

Archival Collections Containing Images from the Revenue Cutter Service

Alfred L. Broadbent, Photograph Albums of the United States Revenue-Cutter Service in Alaska and Siberia, WA Photos 63, Boxes 1 and 2, Yale Collection of Western Americana, Beinecke Rare Book and Manuscript Library, Yale University, New Haven, CT.

Alfred L. Broadbent, Documentary Photographs of Alaska and the Islands off Its Northern Coast, 1894–1995, l, Lot 3198, I. 102. Library of Congress Prints and Photographs Division, Washington, DC.

Carl S. Dentzel Photograph Collection of the American West, 1870s–1990s: Alaska Views, Album photCL 98, Huntington Library, San Marino, CA.

New Bedford Whaling Museum, New Bedford, MA.

Photograph Collection of the Revenue Cutters "Bear" and "Corwin" in Alaska, 1880s–1990s, PhotCL 131 (9 boxes), Huntington Library, San Marino, CA.

Photographs of Activities, Facilities, and Personalities, 1939–1967, Record Group 26: Records of the United States Coast Guard, National Archives and Records Administration, Washington, DC.

U.S. Revenue Cutter Service Photograph Collection, ca. 1905, 1923, 0079 PCA, Alaska State Library, Anchorage.

Bibliography

Subhankar Banerjee, "Subhankar Banerjee Biography," (n.d) https://www.subhankarbanerjee.org/banerjee.html.

John R. Bockstoce, *Whales, Ice and Men: The History of Whaling in the Western Arctic* (Seattle: University of Washington Press, 1986).

Hartson Bodfish, *Chasing the Bowhead* (Cambridge, MA: Harvard University Press, 1936).

William Bradford, *The Arctic Regions* (London: Sampson and Low, 1872).

William Bradford, "Life and Scenery in the Far North," *Journal of the American Geographical Society of New York* 17 (1885), 104.

Committee on the Library, U.S. Congress, *Thanks of Congress to the Officers and Crew of the Steamer Corwin* (H. Rpt 2507) 48th Cong. 2d Sess, February 7, 1885:1–5.

"Crushed Among Icebergs," *Harpers Weekly* 15 (December 2, 1871): 779.

Bathsheba Demuth, *Floating Coast: An Environmental History of the Bering Strait* (New York: W. W. Norton, 2019).

Finis Dunaway, "Reframing the Last Frontier: Subhankar Banerjee and the Visual Politics of the Arctic National Wildlife Refuge," in *A Keener Perception: Ecocritical Studies in American Art History*, ed. Alan Braddock (Tuscaloosa: University of Alabama Press, 2009), 254–274.

Albert Gleaves, "Photography in the Navy," *American Annual of Photography and Photographic Times Almanac* (January 1, 1894): 48.

Capt. M. A. Healy, *Report of the Cruise of the Revenue Marine Steamer Corwin in the Arctic Ocean in the Year 1884* (Washington, DC: Government Printing Office, 1889).

C. L. Hooper, *Report of the Cruise of the Revenue Steamer Corwin in the Arctic Ocean in the Year 1881* (Washington, DC: Government Printing Office, 1884).

Elizabeth Hutchinson, "'Photographic Weather': A Posthumanist Approach to Western Survey Photography," *Panorama: Journal of the Association of Historians of American Art* 6, no. 2 (Fall 2020), https://doi.org/10.24926/24716839.10862.

Tim Ingold, *The Perception of the Environment: Essays on Livelihood, Dwelling and Skill.* New Edition (London: Routledge, 2021).

William Henry Jackson, *William Henry Jackson's "The Pioneer Photographer" Compiled, Edited, and Annotated by Bob Blair; with Original Text from the 1929 Edition by William Henry Jackson in Collaboration with Howard R. Driggs* (Santa Fe: Museum of New Mexico Press, 2005).

"John Carver (Bark) of New Bedford, Mass., mastered by Albert C. Sherman, on voyage from 6 May 1884–11 Nov. 1885" (logbook) New Bedford Whaling Museum, New Bedford, MA.

Robin E. Kelsey, "Viewing the Archive: Timothy O'Sullivan's Photographs for the Wheeler Survey, 1871–74," *The Art Bulletin* 85, no. 4 (2003): 702–723.

J.C.H. King and Henrietta Lidchi, *Imaging the Arctic* (London: British Museum Press, 1998).

Ellen Kittredge Lopp, *Ice Window: Letters from a Bering Strait Village, 1892–1902* (Fairbanks: University of Alaska Press, 2001).

Twila A. Moon, Matthew L. Druckenmiller, and R. L. Thoman (eds), *Arctic Report Card 2024* (Washington, DC: National Oceanic and Atmospheric Administration, 2024), https://doi.org/10.25923/b7c7-6431.

John Muir, "The Corwin's Cruise." *San Franscisco Daily Evening Bulletin* (June 20, 1881): 1.

Peter Nichols, *Oil & Ice,* 2nd ed. (London: Penguin Books, 2010).

Russell Potter, *Arctic Spectacles: The Frozen North in Visual Culture, 1818–1975* (Seattle: University of Washington Press, 2007).

Alton Y. Roppel and S. P. Davey, "Evolution of Fur Seal Management on the Pribilof Islands," *Journal of Wildlife Management* 29, no. 3 (1965): 448–463.

Truman R. Strobridge, *Alaska and the U.S. Revenue Cutter Service, 1867–1915* (Annapolis, Md.: Naval Institute Press, 1999).

United States Treasury Department, *The Cruise of the U.S Revenue Cutter Bear and the Overland Expedition for the Relief of the Whalers in the Arctic Ocean from November 27, 1897 to September 13, 1898* (Washington, DC: Government Printing Office, 1899).

Afterword

Robin E. Kelsey

The Tohono O'odham linguist and poet Ofelia Zepeda has written a poem entitled "Smoke in Our Hair," in which these two lines appear:

> Smoke, like memories, permeates our hair,
> our clothing, our layers of skin.

Zepeda's poem celebrates the smoke of a shared and traditional desert fire, the scent of "mesquite, cedar, piñon, juniper" that settles onto the body, traveling with it, keeping the bearer connected to place and people. In the final five lines of the poem, she writes:

> no matter how far we walk,
> we carry this scent with us.
> New York City, France, Germany—
> we catch the scent of burning wood;
> we are brought home. (Zepada, 2008: 72)

There is smoke and there is smoke.

Each time I get out my sleeping bag to go camping, I smell the wood smoke on its polyester shell, in the down filling, a scent that first sunk into those materials decades ago in the Rocky Mountains. For me, that scent is a Proustian tea-soaked madeleine, unleashing memories of wilder places. It stirs a desire to build a campfire, to sing beside it, and to smell its smoke the next morning in my unwashed hair.

But there is smoke and there is smoke.

The West is burning. The smoke from the 2025 Palisades Fire in Southern California has clung to hair, clothes, burned siding, and whole hillsides. As the fire raged, the acrid effluents of burning plastic and metal mixed with the smoke of burning wood. Chromium, copper, lead, and volatile organic compounds, such as formaldehyde and phosgene, mingled in the air. The lingering scent is not of home or of nights beneath the stars. It is the smell of reckless industrial desire.

As I read the manuscript for this volume, the metaphor of smoke stayed with me. Like smoke, the history of the West clings to us. Whatever roles our ancestors and we have played—or have been forced to play—in that history, we may feel bound to it willy-nilly, bound to its restlessness, its homesickness, its beauty, its subjugations, and its rapacious violence, as if it were a scent we cannot shake.

Photography is part of that scent. As more than one contributor to this volume reminds us, for much of the twentieth century, the dominant take on photography's role in the history of the West was celebratory. Publications and exhibitions exalted practitioners such as Carleton Watkins and Ansel Adams for having divined and captured the aesthetic splendors of the landscape. The association of photography with wilderness conservation, from the establishment of the National Parks to the lobbying of the Sierra Club, bolstered the reputation of photography as the medium of the West *par excellence.* (Bartlett, 1974: 194–210; Dunaway, 2005; Kelsey, 2013a: 11–26)

This exaltation of western landscapes and photography relied on a tacit belief in the benevolence and efficaciousness of boundaries. Around the campfire, Sierra Club-bers may have sung "Don't Fence Me In," but maintaining boundaries was what the photographically celebrated West was all about. Watkins kept miners outside the frame, and Adams did the same with parking lots. Tourists followed suit with their own cameras, understanding that regulating visibility was a prerogative and necessity of the mythic West. The ideological scheme that governed picture making also governed land use. The boundary of the national park or conservation land presumed that those in power could manage the contradictions of the modern West if they just kept industry here and undeveloped spaces there, and more or less maintained the confinement of Native peoples to reservations.

Shifts in attention gradually undermined this faith in boundaries. Native activism and new histories of the West foregrounded the violent taking of traditional lands and the cruel legacies of imperialism. (Brown, 1970; Blackhawk, 2008) Historians of the region developed a keener interest in photographs and prints, putting more emphasis on their instrumental use than on their aesthetic qualities. (Goetzmann, 1967) Photographers such as Stephen Shore, Robert Adams, Richard Misrach, and Lee Friedlander began depicting the West differently, inviting its incongruities and contradictions into the frame. Curators such as Sandra Phillips organized exhibitions addressing land-use issues and impacts. (Philipps, 1996; Best, 2021) Meanwhile, environmentalism began questioning the limits of "protected land." The use of pesticides such as DDT and then climate change became broad public concerns that revealed the limited efficacy of borders as a means of buffering natural places from the effects of industrialization.[1] Authors such as Bill McKibben and William Cronon argued that the framing concepts of "nature" or "wilderness" were flawed or obsolete. (McKibben, 1989; Cronon, 1995) In these many ways, the myth of the West as a limitless space governed by boundaries gave way to the reality of a finite land subject to externalities that no perimeter could contain.

Boundaries mean nothing to smoke. As I write, haze from wildfires in Manitoba, in western Canada, surrounds me in Cambridge, Massachusetts.

Photography suffers the ecological impediment of being a boundary-making machine. In ordinary practice, it cuts out for representation pieces of a continuous visual field.[2] This mode of depiction not only allows a photographer to keep a redwood in the frame while excluding the parking lot beside it, it also keeps the apparatus-that-cuts out of sight. By design, signs of the camera and its operator, and by extension the material underpinnings of the medium, lie outside the view selected for depiction. The view points back to the photographer, but only as a disembodied authorial eye.[3] Photography, in short, thrives on pointing away from itself, on hiding its material basis in the vanishing point.

Over the years, conscientious practitioners and scholars have sought to counteract this tendency. In the West, photographers such as O'Sullivan, Friedlander, and Misrach have found subtle ways of calling attention within the frame to photographic materiality or process.[4] In the last decade or so, a bevy of scholars and practitioners have followed suit, bringing fresh scrutiny to the ecology of photography. Many agree that it is no longer enough to point the camera at the ugly tailing pond instead of the picturesque waterfall. One has to attend to what photography as a material process is extracting from the earth and depositing into the land, water, and air. This shift in focus is part of a larger material turn in the humanities.

Several contributions to this volume take up this call and advance our knowledge of photography's material consequences. Katherine Mintie treats Robert Adams's series "Turning Back" as an occasion to launch an inquiry into the history of photographic paper production. Adams aimed his camera at scenes of timber harvesting in the Pacific Northwest to contemplate and critique deforestation. As Mintie notes, irony haunts this match of medium and message, in light of how much the practice of photography—including Adams's own—has depended, and continues to depend, on paper. Mintie devotes the bulk of her piece to sharing vital details regarding paper production and its environmental effects.

In a similar vein, Alan Braddock provides an account of the use of animal-based gelatin in the production of photographic emulsions. In addition to describing the industrial transformation of animal bones, hides, and other parts into Kodak films, Braddock ruminates on issues of visibility and invisibility, noting how little photography, despite its reputation for exposing industrial ills, has historically exposed its own industrial underpinnings. Braddock's discussion of the 1881 dynamiting of the head of a living mule by military engineers to demonstrate the speed of gelatin emulsions provides us with an emblematic anecdote regarding the callousness to non-human animals that permeates the history of photography.

Attending to materiality can prompt critical self-reflection for scholars as well as photographers. Braddock acknowledges the inspiration he found in the scholarship of Nicole Shukin, whose book *Animal Capital* contains a passage on photographic gelatin. Braddock notes, however, that Shukin "does not mention gelatin's implication in the cover image of her book." As Shakespeare might say, "Ay, there's the rub." If photographers are prone to neglect the material entanglements of their own practices, the same can be said for scholars. We, too, are great users of paper, photography, computers, and digital networks. When we fly to a symposium in New York, France, or Germany, we may carry the scent of home, but our planes soar on the burn of fossil fuels. And when we write about the materiality of photography, we partake of it. (Kelsey, 2013a) To borrow apt words from contributor Alice Cazenave, "How can we talk about the toxicity of photography through a medium that is inherently toxic?"

For the scholars among us, breaking down the boundary between our ecological attentions and the environmental consequences of our professional activities is an uncomfortable exercise. Many of us have been living and breathing the modern industrial life since we were born. Even "getting away from it all" has meant spending time in the wild while extensively outfitted with nylon and polyester. Being cosmopolitan intellectuals, our loves and anchors, our pleasures and cares, our museums and our libraries, our laptops and our conferences are thoroughly implicated in unsustainable industrial processes.

To have our joys and our sorrows so entangled is challenging. Years after I had written a few pieces on the materiality of photography, an eminent retired art historian approached me at a museum event and said with unveiled sarcasm: "So, making photographs requires dangerous chemicals—you're blowing my mind!" Not the most tactful remark, but I understood his complaint. Taking the material turn within art history shifts the conversation in ways troublesome for those in the discipline. In my experience, most art historians did not commit to their scholarly specialty because they bore a special interest or expertise in chemistry or carcinogens. They did so because they found representation, at its best, magical or revelatory. In this respect, my art historian critic and I are in the same boat. If he thinks that my scholarly training and gifts are more evident in my interpretation of pictures *qua* pictures than in my pointing out that Kodak has been a major polluter, he is undoubtedly correct. Where we differ is that I found the tension between photography as a means of celebrating seemingly pristine landscapes and photography as an industrial process too glaring and problematic to ignore. But this shift of attention puts me in a bind that my colleague would evade. I have no good answer to the question: If I am to be ecologically conscientious, what do I do with my love of pictures? This worry, I imagine, lands me in good company.

With respect to this worry, the contributions of Cazenave and Andrés Pardo Piccone to the volume are particularly helpful. By reflecting on historical photographic processes and experimenting with more sensitive or sustainable ones, these writer-practitioners turn to materiality to find a better way forward. Cazenave experiments with developer processes using plants, including the willow tree, that have been affected by industrial silver extraction. In so doing, she crafts a practice that is less toxic and more engaged historically and botanically than the norm. She suggests "the world opens up differently" when we attend more to such modes of ecological engagement and less to the final photograph standing alone.

Pardo Piccone revisits a thirty-year-old recipe for an instant-coffee developer and discusses the challenges of cobbling together a more local developer solution. He muses on the advantages of using spent coffee grounds and takes cues from recipes involving botanical tinctures that William Henry Fox Talbot discovered almost two hundred years ago. Pardo Piccone's precepts—slowing down, walking, doing things yourself, using discarded or locally growing materials, experimenting—undergird for him the possibility of making photographic practice more sustainable.

Meanwhile, Yves Figueiredo suggests that revisiting our priorities can make us better environmental stewards not only as photographers, but also as citizens. If Mintie and Braddock remind us of what prevailing conventions have kept out of the frame, Figueiredo considers the unforeseen consequences of what those conventions have kept within it. In particular, he reflects on the history of exalting forested landscapes in amateur photography and the related effort to cultivate forests in dry California environments. The recent wildfires in the region, he argues, are due in no small part to the application in arid places of a Eurocentric notion that healthy landscapes should feature dense tree cover. His analysis points a way to a more sustainable future and how historical study can inform and support it.[5]

These pieces give me hope. The challenge ahead will be to gather more momentum behind such precepts and practices across a broader population. Of critical importance will be the transfer of habits, values, and expectations from one generation to the next. How can more people experience Pardo Piccone's precepts as he does, as a form of freedom rather than constraint? In the face of so much resistance and inertia, how can practical changes ripple outward at a mass scale?

Woven deeply into this volume's meditation on the ecologies of photography and the American West are concerns regarding the history of colonialism and its suppression of Native voices and cultures. Several contributors, including Will Wilson, Emily Cornish, Jordan Reznik, and Mercedes Dorame, give voice to these concerns and suggest, in words or photographic practice or both, ways of thinking through and beyond the colonial past.

These articulations represent a welcome reckoning in humanities scholarship of recent vintage. In 2007, the year my book *Archive Style* came out, my late friend and colleague François Brunet and I led a workshop on methodological issues in the study of nineteenth-century expedition photography at the Institute for Historical Studies at the University of Michigan. Some American Studies students participating in the workshop took me to task for not attending sufficiently to the historical experiences of Native Americans in my scholarship on Western US surveys. Although they found my critical take on the imperialist project that drove expedition photography clear enough, they felt that my archival research, by remaining focused on the survey as a bureaucratic enterprise, perpetuated an elision of Native perspectives. The criticism stung, but the students were right.

As I reflected on the contributions to this volume that proffer or ponder Native ways and perspectives, two arguments came to mind. One concerns the profound effects in the seventeenth and eighteenth centuries of Native thinking on European political thought. In the midst of the subjugation and displacement of the colonized by violence and coercion, the colonizers had a slew of cultural encounters that changed their own norms and ideals. Absent those encounters, the politics and culture of the United States would be unrecognizable. In their book *The Dawn of Everything: A New History of Humanity*, David Graeber and David Wengrow write: "When it comes to questions of personal freedom, the equality of men and women, sexual mores or popular sovereignty—or even, for that matter, theories of depth psychology—indigenous American attitudes are likely to be far closer to the reader's own than seventeenth-century European ones." (Graeber and Wengrow, 2021: 40–41) In light of this pervasive influence, I wonder if we might frame colonialism as, among other things, an incomplete education.

The contributions to this volume can help us imagine what the next stage of that education might be. It could encompass consideration of Native modes of negotiation between the human and the non-human. (Kimmerer, 2013; Pexa, 2016: 652–667) It could also include consideration of Native ways of contending with world-ending that have emerged in response to colonialism. (Powys White, 2017: 206–215) Reimagining the West as a place of failed boundaries might reinvigorate understandings of negotiation that are premised on circulation and collective responsibility. "This ain't real estate," a skunk says to the character Lipsha about a possible construction site during a vision quest in Louise Erdrich's novel *Bingo Palace*. As we rethink the boundaries that have organized the West, both geopolitically and photographically, we will need the voices of skunks and our other animal relations. We will also need to hold close Robin Wall Kimmerer's reminder that the land belongs to itself; that it is "a gift, not a commodity." (Kimmerer, 2013: 17)

If we regard colonialism and its legacy as an education of the colonizers, how well have the teachers been paid? What thanks have been given for the ideas of freedom and equality to which the nation owes so much? How might we imagine compensation for additional learning of Native precepts and relations? For me, this volume stirred up these questions.

The second argument that came to mind as I read was that encapsulated in Audrey Lorde's famous dictum: "the master's tools will never dismantle the master's house." Like artificial intelligence, photography is often characterized as a tool, suggesting that its essential function is to help us get work done. That is how Talbot represented his invention from the beginning. In letter of 1839, he distinguished his invention from the camera lucida and camera obscura by noting that these "assist the artist in his work" but "do not *work for him*." (emphasis in original; Siegel, 2017: 98) In adopting his mechanism for photogenic drawing, he claims, "you make the powers of nature work for you." (Siegel, 2017: 101) This language aligned photography with the most transformative and disruptive tool of Talbot's day, the steam engine. Today, photographers continue to use the tool of photography in hopes of doing good work in the world. But as John M. Culkin, inspired by the work of Marshall McLuhan, reflected back in 1967, "we shape our tools and thereafter they shape us." (Culkin, 1967: 53) Photography may work for us, but it works upon us, too.

As scholars, I think we are still coming to terms with the work photography has done upon us. Photography is a picture-making device grounded in the graphic optics of perspective. According to a canonical essay by Erwin Panofsky, the perspectival model of space installs the human observer at the center of experience and insists that the noumenal take stage within the phenomenal. We accept only what we observe, and we observe only what we see. In this way, Panofsky says, perspective ushers in the "anthropocracy" of the modern world. (Panofsky, 1997: 71–72) The anthropocratic quality of perspective, its privileging of the human observer as an agent apprehending the world as construed as a visual space, is baked into photography as a technology. Can photography be remade into an instrument honoring a boundaryless ecological reality? Can it become part of a healthy pact between the human and the non-human? What role can the spaces and places of the West play in such a transformation? This volume left me with these questions as well.

These questions, of course, do not arise in a time marked by open inquiry and patient deliberation. So I say one final time, there is smoke and there is smoke.

While I walked recently along a rural road, a pickup truck went by, billowing black exhaust. The owner had evidently modified the truck so that it could engage in "rolling coal," a practice often pursued in protest against environmentalism and its adherents. The fumes I inhaled felt like anger. How could things have come to this?

Americans live in a time of consequence and reckoning, and nowhere are the stakes more palpable than in the West. Greg Grandin, in his book *The End of the Myth: From the Frontier to the Border Wall in the Mind of America*, describes how the frontier mentality has underwritten a national history of violence, racism, and profligacy. (Grandin, 2019) As the frontier myth collapses under its own weight and the empire draws back, a frantic rearguard action seeks to regenerate the national story by reaffirming commitments to burning oil and coal, and to border violence. Even as the nation barricades itself to the south, dreams of taking Canada or Greenland or Mars into our ambit resuscitate the idea that the national destiny lies in owning more—and burning more. The West, once symbolic of an unlimited national future, now seems exhausted, and those seeking to find a new frontier look to the north and to the stars.

As this volume demonstrates abundantly, the West remains a place of photography, a place where light makes us write. May that writing help us to overcome the greed and indulgence that have so marred our relations on this earth!

Notes

1. Rachel Carson famously alerted the country to the toxic effects of DDT (Carson, 1962).
2. Rosalind Krauss has described this operation of photography acutely (Krauss, 1979).
3. I have come to feel keenly this limit on what Blake Stimson and I have called the double index of photography (Kelsey and Stimson, 2008).
4. I have discussed instances elsewhere: on O'Sullivan, see Kelsey, 2013a; on Friedlander, see Kelsey, 2018: 397–398; on Misrach, see Kelsey, 2023.
5. A key source for Figueiredo is Davis, 2025. While I appreciate the skepticism that Davis brings to what she calls arborocentrism, there is a baby in the bathwater: intact or older forests are of vital importance if we are to move toward sustainable carbon cycles. See, for example, Moomaw et al., 2019, especially p. 6, on the issue of fire.

Bibliography

Richard A. Bartlett, *Nature's Yellowstone* (Albuquerque: University of New Mexico Press, 1974).

Makeda Best, *Devour the Land: War and American Landscape Photography since 1970*, exh. cat. (Cambridge, MA: Harvard Art Museums, 2021).

Ned Blackhawk, *Violence over the Land: Indians and Empires in the Early American West* (Cambridge, MA: Harvard University Press, 2008).

Dee Brown, *Bury My Heart at Wounded Knee: An Indian History of the American West* (New York: Holt, Rinehart and Winston, 1970).

Rachel Carson, *Silent Spring* (Boston, MA: Houghton Mifflin, 1962).

William Cronon, "The Trouble with Wilderness; or, Getting Back to the Wrong Nature," in *Uncommon Ground: Rethinking the Human Place in Nature*, ed. William Cronon (New York: Norton, 1995), 69–90.

John M. Culkin, SJ, "A Schoolman's Guide to Marshall McLuhan," *Saturday Review* (March 18, 1967): 53.

Diana K. Davis, "Narrating Nature: Decolonizing Socio-ecological Assemblages," in *Doing Political Ecology*, eds Gregory L. Simon and Kelly Kay (New York: Routledge, 2025), 90–103.

Finis Dunaway, *Natural Visions: The Power of Images in American Environmental Reform* (Chicago: University of Chicago Press, 2005).

William H. Goetzmann, *Exploration and Empire: The Explorer and the Scientist in the Winning of the American West* (New York: Knopf, 1967).

David Graeber and David Wengrow, *The Dawn of Everything: A New History of Humanity* (New York: Farrar, Straus and Giroux, 2021).

Greg Grandin, *The End of the Myth: From the Frontier to the Border Wall in the Mind of America* (New York: Henry Holt/Metropolitan Books, 2019).

Robin Kelsey, "Sierra Club Photography and the Exclusive Property of Vision," in *Eco-Images: Historical Views and Political Strategies*, ed. Gisela Parak (Munich: Rachel Carson Center for Environment and Society, 2013a), 11–26.

Robin Kelsey, "Materiality," *Art Bulletin* 95 (March 2013b): 21–23.

Robin Kelsey, "Photography and the Ecological Imagination," in *Nature's Nation: American Art and Environment*, eds Alan Braddock and Karl Kusserow (Princeton, NJ: Princeton University Press, 2018), 394–405.

Robin Kelsey, "Ecology," in *Art Institute of Chicago Field Guide to Photography and Media*, eds Antawan I. Byrd and Elizabeth Siegel (New Haven, CT: Yale University Press, 2023), 148–155.

Robin Kelsey and Blake Stimson, "Introduction: Photography's Double Index: A Short History in Three Parts," in *The Meaning of Photography*, eds Robin Kelsey and Blake Stimson (Williamstown: Clark Art Institute; New Haven, CT: Yale University Press, 2008), vii-xxxi.

Robin Wall Kimmerer, *Braiding Sweetgrass: Indigenous Wisdom, Scientific Knowledge, and the Teaching of Plants* (Minneapolis: Milkweed Editions, 2013).

Rosalind Krauss, "Stieglitz/Equivalents," *October* 11 (Winter 1979): 133–140.

Bill McKibben, *The End of Nature* (New York: Random House, 1989).

Willliam R. Moomaw, Suan A. Masino, and Edward K. Faison, “Intact Forests in the United States: Proforestation Mitigates Climate Change and Serves the Greatest Good.” *Frontiers in Forests and Global Change* 2, no. 27 (2019): 1–10.

Erwin Panofsky, *Perspective as Symbolic Form*, trans. Christopher S. Wood (New York: Zone Books, 1997).

Christopher J. Pexa, “More than Talking Animals: Charles Alexander Eastman’s Animal Peoples and Their Kinship Critiques of United States Colonialism,” *PMLA* 131, no. 3 (2016): 652–667.

Sandra Phillips, *Crossing the Frontier: Photographs of the Developing West,* exh. cat. (San Francisco: San Francisco Museum of Modern Art, 1996).

Steffen Siegel (ed.), *First Exposures: Writings from the Beginnings of Photography* (Los Angeles: J. Paul Getty Museum, 2017).

Kyle Powys White, “Our Ancestors’ Dystopia Now: Indigenous Conservation and the Anthropocene,” in *The Routledge Companion to the Environmental Humanities*, eds Ursula Heise, Jon Christensen, and Michelle Niemann (New York: Routledge, 2017), 206–215.

Ofelia Zepeda, *Where Clouds are Formed* (Tucson: University of Arizona Press, 2008).

Notes on Contributors

Alan C. Braddock is the Ralph H. Wark Professor of Art History and Environmental Humanities at William & Mary. He is the author of *Thomas Eakins and the Cultures of Modernity* (University of California Press, 2009) and *Implication: An Ecocritical Dictionary for Art History* (Yale University Press, 2023), and coeditor of *A Keener Perception: Ecocritical Studies in American Art History*, with Christoph Irmscher (University of Alabama Press, 2009); *A Greene Country Towne: Philadelphia's Ecology in the Cultural Imagination*, with Laura Turner Igoe (Pennsylvania State University Press, 2016); and *Nature's Nation: American Art and Environment*, with Karl Kusserow (Princeton University Press, 2018), which won three national awards, including the Alfred H. Barr Jr., Award for Excellence from the College Art Association.

Monica Bravo is assistant professor of the history and theory of photography in the Department of Art and Archaeology at Princeton University. She is the author of *Greater American Camera: Making Modernism in Mexico,* published by Yale University Press in 2021. Her next book manuscript, "Silver Pacific: Early American Photography and Its Mining Economy" (under advance contract with Princeton University Press), has been supported by numerous fellowships. Her publications have appeared in the journals *American Art, Art Bulletin, Art History, Art Journal, History of Photography, Modernism/modernity* Print Plus, and *Panorama: Journal of the Association of Historians of American Art*, among other venues.

Alice Cazenave is the 2025–2026 Ansel Adams Research Fellow at the University of Arizona with a PhD in visual anthropology from Goldsmiths, University of London. She is an internationally exhibiting artist and adviser to the Sustainable Darkroom, the world's first charity devising low-toxic methods in film photography. In 2024, she was the artist-in-residence at Hong Kong Photo Festival and the visiting scholar in Art and Sustainability at Rochester Institute of Technology, New York. She has exhibited at the Saatchi (London),

Chappe Contemporary Art Museum (Ekenäs, Finland), BWA Gallery (Bydgoszcz, Poland), Halide Project (Philadelphia), and the London Art Fair, among others. Her research and practice examine links among photographic metals, colonialism, and ecological crises.

Emily Cornish is an assistant professor in the Department of Women, Gender, and Sexuality Studies at The Ohio State University. Her research and teaching examine Pacific Islander women's visual and material culture, with a specific focus on the Hawaiian Kingdom. Her writing on how *ali'i* women mobilized photography to assert and negotiate their sovereignty during the nineteenth century has been published in *Pacific Arts* and the *Journal of New Zealand and Pacific Studies*. Her research has been supported by the Luce/ACLS and the Social Science Research Council, among others.

Binh Danh earned an MFA from Stanford University. His work is represented in the collections of the de Young Museum, the Museum of Fine Arts, Boston, the National Gallery of Art, and the San Francisco Museum of Modern Art, among others. Danh's first monograph, *The Enigma of Belonging*, is published by Radius Books (2023). He teaches photography at San Jose State University.

Mercedes Dorame (Tongva) is a multi-disciplinary artist who calls on her Tongva ancestry to engage the problematics of (in)visibility and ancestral connection to land and sky. Born in Los Angeles, she received her MFA from San Francisco Art Institute (SFAI) and her undergraduate degree from University of California, Los Angeles. She is regular faculty at CalArts. She received the California Community Foundation Fellowship for Visual Arts, the Eiteljorg Contemporary Art Fellowship, and the Creative Capital Award, among others. Dorame's work is in the permanent collections of Allen Memorial Art Museum, the Getty Museum, the Hammer Museum, the Kadist Collection, Los Angeles County Museum of Art (LACMA), Minneapolis Institute of Art, and the San Francisco Museum of Modern Art, among others.

Yves Figueiredo is an Associate Professor of US environmental history at Université Paris Cité (Paris, France). His research focuses on the history of land management in the American West in the long nineteenth century. He has worked on the development of tourism in the region, on the establishment of the first national parks, and on landscape creation and appropriation, notably through photography and pictorial representations. His current research examines the urbanization of Southern California between the 1880s and the 1920s and the role of amateur photography in disseminating environmental imaginaries related to the region's deforestation and reforestation policies.

Carolin Görgen is Associate Professor of American Studies at Sorbonne Université in Paris, France, and junior fellow of the Institut Universitaire de France. She is the author of *The California Camera Club: Collective Visions in the Making of the American West* (University of Oklahoma Press, 2025) and a member of the editorial board of the journal *Photographica*, for which she coedited a special issue on photo albums in 2026. Her research on photo networks across the American West has received support from the Terra Foundation for American Art, the Huntington Library, Yale University, the Amon Carter Museum, and the Thomas Mann House, among others. She was a 2025 Fulbright visiting scholar at the University of California, Los Angeles.

Elizabeth Hutchinson is Associate Professor of Art History at Barnard College, Columbia University. She is the author of *The Indian Craze: Primitivism, Modernism, and Transculturation in American Art, 1890–1915* (Duke University Press, 2009) and co-curator of *Shifting Shorelines: Art, Industry, and Ecology along the Hudson River* (Columbia University Press, 2024). Her numerous publications include articles and exhibition catalog essays that incorporate transcultural and ecocritical approaches to landscape representation in the American West, including studies of the work of Carleton Watkins, William Henry Jackson, and Albert Bierstadt. She is at work on a book about Eadweard Muybridge's Pacific coast photography. Building on her first book, Dr. Hutchinson has an ongoing interest in the collecting and display of Indigenous art of the United States and is currently researching the provenance history of *Nuevomexicano* paintings of saints.

Robin Kelsey is Shirley Carter Burden Professor in the Department of History of Art and Architecture at Harvard University, where he served as Dean of Arts and Humanities from 2016 to 2024. He is the author of *Photography and the Art of Chance* (Cambridge, MA: Belknap Press of Harvard University Press, 2015) and *Archive Style: Photographs and Illustrations for U.S. Surveys, 1850–1990* (Berkeley and Los Angeles: University of California Press, 2007), as well as numerous articles and essays. With Blake Stimson, he coedited the anthology *The Meaning of Photography* (Williamstown, MA: Sterling and Francine Clark Art Institute, 2008). He holds a PhD from Harvard and a JD from Yale Law School and has practiced law in California.

Emilia Mickevicius is the Norton Family Assistant Curator of Photography, a joint appointment between Phoenix Art Museum and the Center for Creative Photography at the University of Arizona. Her exhibition projects have included *Funny Business: Photography and Humor* (2025) and *Muscle Memory: Lens on the Body* (2026). From 2019 to 2023, she worked in the photography department at the San Francisco Museum of Modern Art. In 2019, she received her PhD from Brown University, where her research was supported by the Henry Luce Foundation/American Council of Learned Societies, the Crystal Bridges Museum of American Art, and the Getty Research Institute.

Katherine Mintie is Head of Collections at the Center for Creative Photography at the University of Arizona. Her writing on the material history of photography has appeared in several journals, including *The Art Bulletin*, *Transbordeur: Photographie, Histoire, Société*, and *Panorama: Journal of the Association of Historians of American Art*.

Andrés Pardo Piccone is an experimental filmmaker, editor, professor, and researcher based in Montevideo, Uruguay. With more than twenty-five years of experience in film production and postproduction, he works across documentary, fiction, and experimental cinema. He founded CuriosoLab, a platform for sustainable photography and low-impact image making rooted in everyday environments. His work explores historical photo processes, alternative chemistry, and the use of local or discarded materials. Author of *Back to Basics: A Guide to Ecological Photochemistry* (The Sustainable Darkroom, 2024), he is currently researching Edward Bach solarization and energy-driven developers, investigating how analog practices can rethink our relationship with capitalism, technology, nature, and time.

Jordan Reznick is Assistant Professor of American Studies and Film and Media Studies at Grinnell College, Iowa. His current book project, *Landing the Camera: California Indian Ecologies and Colonial Photography, 1848–1890* has been supported by the Getty Research Institute, the California State Library, the National Endowment for the Humanities, and the Peter E. Palmquist Memorial Fund. His articles have appeared in *American Art*, *Art Journal*, and *California History*.

Martha A. Sandweiss is Professor of History, Emerita, Princeton University. She began her career as Curator of Photographs at the Amon Carter Museum in 1979 and taught for twenty years at Amherst College before moving to Princeton. She is the prize-winning author of numerous books on western history and photography including: *The Girl in the Middle: A Recovered History of the American West* (Princeton University Press, 2025); *Passing Strange: A Gilded Age Tale of Love and Deception across the Color Line* (Penguin Press, 2009); and *Print the Legend: Photography and the American West* (Yale University Press, 2002). She is coauthor of *Eyewitness to War: Prints and Daguerreotypes of the Mexican War* (Amon Carter Museum and Smithsonian Institution Press, 1989). She is also the editor of *Photography in Nineteenth-Century America* (Amon Carter Museum and Harry N. Abrams, Inc., 1991) and coeditor of *The Oxford History of the American West* (Oxford University Press, 1994).

Jennifer Tucker is Professor of History of Science and Visual Culture in the History Department at Wesleyan University, Middletown, Connecticut, and Founding Director of Wesleyan's Center for the Study of Guns and Society. She is the author of the book *Nature Exposed: Photography as Eyewitness in Victorian Science* (2005) and of numerous articles, book chapters, and media essays about the culture of photographic evidence in science, law, and popular culture. Previous projects include studies of photography and alkali pollution in a nineteenth-century northern English chemical town, in *International Labour and Working Class History*; "Gasworks, Ballooning, and the Victorian Gas Field," in *British Art Studies* (2022); and "Diseases in the Darkroom: Morality and Mortality in Nineteenth-Century Chemical Work," in *PhotoResearcher* (2024). She is coeditor of Routledge's Photography/History series and a Fellow of the Royal Historical Society, UK.

Will Wilson (Diné) is Associate Professor of Photography and Media at the University of Texas at Austin. His work centers Indigenous ecological knowledge, environmental justice, and visual sovereignty through long-term projects including *Auto Immune Response*; *Critical Indigenous Photographic Exchange* (CIPX); and *Connecting the Dots for a Just Transition*. He is coeditor of *Speaking with Light: Contemporary Indigenous Photography* (Radius Books, 2023), and his writing has appeared in *Devour the Land* (Harvard Art Museums, 2021) and *Out of Site: Survey Science and the Hidden West* (Autry Museum, 2024). Wilson's research examines uranium extraction on Dinétah through critical Indigenous cartography and land-based storytelling.

Colophon

Lieven Gevaert Research Centre for Photography, Art and Visual Culture
Arts Faculty KU Leuven
Blijde-Inkomststraat 21 box 3313
B-3000 Leuven
Belgium

Editors: Monica Bravo and Carolin Görgen
Layout and cover design: Theo van Beurden

GPRC
Guaranteed
Peer Reviewed
Content
www.gprc.be

ISBN 978 94 6270 523 4
eISBN 978 94 6166 766 3 (epdf)
eISBN 978 94 6166 767 0 (epub)
https://doi.org/10.11116/9789461667670
D/2026/1869/34
NUR: 652, 654

This publication is made possible in part by the Barr Ferree Foundation Fund for Publications, Department of Art and Archaeology, Princeton University

Previously published in the Lieven Gevaert Series

Volume 35
Valérie Mavridorakis, Alexander Streitberger, Hilde Van Gelder, and Erik Verhagen (eds)
Entangled Art Histories: The United States and the Two Germanies 1960–1990 (2025)

Volume 34
Jonathan Stafford, Henning Trüper, and Burkhardt Wolf (eds)
Moral Seascapes: On the Ethics and Aesthetics of Maritime Emergency (2024)

Volume 33
Herman Asselberghs, Robbrecht Desmet, Bart Meuleman, and Peter Jan Perquy (eds)
Dirk Lauwaert. Selected Writings, 1983–2008 (2023)

Volume 32
Olga Smith
Contemporary Photography in France (2022)

Volume 31
Jacob W. Lewis and Kyle Parry (eds)
Ubiquity. Photography's Multitudes (2021)

Volume 30
Hilde Van Gelder
Ground Sea. Photography and the Right to Be Reborn (2021)

Volume 29
Alexandra Moschovi
A Gust of Photo-Philia. Photography in the Art Museum (2020)

Volume 28
Alexander Streitberger
Psychical Realism. The Work of Victor Burgin (2020)

Volume 27
Alexander Streitberger and Hilde Van Gelder (eds)
"Disassembled" Images. Allan Sekula and Contemporary Art (2019)

Volume 26
Mieke Bleyen and Liesbeth Decan (eds)
Photography Performing Humor (2019)

Volume 25
Jorge Ribalta
Renaissance. Scenes of Industrial Reconversion in the Nord–Pas-de-Calais Coalfield (2018)

Volume 24
Maren Polte
A Class of Their Own. The Düsseldorf School of Photography (2017)

Volume 23
Stéphane Symons and Hilde Van Gelder (eds)
Victor Burgin's Parzival *in Leuven. Reflections on the "Uncinematic"* (2017)

Volume 22
Liesbeth Decan
Conceptual, Surrealist, Pictorial. Photobased Art in Belgium (1960s – early 1990s) (2016)

Volume 21
Brianne Cohen and Alexander Streitberger (eds)
The Photofilmic. Entangled Images in Contemporary Art and Visual Culture (2016)

Volume 20
Katarzyna Ruchel-Stockmans
Images Performing History. Photography and Representations of the Past in European Art after 1989 (2015)

Volume 19
Hilde Van Gelder (ed.)
Allan Sekula. Ship of Fools / The Dockers' Museum (2015)

Volume 18
Erik Verhagen
Jan Dibbets. The Photographic Work (2014)

Volume 17
Mieke Bleyen
Minor Aesthetics. The Photographic Work of Marcel Mariën (2014)

Volume 16
Ariella Azoulay
Aïm Deüelle Lüski and Horizontal Photography (2014)

Volume 15
Raphaël Pirenne and Alexander Streitberger (eds)
Heterogeneous Objects. Intermedia and Photography after Modernism (2013)

Volume 14
T.J. Demos and Hilde Van Gelder (eds)
In and Out of Brussels. Figuring Postcolonial Africa and Europe in the Films of Herman Asselberghs, Sven Augustijnen, Renzo Martens, and Els Opsomer (2012)

Volume 13
Mieke Bleyen (ed.)
Minor Photography. Connecting Deleuze and Guattari to Photography Theory (2012)

Volume 12
Alexander Streitberger
Shifting Places. Peter Downsbrough, the Photographs (2011)

Volume 11
Maarten Vanvolsem
The Art of Strip Photography. Making Still Images with a Moving Camera (2011)

Volume 10
Jan Baetens, Alexander Streitberger, and Hilde Van Gelder (eds)
Time and Photography (2010)

Volume 9
Alexander Streitberger (ed.)
Situational Aesthetics. Selected Writings by Victor Burgin (2009)

Volume 8
Barbara Baert (ed.)
Fluid Flesh. The Body, Religion and the Visual Arts (2009)

Volume 7
Hilde Van Gelder and Helen Westgeest (eds)
Photography between Poetry and Politics (2008)

Volume 6
Henri Van Lier
Philosophy of Photography (2007)

Volume 5
Patricia Allmer and Hilde Van Gelder (eds)
Collective Inventions: Surrealism in Belgium (2007)

Volume 4
Jan Baetens and Hilde Van Gelder (eds)
Critical Realism in Contemporary Art.Around Allan Sekula's Photography (2006)

Volume 3
Hilde Van Gelder (ed.)
In the Name of Mozart. Photographs by Malou Swinnen (2006)

Volume 2
Hilde Van Gelder (ed.)
Constantin Meunier. A Dialogue with Allan Sekula (2005)

Volume 1
Els Vanden Meersch.
Paranoid Obstructions (2004)

Cover image

Binh Danh, *Yosemite Falls, Yosemite, CA* (April 15, 2012), 2012, Daguerreotype, 8.5×6.5 in.,
The Huntington Library, Art Museum, and Botanical Gardens, Courtesy of the artist.